Design and Construction of Steel Structures

This book presents the design of steel structures and defines the approaches for various design codes, including AISC, BS, and EC3. It also discusses the theoretical background for the design of different structural members and provides illustrative examples of each, as well as structural connections, base plates with anchor bolt designs, and more. In addition, best practices for on-site construction methods, including receipt of materials, quality control and assurance, and inspection, are also presented. Non-destructive testing methods are discussed.

Features:

- Provides coverage of American (AISC) and European (EC3) structural codes.
- Examines various types of structural loading, including dead loads, live loads, wind, seismic loads/earthquakes, snow, and more.
- Includes numerous practical examples as well as ancillary Excel worksheets to aid in design calculations.
- Welding process, quality control and construction method statement by a case study.

Design and Construction of Steel Structures

Mohamed A. El-Reedy, Ph.D.

CRC Press
Taylor & Francis Group
Boca Raton London New York

CRC Press is an imprint of the
Taylor & Francis Group, an **informa** business

Designed cover image: Shutterstock

First edition published 2025
by CRC Press
2385 NW Executive Center Drive, Suite 320, Boca Raton FL 33431

and by CRC Press
4 Park Square, Milton Park, Abingdon, Oxon, OX14 4RN

CRC Press is an imprint of Taylor & Francis Group, LLC

ISBN: 978-1-032-29524-4 (hbk)
ISBN: 978-1-032-30929-3 (pbk)
ISBN: 978-1-003-30730-3 (ebk)

DOI: 10.1201/9781003307303

Typeset in Times
by codeMantra

This book is dedicated to the spirits of my mother and father, my wife, my son Hisham, and my daughters, Maey and Mayar

Contents

Preface

Steel structures are used worldwide for high-rise buildings, workshops, and various other applications. There has been significant development in steel structure design theory over time, especially with the introduction of high-strength bolts after World War II. To save costs, considerable research and development have been undertaken, transitioning from allowable strength design developed over decades to load resistance factor design (LRFD).

In many projects globally, structural engineers handle the design of steel structures, while mechanical engineers oversee the construction process. However, designers should be familiar with construction methods to create realistic designs, and on-site engineers should understand the design concepts to make informed during construction decisions.

Therefore, the objective of this book is to comprehensively address both design and construction aspects, serving as a valuable handbook for engineers engaged in steel structure projects. This book will cover both the European code EC3 and the American code AISC.

In summary, this book presents the design of steel structure members in a straightforward manner, describing the welding process, its characteristics, the required inspections during fabrication and erection, as well as the steps involved in constructing a complete steel structure building.

The European design code is generally newer than the American Institute of Steel Construction (AISC) design code. However, the European code is growing rapidly as many countries are transitioning to this standard. The concept of this code is that it can be used in any country within the European Union, with each country having its own Annex to incorporate parameters that align with their specific environment, workmanship, and other factors.

The European code directly addresses limit states design using the LRFD approach. Its comprehensive nature aims to cover all modes of failure from a strength or load perspective. Therefore, it delves theoretically into greater depth, as described in many textbooks. This book aims to swiftly guide readers to the practical application of this code for designing structures without requiring them to navigate complex equations and research. Each design element will be covered by an Excel sheet provided to the readers.

This book will explore steel structure design according to the European code while also incorporating the AISC code to enhance clarity on design concepts, owing to the latter's greater maturity in the field.

The best practical design aims to optimize materials and reduce the tonnage of steel, necessitating a realistic and efficient approach. This book outlines the method for calculating loads on steel structures, detailing all loads that impact buildings, including dead load, live load, wind, seismic, snow, and others, in accordance with EC3 and ASCE. It will present key aspects of steel structure design for various members, including tension members, beams, columns, and beam–columns. The design will be based on EC3 as well as AISC, accompanied by examples. Additionally, the behaviour of steel structural members will be discussed, covering theoretical backgrounds such as buckling and lateral–torsional buckling for beams, as well as other parameters and behaviours that can influence steel structure design based on the European code EC3 and AISC.

A crucial aspect of steel structure design is connection design. This course will cover the design of fixed and shear connections. The design of anchor bolts will be illustrated and exemplified using Excel sheets and design tools. The design examples performed in the Excel sheets aim to make it easier for readers to follow.

The construction methods will be discussed, including types of bolted and welded connections. Non-destructive testing for welding and raw steel will be a mandatory topic in this book, along with various quality control measures for steel structures, particularly concerning bolted and welded connections.

The target of this book is to encompass everything from the design to the construction of steel structures, as per EC3 and AISC. This book consists of 11 chapters: Chapter 1 introduces the European code, while Chapter 2 covers the actions that affect the structures. Chapters 3 to 6 focus on the design of the main steel structure elements, and Chapter 7 addresses the design of composite sections. The design of connections, whether welded or bolted, is illustrated in Chapters 8 and 9, which also discuss the quality of welding and bolted connections, as well as general quality during construction. Chapter 10 presents the non-destructive testing methods that will be used on-site, and Chapter 11 outlines the steps involved in the fabrication and erection of steel structure projects.

I hope this book serves as a valuable guide for engineers involved in both the design and construction of steel structures.

Mohamed A. E-Reedy, Ph.D.
elreedyma@gmail.com
www.elreedyman.com

Author

Dr. Mohamed Abdallah El-Reedy has a strong background in structural engineering, with a primary focus on the reliability of concrete and steel structures. He has provided consultation to various oil and gas industries in Egypt, as well as to international companies such as the International Egyptian Oil Company and British Petroleum (bp). Moreover, he offers design packages for concrete and steel structures for residential buildings, warehouses, and telecommunication towers. He has participated in liquefied natural gas and natural gas liquid projects alongside international engineering firms. Currently, he oversees reliability, inspection, and maintenance strategies for onshore concrete structures and offshore steel platforms. He has completed these tasks for hundreds of structures in the Gulf of Suez in the Red Sea. He also consulted with and trained executives from numerous organizations, including the Arabian American Oil Company, BP, Apache, Shell, Abu Dhabi Marine Operating Company, the Abu Dhabi National Oil Company, the Ministry of Interior of Saudi Arabia, Qatar Telecom, the Egyptian General Petroleum Corporation, Saudi Arabia Basic Industries Corporation, the Kuwait Petroleum Corporation, Qatar Petrochemical Company, Petroleum Development Oman, and the PETRONAS Malaysian oil company. He has taught technical courses on the repair and maintenance of reinforced concrete and steel structures worldwide, particularly in the Middle and Far East. He has written numerous publications and presented many papers at local and international conferences sponsored by the American Society of Mechanical Engineers, the American Concrete Institute, the American Society for Testing and Materials, and the American Petroleum Institute. He has published many research papers in international technical journals and has authored 11 books on topics such as offshore structures, asset integrity management, total quality management, economic management for engineering projects, repair and protection of reinforced concrete structures, advanced concrete materials, onshore structure design in petroleum facilities, and concrete structure reliability. He received his bachelor's degree from Cairo University in 1990, his master's degree in 1995, and his PhD from Cairo University in 2000.

Symbols

The convention for member axes is given below:

x-x	along the member
y-y	axis of the cross section
z-z	axis of the cross section

For steel members, the conventions used for cross-sectional axes are as follows:

y-y	cross-sectional axis parallel to the flanges
z-z	cross-sectional axis perpendicular to the flanges

For angle sections:

y-y	axis parallel to the smaller leg
z-z	axis perpendicular to the smaller leg

Where necessary:

u-u	major principal axis (where this does not coincide with the y-y axis)
v-v	minor principal axis (where this does not coincide with the z-z axis)
b	width of a cross section
h	depth of a cross section
d	depth of a straight portion of a web
tw	web thickness
tf	flange thickness
r	radius of root fillet
r_1	radius of root fillet
r_2	toe radius
t	thickness
P_k	nominal value of the effect of prestressing imposed during erection
Gk	nominal value of the effect of permanent actions
XK	characteristic values of material property
X_n	nominal values of material property
R_d	design value of resistance
R_k	characteristic value of resistance
γM	general partial factor
γMi	particular partial factor
γMf	partial factor for fatigue
η	conversion factor
a_d	design value of geometrical data
fy	yield strength
fu	ultimate strength
R_{eh}	yield strength to product standards
R_m	ultimate strength to product standards
A_0	original cross-sectional area
εy	yield strain
ε_u	ultimate strain
ZEd	required design Z-value resulting from the magnitude of strains from restrained metal shrinkage under the weld beads.

ZRd	available design Z-value
E	modulus of elasticity
G	shear modulus
v	Poisson's ratio in elastic stage
α	coefficient of linear thermal expansion
α_{cr}	factor by which the design loads would have to be increased to cause elastic instability in a global mode
FEd	design loading on the structure
Fcr	elastic critical buckling load for global instability mode based on initial elastic stiffnesses
HEd	design value of the horizontal reaction at the bottom of the storey to the horizontal loads and fictitious horizontal loads
V_{Ed}	total design vertical load on the structure on the bottom of the storey
$\delta_{H,Ed}$	horizontal displacement at the top of the storey, relative to the bottom of the storey
h	storey height
λ	non-dimensional slenderness
N_{Ed}	design value of the axial force
φ	global initial sway imperfection
φ_0	basic value for global initial sway imperfection
αh	reduction factor for height h applicable to columns
h	height of the structure
α_m	reduction factor for the number of columns in a row
m	number of columns in a row
$e0$	maximum amplitude of a member imperfection
L	member length
$\eta init$	amplitude of the elastic critical buckling mode
ηcr	shape of the elastic critical buckling mode
$e_{0,d}$	design value of the maximum amplitude of an imperfection
MRk	characteristic moment resistance of the critical cross section
NRk	characteristic resistance to normal force of the critical cross section
α	imperfection factor
$cr\ EI\ \eta$	bending moment due to η_{cr} at the critical cross section
χ	reduction factor for the relevant buckling curve
$\alpha_{ult,k}$	minimum force amplifier to reach the characteristic resistance without taking buckling into account
α_{cr}	minimum force amplifier to reach elastic critical buckling
q	equivalent force per unit length
δ_q	in-plane deflection of a bracing system
q_d	equivalent design force per unit length
M_{Ed}	design bending moment
k	factor for $e_{0,d}$
ε	strain
σ	stress
$\sigma com,Ed$	maximum design compressive stress in an element
ε	coefficient depending on f_y

c	width or depth of a part of a cross section
α	portion of a part of a cross section in compression
ψ	stress or strain ratio
$k\sigma$	plate buckling coefficient
d	outer diameter of circular tubular sections
$\gamma M0$	partial factor for resistance of cross sections whatever the class is
$\gamma M1$	partial factor for resistance of members to instability assessed by member checks
$\gamma M2$	partial factor for resistance of cross sections in tension to fracture
$\sigma x,Ed$	design value of local longitudinal stress
$\sigma z,Ed$	design value of local transverse stress
τEd	design value of local shear stress
NEd	design normal force
My,Ed	design bending moment, y-y axis
Mz,Ed	design bending moment, z-z axis
NRd	design values of resistance to normal forces
My,Rd	design values of resistance to bending moments, y-y axis
Mz,Rd	design values of resistance to bending moments, z-z axis
s	staggered pitch, the spacing of the centres of two consecutive holes in the chain measured parallel to the member axis
p	spacing of the centres of the same two holes measured perpendicular to the member axis
n	number of holes extending in any diagonal or zig-zag line progressively across the member or part of the member d_0 diameter of the hole
eN	shift of the centroid of the effective area A_{eff} relative to the centre of gravity of the gross cross section
ΔMEd	additional moment from shift of the centroid of the effective area A_{eff} relative to the centre of gravity of the gross cross section
$Aeff$	effective area of the cross section
Nt,Rd	design values of resistance to tension forces
$N_{pl,Rd}$	design plastic resistance to normal forces of the gross cross section
$N_{u,Rd}$	design ultimate resistance to normal forces of the net cross section at holes for fasteners
A_{net}	net area of the cross section
$N_{net,Rd}$	design plastic resistance to normal forces of the net cross section
$N_{c,Rd}$	design resistance to normal forces of the cross section for uniform compression
$M_{c,Rd}$	design resistance for bending about one principal axis of the cross section
W_{pl}	plastic section modulus
Wel,min	minimum elastic section modulus
$Weff,\mathrm{min}$	minimum effective section modulus
A_f	area of the tension flange
$A_{f,net}$	net area of the tension flange
VEd	design shear force

$V_{c,Rd}$	design shear resistance
$V_{pl,Rd}$	plastic design shear resistance
A_v	shear area
η	shear area factor
S	first moment of area
I	second moment of area
Aw	area of a web
Af	area of one flange
TEd	design value of total torsional moments
TRd	design resistance to torsional moments
$T_{t,Ed}$	design value of internal St. Venant torsion
$T_{w,Ed}$	design value of internal warping torsion
$\tau t,Ed$	design shear stresses due to St. Venant torsion
$\tau w,Ed$	design shear stresses due to warping torsion
$\sigma w,Ed$	design direct stresses due to bimoment B_{Ed}
BEd	bimoment
$V_{pl,T,Rd}$	reduced design plastic shear resistance making allowance for the presence of a torsional moment
ρ	reduction factor to determine reduced design values of resistance to bending moments making allowance for the presence of shear forces
MV,Rd	reduced design values of resistance to bending moments making allowance for the presence of shear forces
MN,Rd	reduced design values of resistance to bending moments making allowance for the presence of normal forces
n	ratio of design normal force to design plastic resistance to normal forces of the gross cross section
a	ratio of web area to gross area
α	parameter introducing the effect of biaxial bending
β	parameter introducing the effect of biaxial bending
eN,y	shift of the centroid of the effective area A_{eff} relative to the centre of gravity of the gross cross section (y-y axis)
eN,z	shift of the centroid of the effective area A_{eff} relative to the centre of gravity of the gross cross section (z-z axis)
$W_{eff,min}$	minimum effective section modulus
Nb,Rd	design buckling resistance of a compression member
χ	reduction factor for the relevant buckling mode
Φ	value to determine the reduction factor χ
a_0, a, b, c, d	class indexes for buckling curves
N_{cr}	elastic critical force for the relevant buckling mode based on the gross cross-sectional properties
i	radius of gyration about the relevant axis, determined using the properties of the gross cross section
λ_1	slenderness value to determine the relative slenderness
λ_T	relative slenderness for torsional or torsional–flexural buckling
$N_{cr,TF}$	elastic torsional–flexural buckling force
$N_{cr,T}$	elastic torsional buckling force

$M_{b,Rd}$	design buckling resistance moment
χ_{LT}	reduction factor for lateral–torsional buckling
Φ_{LT}	value to determine the reduction factor χ_{LT}
α_{LT}	imperfection factor
λ_{LT}	non-dimensional slenderness for lateral–torsional buckling
M_{cr}	elastic critical moment for lateral–torsional buckling
$\lambda_{LT,0}$	plateau length of the lateral–torsional buckling curves for rolled sections
β	correction factor for the lateral–torsional buckling curves for rolled sections
$\chi_{LT,mod}$	modified reduction factor for lateral–torsional buckling
f	modification factor for χ_{LT}
k_c	correction factor for moment distribution
ψ	ratio of moments in segment
L_c	length between lateral restraints
Λ_f	equivalent compression flange slenderness
I_{fz}	radius of gyration of compression flange about the minor axis of the section
$I_{eff,f}$	effective second moment of area of compression flange about the minor axis of the section
$Aeff,f$	effective area of compression flange
$Aeff,w,c$	effective area of compressed part of web
λ_{c0}	slenderness parameter
$fl\ k$	modification factor
Δ_{My}	moments due to the shift of the centroidal y-y axis
Δ_{Mz}	moments due to the shift of the centroidal z-z axis
χy	reduction factor due to flexural buckling (y-y axis)
χz	reduction factor due to flexural buckling (z-z axis)
k_{yy}	interaction factor
k_{yz}	interaction factor
kzy	interaction factor
kzz	interaction factor
λ_{op}	global non-dimensional slenderness of a structural component for out-of-plane buckling op χ reduction factor for the non-dimensional slenderness λ_{op}
$\alpha_{ult,k}$	minimum load amplifier of design loads to reach the characteristic resistance of the most critical cross section
$\alpha_{cr,op}$	minimum amplifier for the in-plane design loads to reach elastic critical resistance with regard to lateral or lateral–torsional buckling
NRk	characteristic value of resistance to compression
My,Rk	characteristic value of resistance to bending moments about the y-y axis
Mz,Rk	characteristic value of resistance to bending moments about the z-z axis
Q_m	local force applied at each stabilized member at the plastic hinge locations

L_{stable}	stable length of segment
L_{ch}	buckling length of chord
$h0$	distance of centrelines of chords of a built-up column a distance between restraints of chords
α	angle between axes of chord and lacings
$i\text{min}$	minimum radius of gyration of single angles
A_{ch}	area of one chord of a built-up column
$N_{ch,Ed}$	design chord force in the middle of a built-up member
$I_{Ed\,M}$	design value of the maximum moment in the middle of the built-up member
I_{eff}	effective second moment of area of the built-up member
S_v	shear stiffness of built-up member from the lacings or battened panel
n	number of planes of lacings
A_d	area of one diagonal of a built-up column
d	length of a diagonal of a built-up column
A_V	area of one post (or transverse element) of a built-up column
I_{ch}	in-plane second moment of area of a chord
I_b	in-plane second moment of area of a batten
μ	efficiency factor
Cmy	equivalent uniform moment factor
Cmz	equivalent uniform moment factor
$CmLT$	equivalent uniform moment factor
μ_y	factor
μ_z	factor
$N_{cr,y}$	elastic flexural buckling force about the y-y axis
$N_{cr,z}$	elastic flexural buckling force about the z-z axis
C_{yy}	factor
C_{yz}	factor
C_{zy}	factor
C_{zz}	factor
w_y	factor
w_z	factor
n_{pl}	factor
$\lambda_{\max}$	maximum of λ_y and λ_z
b_{LT}	factor
c_{LT}	factor
d_{LT}	factor
e_{LT}	factor
ψ_y	ratio of end moments (y-y axis)
$C_{my,0}$	factor
$C_{mz,0}$	factor
aLT	factor
IT	St. Venant torsional constant
I_y	second moment of area about y-y axis
$M_{i,Ed(x)}$	maximum first-order moment
$\lvert\delta x\rvert$	maximum member displacement along the member

α_s	factor
α_h	factor
C_m	equivalent uniform moment factor
γ_G	partial factor for permanent loads
G_k	characteristic value of permanent loads
γ_Q	partial factor for variable loads
Q_k	characteristic value of variable loads
$\lambda_{eff,v}$	effective slenderness ratio for buckling about the v-v axis
$\lambda_{eff,y}$	effective slenderness ratio for buckling about the y-y axis
$\lambda_{eff,z}$	effective slenderness ratio for buckling about the z-z axis
L	system length
L_{cr}	buckling length
S	shear stiffness provided by sheeting
Iw	warping constant
$C_{\vartheta,k}$	rotational stiffness provided by stabilizing continuum and connections
K_{υ}	factor for considering the type of analysis
K_{ϑ}	factor for considering the moment distribution and the type of restraint
$C_{\vartheta R,k}$	rotational stiffness provided by the stabilizing continuum to the beam assuming a stiff connection to the member
$C_{\vartheta C,k}$	rotational stiffness of the connection between the beam and the stabilizing continuum
$C_{\vartheta D,k}$	rotational stiffness deduced from an analysis of the distortional deformations of the beam cross sections
L_m	stable length between adjacent lateral restraints
L_k	stable length between adjacent torsional restraints
L_s	stable length between a plastic hinge location and an adjacent torsional restraint
C_1	modification factor for moment distribution
C_m	modification factor for linear moment gradient
C_n	modification factor for the non-linear moment gradient a distance between the centroid of the member with the plastic hinge and the centroid of the restraint members
B_0	factor
B_1	factor
B_2	factor
η	ratio of critical values of axial forces to the radius of gyration related to the centroid of the restraining member
βt	ratio of the algebraically smaller end moment to the larger end moment
R_1	moment at a specific location of a member
R_2	moment at a specific location of a member
R_3	moment at a specific location of a member
R_4	moment at a specific location of a member
R_5	moment at a specific location of a member
R_E	maximum of R_1 or R_5
R_s	maximum value of bending moment anywhere in the length L_y

C	taper factor
hh	additional depth of the haunch or taper
$h$max	maximum depth of the cross section within the length L_y
$h$min	minimum depth of the cross section within the length L_y
h_s	vertical depth of the unhaunched section
Lh	length of haunch within the length L_y
L_y	length between restraints

1 Introduction to Steel Structure

The steel structure is the most popular choice for industrial projects, as the advantages of steel align well with the requirements of such projects. Below is a sample of its advantages:

- High strength in tension and compression allows for smaller cross sections, making it more suitable for long spans.
- Elasticity: Steel is a prime example of a material that adheres to Hooke's law.
- Ductility: Steel exhibits significant elongation before breaking, allowing it to withstand overloads without sudden failure.
- It can be easily strengthened without the need for advanced technology.
- Steel typically demonstrates excellent quality control, as rolled cross sections are produced in factories that operate under stringent quality control systems.
- The erection process is easier and faster, as it can be completed in a short timeframe.
- Steel can be recovered from scrap.
- It retains value, which is beneficial if the structure needs to be demolished.

However, steel does have disadvantages, notably the need for periodic maintenance. This aligns with the needs of the industrial perspective, where maintenance is budgeted and handled by competent employee.

1.1 STEEL STRUCTURAL PROJECTS

There are various ways to classify the phases of a structural project, as projects can vary significantly in scope and execution methods. The classification methods can be all-inclusive or, at the very least, partially satisfactory. For civil engineering structures, the following phases may be helpful:

1. Feasibility study
2. Preliminary design
3. Detailed design
   A. Analysis
   B. Proportioning of elements
   C. Drawings, specifications, and other contract documents
4. Contract
5. Fabrication and construction
6. Operation and maintenance

DOI: 10.1201/9781003307303-1

1.1.1 Feasibility Study

This phase is crucial, with the primary focus on economic analysis, although structural engineers provide some minor assistance. During this phase, the engineer acts as a consultant to the prospective owner, builder, or financier. The engineer must not only recognize the standard structural systems worthy of investigation for the specific case but also devise new or modified systems to meet special demands. Very rough designs are typically made, often relying on judgement and experience with minimal computations.

The approximate costs of the promising schemes are estimated and compared with the tentative budgets. Time schedules are prepared roughly in this phase. If the project appears reasonable to those immediately concerned, their next task is to convince others of its value. The power of decision may rest with an individual builder, a corporation executive, or a board.

If a site has not been predetermined, this phase may also involve the structural engineer in defining a suitable site location. It is also true that the ideas generated and the course set in this early work have a significant impact on the project's success.

Therefore, the experience of the engineer or the engineering office is essential.

1.1.2 Preliminary Design Phase

In this phase, much is required from a competent engineering office, as the engineer will decide on the type of structure that will provide the function at the lowest cost. During this stage, the designer chooses between different structural system alternatives, such as using a truss, frame, main girder, or joist only. In the case of high-rise buildings, the engineer decides on the type of structural system that will resist lateral loads from wind and earthquakes; these systems may include shear walls, tube systems, frames with shear walls, and X-bracing or K-bracing, as shown in Figures 1.1 and 1.2 along the building Figure 1.3 presents the relation between building story and sway for different structure systems like K bracing system and moment release or combined frame system.

FIGURE 1.1 X-bracing.

FIGURE 1.2 K-bracing.

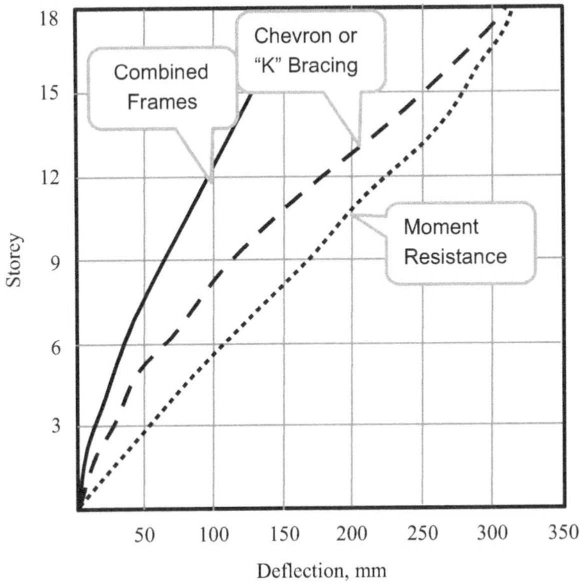

FIGURE 1.3 Relation between horizontal deflections and the number of storeys with different structural systems.

1.1.3 Detail Design

By the end of this phase, the owner should have a complete drawing set, specifications, material take-off, and in general, a complete tender package.

Refinements and alterations are made throughout the detail design, but generally, the basic structure has been defined by the time this phase starts. What remains is the detailed structural analysis and proportioning of elements. Primarily in statically indeterminate systems, though also in determinate ones, this may involve successive approximations in the search for dimensions. Up-to-date technical competence and alertness are prime requisites. The tender package and contract documents are generally prepared under the supervision of the structural engineer, or, in the case of legal documents, with their assistance or cooperation. They may also participate in the preparation of the engineer's final cost estimation, which is used as the standard for evaluating construction bids.

1.1.4 Contract

The structural engineer usually participates in negotiations with the contractors or in the review of competitive bids (contracts). Frequently, contractors raise technical questions or propose design alternatives that require the structural engineer's judgements.

1.1.5 Fabrications and Construction

Having undertaken construction, the detailed shop and construction drawings from the steel fabricator and other manufacturers are submitted to the engineer for approval (fabrications). Generally, a representative of the engineer remains on site for the duration of the project or visits periodically for smaller projects (construction). His primary duties are to inspect the work and ensure that construction complies with the plans and specifications. He is also responsible for project scheduling and progress. In addition, on almost all jobs, questions regarding interpretations of the design arise, requiring his attention.

1.1.6 Operations and Maintenance

Once the job is completed, operations and maintenance pass to the operations and maintenance staff. However, new construction frequently undergoes a shakedown period during which minor alterations are found to be necessary. This is particularly true in novel structures or those incorporating some innovative details. During this period, the engineer remains intimately involved in the projects.

Several aspects of the above breakdown require comment. In practice, the dividing lines between phases are rarely as clearly defined as discussed here; different phases are often combined or occur simultaneously. Some phases may even be dispensed with entirely. For instance, if the practicability of a project is so obvious, a feasibility study may be unnecessary.

In other cases, if the choice of a structural system is clear, final design can begin almost immediately, even in the case of traditional structures such as workshops and warehouses.

1.2 ULTIMATE LIMIT STATES

1.2.1 General Partial Safety Factors

The philosophy of the European code aligns with the British standard as it takes partial safety factors into consideration.

The development of steel and concrete structures accelerated rapidly after World War II, resulting in many structures worldwide and, consequently, a vast database. This database includes numerous studies discussing the variability and uncertainty in loads and strength in steel structures.

As a result of these extensive studies, specifications are increasingly oriented towards ultimate limit states design, which can also be called load resistance factor design. We use many factors for strength and load to address the variability in both and to safely load the materials up to their ultimate limit.

Instead of the traditional single factor of safety used in allowable stress design, limit state design provides for several partial safety factors that relate the characteristic values, such as loads and strength, to design values. In principle, the magnitude of a partial safety factor should be linked to the degree of uncertainty or variability of a particular quantity related to actions (loads) or material properties, as determined statistically. In practice, while this appears to be the case, the actual values of the partial safety factors used incorporate significant elements of the global safety factor and do not represent a rigorous probabilistic treatment of the uncertainties.

The principal sources of variability are associated with actions, material properties, and physical geometry. These are formally accounted for by applying appropriate partial safety factors to characteristic values.

These values for resistance and loads are derived from the reliability index, *b*, which represents the distance between the mean value of strength and the mean value of the load. It is important to note the relationship between the reliability index and the probability of failure. The probability of failure increases when the reliability index decreases, as shown in Figure 1.4.

The European code is uniquely structured to serve the various countries in the European Union. At certain points in the code, reference is made to the national annex; this implies that each country should have its own national annex. For example, wind maps should be unique to each country, with wind speed values distributed according to the respective country maps.

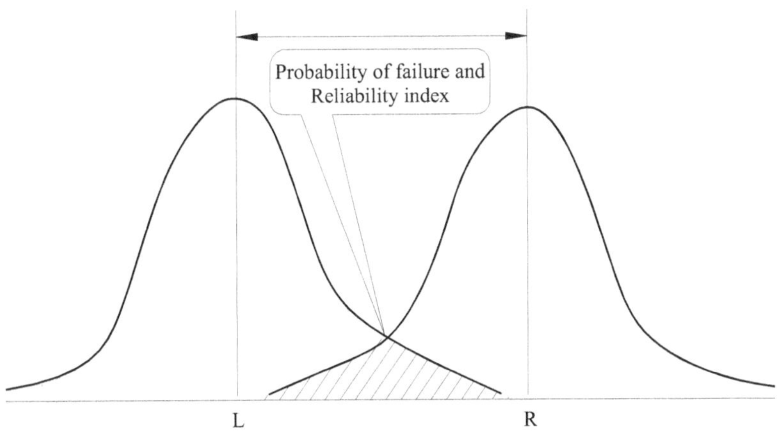

FIGURE 1.4 Relation between the probability of failure and the reliability index.

Based on the European Code (EC), the characteristic actions (F_k) are multiplied by the partial safety factors on actions (γ_F) to obtain the design loads (F_d):

$$F_d = \gamma_F F_k$$

The effects of applying the design loads to the structure – such as bending moment, shear force, and others are termed the "design effects" E_d.

The design resistance R_d is obtained by dividing the characteristic strengths R_k by the partial safety factors on materials γ_M, including appropriate modifications to account for other considerations, such as buckling. For a satisfactory design, the design resistance should be greater than the "design effect."

1.2.2 Variability of Actions (Loadings)

There is considerable variability in the magnitude of applied actions. The values of the partial safety factors reflect this degree of variability. For example, dead loads, which can be quantified with some confidence, attract a lower factor than imposed loads, which are related to a broad category of building use.

The main philosophy of the EC is that, where different actions are considered in combination, lower values of partial safety factors may apply. This reflects the reduced probability of the simultaneous occurrence of maximum actions of different types.

Different partial safety factors may also apply if a particular action has a beneficial effect.

The number of different cases that may need to be considered can become very high; however, in many situations, a small number of potentially critical combinations can be identified.

1.2.3 Variability of Material Properties

1.2.3.1 Strength

The variability of structural materials is reflected in variations in the strength of structural components. For structural steel, the most important property in this context is the yield strength. The characteristic yield strength is normally defined as the value below which only a small proportion of all values would be expected to fall. For practical reasons, a nominal value, which typically corresponds to the specified minimum yield strength, is generally used as the characteristic value for structural design purposes. This practice is exemplified in Eurocode 3, which tabulates nominal values of yield strength for different grades of steel.

The design value for the strength of steel is defined as the characteristic value divided by the appropriate partial safety factor.

1.2.3.2 Stress–Strain Behaviours of Structural Steel

Structural steel is an important construction material. It possesses attributes such as strength, stiffness, toughness, and ductility, which are highly desirable in modern construction.

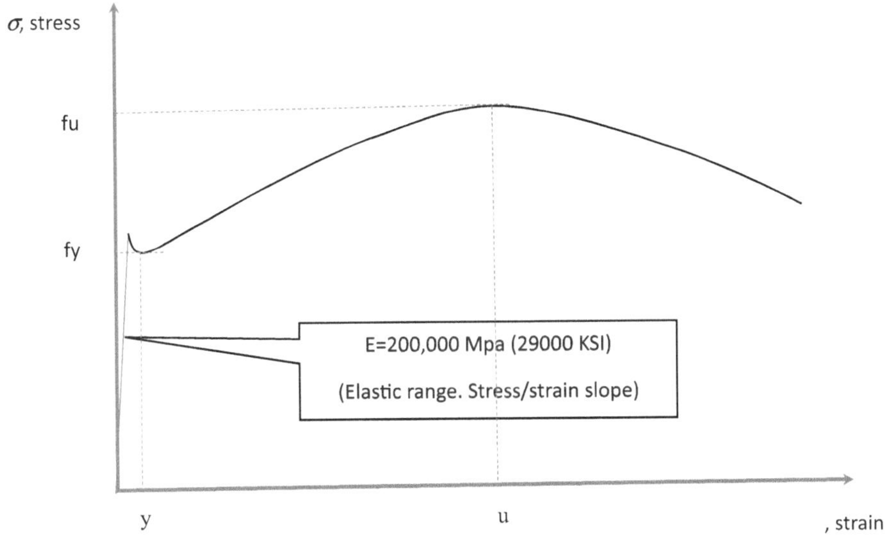

FIGURE 1.5 Uniaxial stress–strain behaviour of steel.

Strength is the ability of a material to resist stresses and is measured in terms of the material's yield strength, F_y, and ultimate or tensile strength, F_u. For steel, the ranges of F_y and F_u ordinarily used in construction are 248 to 345 MPa until 400 to 483 MPa, respectively, although higher strength steels are becoming more common. Stiffness is the ability of a material to resist deformation and is measured as the slope of the material's stress–strain curve. Referring to Figure 1.5, which shows uniaxial engineering stress–strain curves obtained from tests for various grades of steel, it can be seen that the modulus of elasticity, E, does not vary appreciably for the different steel grades. Therefore, a value of 200 GPa is often used for design purposes.

The definition of toughness as its the ability of a material to absorb energy before failure and is measured as the area under the material's stress–strain curve. Lower grade steels possess high toughness, making them suitable for both static and seismic applications. Ductility is the ability of a material to undergo significant inelastic, or plastic, deformation before failure and is measured in terms of percent elongation or percent reduction in area of the specimen tested in uniaxial tension.

Nominal values of yield strength F_y and ultimate tensile strength for hot rolled sections in EC are presented in Table 1.1.

Now we will discuss the principles of designing the main component.

1.2.3.3 Other Material Properties

Other material properties, notably modulus of elasticity, shear modulus, Poisson's ratio, coefficient of linear thermal expansion, and density, are much less variable than strength, and their design values are typically quoted as deterministic, with no partial safety factor applied.

In addition to the quantified values used directly in structural design, certain other material properties are normally specified to ensure the validity of the design

TABLE 1.1
Nominal Values of Yield Strength f_y and Ultimate Strength f_u for Hot Rolled Structural Steel

Standard and Steel Grade	Nominal Thickness of the Element t (mm)			
	$t \leq 40$mm		40 mm $< t \leq 80$ mm	
	f_y (N/mm²)	f_u (N/mm²)	f_y (N/mm²)	f_u (N/mm²)
		En 10025-2		
S 235	235	360	215	360
S 275	275	430	255	410
S 355	355	510	335	470
S 450	440	550	410	550
		En 10025-3		
S 275 N/NL	275	390	255	370
S 355 N/NL	355	490	335	470
S 420 N/NL	420	520	390	520
S 460 N/NL	460	540	430	540
		En 10025-4		
S 275 N/NL	275	370	255	360
S 355 N/NL	355	470	335	450
S 420 N/NL	420	520	390	500
S 460 N/NL	460	540	430	530
		En 10025-5		
S 235 W	235	360	215	340
S 355 W	355	510	335	490
		En 10025-6		
S 460 Q/QL/QL 1	460	570	440	550
		En 10210-1		
S 235 H	235	360	215	340
S 275 H	275	430	255	410
S 355 H	355	510	335	490
S 275 NH/NLH	275	390	255	370
S 355 NH/NLH	355	490	335	470
S 420 NH/NHL	420	540	390	520
S 460 NH/NLH	460	560	430	550
		En 10219-1		
S 235 H	275	390		
S 275 H	355	490		
S 355 H	420	520		

(Continued)

TABLE 1.1 (*Continued*)
Nominal Values of Yield Strength f_y and Ultimate Strength f_u for Hot Rolled Structural Steel

Standard and Steel Grade	Nominal Thickness of the Element t (mm)			
	$t \leq 40$mm		40 mm $< t \leq 80$ mm	
	f_y (N/mm²)	f_u (N/mm²)	f_y (N/mm²)	f_u (N/mm²)
S 275 NH/NLH	275	370		
S 355 NH/NLH	355	470		
S 460 NH/NLH	460	550		
S 275 MH/MLH	275	360		
S 355 MH/MLH	355	470		
S 420 MH/MLH	420	500		
S 460 MH/MLH	460	530		

procedures included within codified rules. For instance, Eurocode 3 stipulates minimum requirements for the ratio of ultimate to yield strength, elongation at failure, and ultimate strain if plastic analysis is to be used.

The material coefficients to be adopted in calculations for the structural steels covered by this Eurocode should be taken as follows:

- Modulus of elasticity $E = 210{,}000$ N/mm^2
- Shear modulus 81,000 N/mm^2
- Poisson's ratio in the elastic stage $\nu = 0{,}3$
- Coefficient of linear thermal expansion $\alpha = 12 \times 10^{-6}$ per K (for $T \leq 100°C$)

It is worth mentioning that in the case of calculating the structural effects of unequal temperatures in composite concrete–steel structures to EN 1994, the coefficient of linear thermal expansion is taken as $\alpha = 10 \times 10^{-6}$ per K.

1.2.3.4 Variability of Geometry

Geometrical data are generally represented by their nominal values, which are the values to be used for design purposes. Variability, for instance in cross-sectional dimensions, is accounted for in the partial safety factors applied elsewhere. Other imperfections, such as lack of verticality, lack of straightness, lack of fit, and unavoidable minor eccentricities present in practical connections, should also be allowed for. These may influence the global structural analysis, the analysis of the bracing system, or the design of individual structural elements, and are generally accounted for in the design rules themselves.

The geometry parameters and axes for hot rolled sections in EC3 are presented in Figure 1.6.

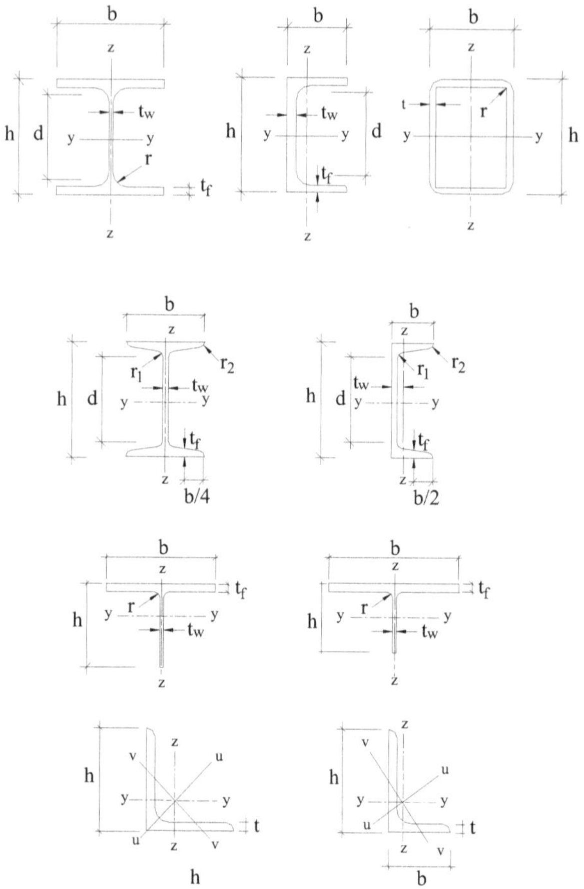

FIGURE 1.6 Geometry for rolled sections based on EC.

TABLE 1.2
Differences in Axis between AISC and Eurocode

Axes	AISC	Eurocodes
Along the member		X
Major axis	X	Y
Minor axis	Y	Z

There are differences in the axes between AISC and the Eurocodes, as shown in Table 1.2, and this is clearly presented in the section tables as shown in Appendix A. The comparison between the BS and Eurocode for the frequently used symbols is presented in Table 1.3.

TABLE 1.3
Comparison of Frequency Used in BS5950-1 and Eurocodes

	AISC	Eurocodes
Area	A	A
Elastic modulus	Z	W_{ef}
Plastic modulus	S	W_{pt}
Inertia about major axis	l_x	l_y
Inertia about minor axis	l_y	l_z
Warping constant	H	l_w
Torsion constant	J	l_t
Radius of gyration	r	L
Applied axial force	F	N
Resistance to axial force	P	N_{Rd}
Bending moment	M	M
Applied shear force	F_v	V
Shear resistance	P_v	V_{Rd}
Yield stress	P_y	f_y
Bending strength	P_b	$X_{LT}{}^{f}{}_{y}$
Compressive strength	P_c	$X^{f}{}_{y}$

1.2.3.4.1 Ductility Requirements

For steels, a minimum ductility is required, which should be expressed in terms of limits for

- the ratio f_u/f_y of the specified minimum ultimate tensile strength f_u to the specified minimum yield strength f_y;
- the elongation at failure on a gauge length of 5.65 o A (where A_0 is the original cross-sectional area);
- the ultimate strain ε_u, where ε_u corresponds to the ultimate strength f_u. The limiting values of the ratio f_u/f_y, the elongation at failure, and the ultimate strain ε_u may be defined in the National Annex. The following values are recommended:
- $f_u/f_y \geq 1{,}10$;
- elongation at failure not less than 15%;
- $\varepsilon_u \geq 15\varepsilon_y$, where ε_y is the yield strain ($\varepsilon_y = f_y/E$).

Steel that conforms to one of the steel grades listed in Table 1.1 should be accepted as meeting these requirements.

1.3 DESIGN SITUATIONS

The relevant design situations based on European code EC3 shall be selected, taking into account the circumstances under which the structure is required to fulfil its function.

Design situations shall be classified as follows:
Persistent design situations, which refer to the conditions of normal use;

- Transient design situations, which refer to temporary conditions applicable to the structure, during execution or repair;
- Accidental design situations, which refer to exceptional conditions applicable to the structure or to its exposure to fire, explosion, impact, or the consequences of localized failure.
- Seismic design situations, which refer to conditions applicable to the structure when subjected to seismic events.

The selected design situations shall be sufficiently severe and varied to encompass all conditions that can reasonably be foreseen to occur during the execution or throughout the structure's lifetime.

1.3.1 Design Procedure

Normally, we start with the preliminary design, as it is a rapid approximate manual method of designing a structure, as opposed to carrying out rigorous analysis and detailed design. The overall aim for a given structure is to identify critical loads, estimate design actions, and select sections. The problem is bound up with conceptual design, alternative systems, idealization, identification of critical members, and rationalization. The process depends greatly on the designer's experience and the use of appropriate design aids. The term is also often applied to manual design. Preliminary design is needed for the following reasons:

- For obtaining sections and weights for cost estimation;
- To compare alternative proposals;
- For obtaining initial sections for computer analysis;
- To check a final design.

The need for approximate manual methods is more important than ever. It is necessary to know if the output is right, wrong, or complete nonsense.

Methods of preliminary analysis that is not dependent on member sizes are set out for both elastic and plastic theories.

Redundant structures are treated as statically determinate by approximately locating points of contraflexure using sub-frames or by assigning values of actions at critical positions.

After defining the structure system in the preliminary design, we proceed to the detailed engineering by building the computer model and then select the safe member. You are now ready to deliver the complete engineering tender package.

FURTHER READING

British Standards Institution. (2005). *Eurocode 3: Design of Steel Structures: Part 1.1 General Rules and Rules for Buildings*, BS EN 1993-1-1, BSI, London.

Elreedy, MA. (2012). *Reinforced Concrete structure reliability*, CRC press, USA.

Simoes da Silva, L., Simoes, R., and Gervasio, H. (2010). *ECCS Eurocode Design Manuals*, Ernst & Sohn, A Wiley Company.

2 Loads Calculation

2.1 INTRODUCTION

The load is the primary item to be defined in the design process. The European code is a pioneer in referring to loads, which are termed "actions" in AISC or BS to be more generic. The European code covers all the countries within the European Union, so it must be broad enough to address all modes of failure and all types of loads or actions that may affect the structure throughout its lifetime. The terminology used for all loads or actions is consistent, as it is based on statistical data and the probability of occurrence. The European code presents this clearly compared to other codes or standards. This chapter also presents load calculations based on EC and AISC.

2.2 CLASSIFICATION OF ACTIONS (LOADS)

In general, the concept remains the same, so this book may refer to them as loads or actions.

Actions shall be classified by their variation over time as follows:

- Permanent loads (*G*), such as the self-weight of structures, fixed equipment, and road surfacing.
- Indirect actions caused by shrinkage and uneven settlements.
- Variable loads *(Q)*, such as. imposed loads on building floors, beams, and roofs, wind actions, or snow loads.
- Accidental actions *(A)*, such as explosions or impacts from vehicles.

Indirect actions caused by imposed deformations can be either permanent or variable.

Certain loads, such as seismic loads and snow loads, may be considered either accidental or variable actions, depending on the location of the site.

Actions caused by water may be considered as permanent and/or variable actions depending on the variation of their magnitude over time.

Actions shall also be classified as follows:

- By their origin, as direct or indirect
- By their spatial variation, as fixed or free.
- By their nature or the structural response, as static or dynamic.

2.2.1 Dead Load

The dead load is considered the structure's self-weight, which can be defined by knowing the dimensions of the structural elements and the density of the materials from which these members are made, such as concrete, steel, or wood. Therefore, in

DOI: 10.1201/9781003307303-2

the first step, it is necessary to define the member dimensions preliminarily, in addition to identifying the material from which the member will be constructed.

The dead load also includes the weight of finishes such as plastering, tiles, or walls made from bricks and other finishing materials that are not considered the main structural components.

It is very important to note that the dead load constitutes a significant portion of the total load during design; consequently, the weight of the building members must be calculated accurately, especially in the case of reinforced concrete structures.

The densities of various materials commonly used in construction are presented in Table 2.1, which should be considered when calculating the dead load.

TABLE 2.1
Nominal Weight for Different Materials Based on Eurocodes

Materials	ASCE (lb/ft³)	Density γ (kN/m³)
Concrete		
Lightweight (varies with density class)	8 per inch	
Density class LC 1.0	105	
Density class LC 1.2		9.0 to 10.0
Density class LC 1.4		10.0 to 12.0
Density class LC 1.6		12.0 to 14.0
Density class LC 1.8		14.0 to 16.0
Density class LC 2.0		16.0 to 18.0
Normal weight		18.0 to 20.0
Heavyweight		*24
For reinforced and prestressed concrete;		>28
unhardened concrete		Above + 1
Mortar		
Cement mortar		19–23
Gypsum mortar; lime mortar	20.4	12–18
Lime-cement mortar	20.4	18–20
Masonry Units		
Dense limestone	147	20–29
Granite	153	27–30
Sandstone	137	21–27
Marble	156	8
Glass blocks, hollow	120	21
Terra cotta, solid		
Densities are not quoted for concrete, aerated concrete, calcium silicate, clay, or manufactured stone		
Metals		
Aluminium	170	27
Copper	556	87–89
Steel	492	77–78.5
Zinc	449	71–72

(Continued)

TABLE 2.1 (Continued)
Nominal Weight for Different Materials Based on Eurocodes

Materials	ASCE (lb/ft³)	Density γ (kN/m³)
Wood		
Timber (depending on strength class C14 – C17; see pr EN 338)		
Plywood:		
Raw plywood (softwood and birch)	45	2.9–9.0
Laminboard and block board		6
Particleboards:		
Chipboard		4
Cement-boarded particleboard		8
Fiber building board:		12
Hardboard, standard and tempered		10
Medium density fibreboard		8
Soft board		4
Other Materials		
Glass, in sheets		
Plastics	0.34 (5 mm)	25
Acrylic sheet	0.48 (6 mm)	12
Polystyrene, expanded, granules		0.25
Slate		29

Products	Density γ (kN/m³)	Angle of Repose ø (°)
Eggs, in stands	4.0–5.0	–
Flour		
Bulk	6.0	25
Bagged	5.0	–
Fruit		
Apples	8.3	30
• Loose	6.5	–
• Boxed	7.8	–
Cherries	5.9	–
Pears	2.0	–
Raspberries, in trays	1.2	–
Strawberries, in trays	6.8	–
Tomatoes		
Sugar		
Loose, piled	7.5–10.0	35
Dense and bagged	16.0	–
Vegetables, Green		
Cabbages	4.0	–
Lettuce	5.0	–

(Continued)

Products	Density γ (kN/m³)	Angle of Repose ø (°)
Vegetables, legumes		
Beans		
• General	8.1	35
• Soya	7.4	30
Peas	7.8	–
Vegetables, root		
General	8.8	–
Beetroot	7.4	40
Carrots	7.8	35
Onions	7	35
Turnips	7	35
Potatoes		
In bulk	7.6	35
In boxes	4.4	–
Sugar beet		
Dried and chopped	2.9	35
Raw	7.6	–
Wet shreds	10.0	–

Nowadays, computer software is used in structural analysis to calculate the self-weight of structural members based on their materials and dimensions. However, the finishing materials and the dead load affecting the building must be input manually into the program.

Recently, advancements in structural analysis software allow for the calculation of the self-weight of structural members after defining their materials and dimensions. However, the finishing materials and the dead load that occupy the building must still be entered manually into the program.

2.2.2 Live Load (Imposed Load)

In the last decade, there have been significant advances in the application of structural reliability theory (El-Reedy et al., 2000). One area of focus has been live loads in buildings, including the development of realistic stochastic models. Rapid advances in the probabilistic modelling of live loads in recent years can be attributed to a growing awareness of designers' uncertainty regarding the loads acting on a structure, as well as the acceptance of a probabilistic assessment of these loads.

In general, the live load is composed of two parts: the sustained load and the extraordinary load. The summation of these two types of loads is required to obtain the probabilistic model of the total live load on a residential building throughout the lifetime of the structure. The European code for actions primarily depends on statistics and reliability approaches.

2.2.2.1 Live Load as per European Code

The imposed load values according to the EC depend on categorizing buildings based on their use. The categories of the buildings are shown in Table 2.2.

Values for the uniform load (q_k) and the concentrated imposed load (Q_k) are given in Table 2.3 below. Where a range is given in this table, the value may be specified by the National Annex. The recommended values, intended for separate application, are underlined. q_k is intended for determining general effects, while Q_k is for local effects. The National Annex may define different conditions for using Table 2.3.

The following considerations should be taken into account:

1. Where necessary q_k and Q_k should be increased in the design for stairs and balconies, depending on occupancy and dimensions.

TABLE 2.2
Categories as per EC

Category	Specific Use	Example
A	Areas for domestic and residential activities	Rooms in residential buildings and houses; bedrooms and wards in hospitals; bedrooms in hotels and hostels kitchens and toilets.
B	Office areas	
C	Areas where people may congregate (with the exception of areas defined under category A, B, and D (NOTE 1)[1))	C1: Areas with tables, etc. e.g. areas in schools, cafes, restaurants, dining halls, reading rooms, and receptions. C2: Areas with fixed seats, e.g. areas in churches, theatres or cinemas, conference rooms, lecture halls, assembly halls, waiting rooms, and railway waiting rooms. C3: Areas without obstacles for moving people, e.g. areas in museums and exhibition rooms And access areas in public and administration buildings, hotels, hospitals, railway station forecourts. C4: Areas with possible physical activities, e.g. dance halls, gymnastic rooms, and stages. C5: Areas susceptible to large crowds, e.g. in buildings for public events like concert halls, sports halls including stands, terrace, and access areas and railway platforms.
D	Shopping areas	D1: Areas in general retail shops. D2: Areas in department storeys.

Attention is drawn to 6.3.1.1(2), in particular for C4 and C5. See EN 1990 when dynamic effects need to be considered. For category E, see Table 6.3.

Note 1: Depending on their anticipated uses, areas likely to be categorized as C2, C3, and C4 may be categorized as C5 by decision of the client and/or National annex.

Note 2: The National annex may provide sub categories to A, B, C1 to C5, D1 and D2.

Note 3: See 6.3.2 for storage or industrial activity.

TABLE 2.3
Imposed Loads on Floors, Balconies, and Stairs in Buildings

Categories of Loaded Areas	q_k(kN/m^2)	Q_k(kN)
Category A		
• Floors	1.5 to **2.0**	2.0 to 3.0
• Stairs	**2.0 to** 4.0	2.0 to 4.0
• Balconies	**2.5 to** 4.0	2.0 to 3.0
Category B	2.0 to **3.0**	1.5 to 4.5
Category C	2.0 to **3.0**	3.0 to **4.0**
• C1	3.0 to **4.0**	2.5 to 7.0 (**4.0**)
• C2	3.0 to **5.0**	**4.0** to 7.0
• C3	4.5 to **5.0**	3.5 to **7.0**
• C4	**5.0** to 7.5	3.5 to **4.5**
• C5	4.0 to 5.0	3.5 to 7.0 (**4.0**)
Category D	4.0 to 5.0	3.5 to **7.0**
• D1		
• D2		

2. For local verifications, a concentrated load Q_k acting alone should be considered.
3. For concentrated loads from storage racks or lifting equipment, Q_k should be determined for the individual case.
4. The concentrated load shall be considered to act at any point on the floor, balcony, or stairs over an area shaped appropriately for the use and form of the floor. It is noted that the shape may typically be assumed to be a square with a width of 50 mm.
5. The vertical loads on floors due to the traffic of forklifts shall be taken into account.
6. If floors are subjected to multiple uses, they shall be designed for the most unfavourable loading category that produces the highest effects of actions on the member under consideration.
7. Provided that a floor allows lateral distribution of loads, the self-weight of movable partitions may be accounted for by a uniformly distributed load q_k, which should be added to the imposed loads of the floors obtained from Table 2.6. As per EC, the uniformly distributed load for the partitions is dependent on their self-weight as follows:

   a. For movable partitions with a self-weight of 1.0 kN/m wall length: q_k =0.5 kN/m^2;
   b. For movable partitions with a self-weight of 2.0 kN/m wall length: q_k =0.8 kN/m^2;

c. For movable partitions with a self-weight of 3.0 kN/m wall length: q_k =1.2 kN/m^2.

8. Heavier partitions should be considered in the design, taking into account:
   - The locations and directions of the partitions;
   - The structural form of the floors.

2.2.2.2 Loads on the Roofs

According to the EC, roofs shall be categorized according to their accessibility into three categories, as shown in Table 2.4.

Imposed loads for roofs of category H should be those given in the following section; however, imposed loads for roofs of category I are given in Table 2.6.

The loads for roofs of category K, which provide areas for helicopter landing, should adhere to the helicopter classes HC, as shown in Table 2.8.

2.2.2.3 Values of Actions

For roofs of category H, the minimum characteristic values Q_k and q_k should be used as presented in Table 2.3. These values are related to the projected area of the roof under consideration.

2.2.2.3.1 In Case of Roof of Category H

For category H, q_k may be selected within the range of 0.00–1.0 kN/m^2, and Q_k may be selected within the range of 0.9–1.5 kN.

Where a range is given, the values may be specified by the National Annex. The recommended values are as follows:

q_k=0.4 kN/m^2, Q_k= 1.0kN

q_k may be varied by the National Annex depending on the roof slope.

q_k may be assumed to act on an area A, which may be defined by the National Annex.

The recommended value for A is 10 m^2, within the range from zero to the total area of the roof.

TABLE 2.4
Categories of the Roofs

Categories of Loaded Area	Specific Use
H	Roofs not accessible except for normal maintenance and repair.
I	Roofs accessible with occupancy according to categories A to D.
K	Roof accessible for special services, such as helicopter landing areas.

The minimum values given in Table 6.10 do not account for uncontrolled accumulations of construction materials that may occur during maintenance.

For roofs, separate verifications shall be performed for the concentrated load Q_k and the uniformly distributed load q_k, treating them as independent actions.

Roofs, other than those with roof sheeting, should be designed to resist 1.5 kN on an area based on a square with sides of 50 mm. Roof elements with a profiled or discontinuously laid surface should be designed so that the concentrated load Q_k acts over the effective area provided by load-spreading arrangements.

2.2.2.3.2 In Case of Roof of Category K

For roofs of category K, the actions from helicopters on landing areas should be determined in accordance with Table 2.5.

Therefore, in the case of loads due to helicopters, there will be an impact load effect. The dynamic factor to be applied to the take-off load Q_k, to account for impact effects, may be taken as 1.40.

Access ladders and walkways should be assumed to be loaded according to category H for roof slopes $< 20°$.

For walkways that are part of a designated escape route, q_k should be according to Table 2.6. For service walkways, a minimum characteristic value Q_k of 1.5 kN should be applied.

2.2.2.3.3 Imposed Load Reduction Factor

As the number of storeys in a building increases, the likelihood that all floors will be loaded to the full design level decreases. Similarly, large floor areas will seldom be subjected to the full design loading uniformly. To reflect this, reduction factors for imposed loads may be applied for the design of floors, beams, roofs, columns, and walls.

For the design of individual floors, beams, and roofs, the area reduction factor αA may be applied. For the design of columns and walls, the reduction factor αn for the number of storeys may be applied. The reduction factor αn relates to the number of floors supported by the column section under consideration and may be applied to the total imposed load being carried. If, for a given column or wall, $\alpha_A < \alpha_n$, then α_A may be used in place of α_n, but α_A and α_n may not be used together.

The following reduction factor α_A, based on the following equation, may be applied to the q_k values for imposed loads for floors in Table 2.6 and accessible roofs, Category I.

TABLE 2.5
Imposed Loads on Roofs of Category, K, for Helicopter

Class of Helicopter	Take-off Load *Q* of Helicopter	Take-off Load Q_k	Dimension of the Loaded Area (*m*×*m*)
HC1	$Q \leq 20$ kN	$Q_k = 20$ kN	0.2×0.2
HC2	20 kN $< Q \leq 60$ kN	$Q_k = 60$ kN	0.3×0.3

TABLE 2.6
Recommended Values of ψ Factors for Buildings

Action	Ψ_0	Ψ_1	Ψ_2
Imposed loads in buildings, category (see EN 1991-1-1)			
Category A: domestic, residential areas	0.7	0.5	0.3
Category B: office areas	0.7	0.5	0.3
Category C: congregation areas	0.7	0.7	0.6
Category D: shopping areas	0.7	0.7	0.6
Category E: storage areas	1.0	0.9	0.8
Category F: traffic area, vehicle weight ≤30 kN	0.7	0.7	0.6
Category G: traffic area, 30 kN < vehicle weight	0.7	0.5	0.3
≤160 kN	0	0	0
Category H: roofs			
Snow loads on buildings (see EN 1991-1-3)[a]			
Finland, Iceland, Norway, Sweden	0.70	0.50	0.20
Remainder of CEN Member States, for sites located at	0.70	0.50	0.20
altitude H > 1,000 m a.s.l.	0.50	0.20	0
Remainder of CEN Member States, for sites located at			
altitude H ≤ 1,000 m a.s.l.			
Wind loads on buildings (see EN 1991-1-4)	0.6	0.2	0
Temperature (non-fire) in buildings (see EN 1991-1-5)	0.6	0.5	0

Note: The Ψ values may be set by the National annex.

[a] For countries not mentioned below, see relevant local conditions.

$$\alpha_A = \frac{5}{7}\psi_0 + \frac{A_0}{A} \leq 1,0$$

With the restriction for categories C and D,

$$\alpha_A \geq 0.6$$

where

ψ_0 is the factor according to EN 1990 as in Table 2.6
A_0 is 10,0 m^2
A is the loaded area

It is worth mentioning that a revised expression for the reduction factor αA for imposed loads on floors and accessible roofs is provided in Clause NA.2.5 of the UK National Annex to EN 1991-1-1 (see equation 2.1)2.2 and replaces those given in Clause 6.3.1.2(10) of EN 1991-1-1.

$$\alpha A = 1.0 - A/1,000 \geq 0.75 \qquad (2.1)$$

where A is the area (m^2) supported.

There is another reduction due to the number of storeys. It is stipulated that the area is classified according to Table 2.6 into categories A to D. For columns and walls, the total imposed loads from several storeys may be multiplied by the reduction factor α_n.

The recommended values for α_n are obtained from the following equation:

$$\alpha_n = 2 + (n-2)\psi_0 / n$$

where

- n is the number of storeys (>2) above the loaded structural elements from the same category.
- ψ_0 is in accordance with EN1990 values as in Table 2.6.

where

- Ψ_0 is the factor for combination value of a variable action
- Ψ_1 is the factor for frequent value of a variable action
- Ψ_2 is the factor for quasi-permanent value of a variable action

Revised expressions are provided in the UK National Annex (Clause NA.2.6 and Equation NA.2), as given by equations (2.2–2.4) below. These reduction factors may be applied to the total imposed load experienced by a given column but may only be employed when the imposed load is the leading variable action in a load combination. When the imposed load is an accompanying action, either ψ_0 or α_n may be applied, but not both.

$$\alpha_n = 1.1 - n/10 \text{ for } 1 \le n \le 5 \tag{2.2}$$

$$\alpha_n = 0.6 \text{ for } 5 < n \le 10 \tag{2.3}$$

$$\alpha_n = 0.5 \text{ for } n > 10 \tag{2.4}$$

2.2.3 Wind Load

Wind load governs the design based on the building's geographic location and its shape. Therefore, the parameters that influence wind load are approximately similar across different codes, as all are based on aerodynamic theory.

The concept of wind load is consistent among ASCE, BS standards, the European code, and others. Traditionally, wind loads acting on structures have been accounted for by using static force equivalents. The first step is usually to determine a basic wind speed for the general region in which the structure is to be built, using information derived from meteorological studies. This basic wind speed may represent an extreme velocity measured at a height of 10 m and averaged over a period of 3 seconds, with a return period of 50 years (i.e., a velocity that will, on average, be reached or exceeded once in 50 years, or have a probability of being exceeded of 1/50).

2.2.3.1 Wind Load in EC-1

Two approaches are presented in ENV 1991-2-4. The simple method is applicable where structures are unlikely to be susceptible to dynamic excitation. In practice, steel-framed buildings less than 200 m tall should satisfy this condition, but for very slender or unusual structures, more detailed calculations may be required for justification. This section focusses on the simplified approach and applies to "permanent" structures. In general, temporary structures may be designed for lower wind pressures.

Wind loads are treated as quasi-static pressures acting normal to the surface of the building (except in special circumstances, such as tangential friction forces due to wind blowing over large flat surfaces). The characteristic wind pressure is the reference mean wind velocity pressure, q_b, derived from the corresponding velocity ($v_{b,0}$). The reference wind velocity ($v_{b,0}$) is determined from wind maps for each member state, and each country has its own wind speed map.

Exceptionally, the value of $v_{ref,0}$ may need to be modified to account for altitude or wind direction in some countries.

The reference wind pressure is then determined directly:

$$q_b = 0.5\rho(v_b)^2$$

where ρ is the density of air, taken as 1.25 kg/m^3 unless otherwise specified.

This is modified by an exposure coefficient $c_e(z)$, which accounts for terrain roughness according to Table 2.7, topography, and height above ground. For "flat" terrain, the exposure coefficient $c_e(z)$ can be determined from Figure 2.1, which relates height to terrain category. Terrain is generally considered "flat" except for locations close to isolated hills and escarpments.

The parameters of this table are calibrated to obtain the best fit of available data for the values k_r, z_o, and z_{min}.

TABLE 2.7
Terrain Categories and Related Parameters

Terrain Category	k_r	z_o (m)	z_{min} (m)	ε
0. Sea or coastal area exposed to the open sea	0.16	0.003	1	
I. Rough open sea, lakes with at least 5 km fetch upwind and smooth flat country without obstacles	0.17	0.01	2	(0.13)
II. Farmland with boundary hedges, occasional small farm structures, houses, or trees	0.19	0.05	4	(0.26)
III. Suburban or industrial areas and permanent forests	0.22	0.3	8	(0.37)
IV. Urban areas in which at least 15% of the surface is covered with buildings and their average height exceeds 15 m	0.24	1	16	(0.46)

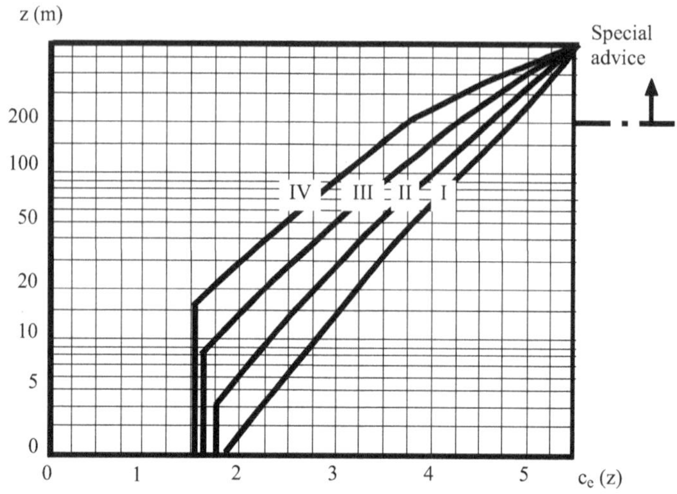

FIGURE 2.1 Exposure coefficient $c_e(z)$ as a function of height z above ground.

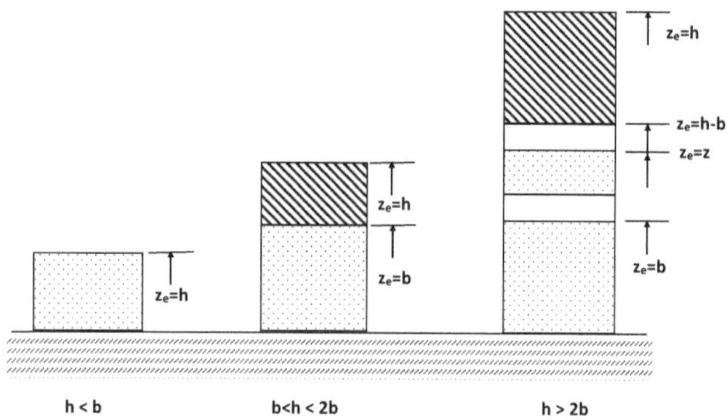

FIGURE 2.2 Reference height z_e as a function of the breadth and actual building height.

For tall buildings, sections may be defined with appropriate height ranges, and wind pressures calculated for each section according to the corresponding "reference height," as shown in Figure 2.2.

2.2.3.1.1 Wind Pressure Distribution

The net pressure on a surface is the algebraic sum of the internal and external pressures, as indicated in Figure 2.2

Wind pressure coefficients c_{pe} and c_{pi} are used to determine the distribution of external and internal pressures, respectively (Figure 2.3).

Values are tabulated for

- vertical walls of rectangular buildings
- flat roofs

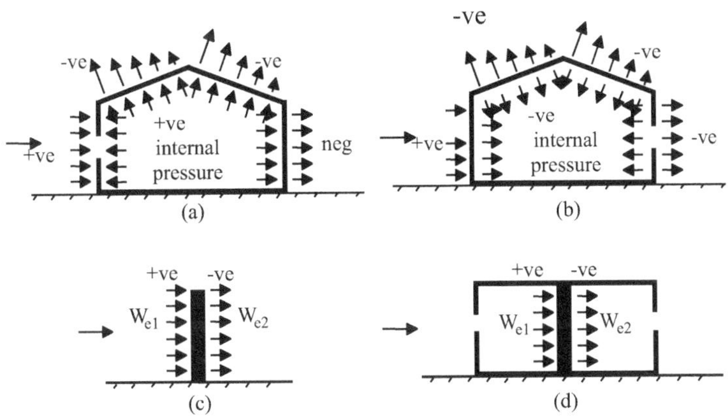

FIGURE 2.3 Pressure on surfaces of buildings.

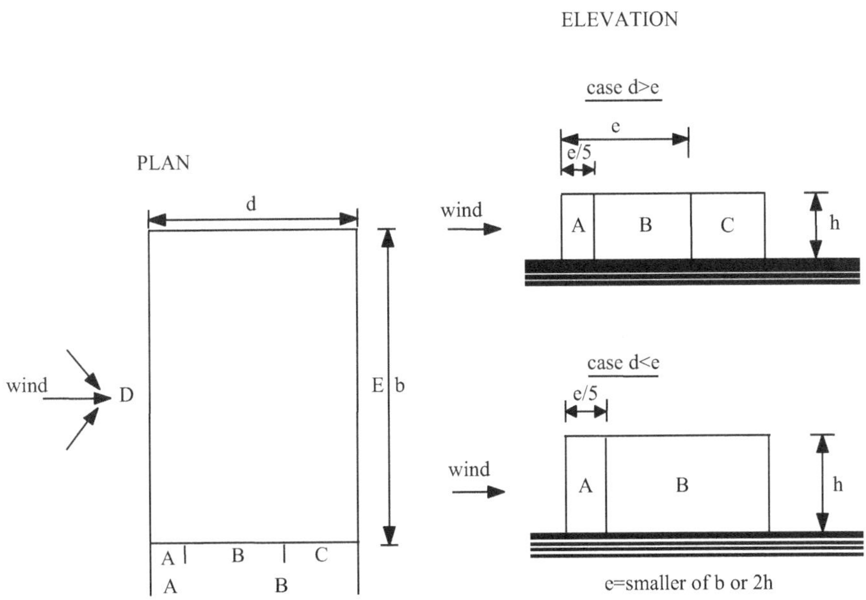

FIGURE 2.4 External pressure coefficients for vertical walls of rectangular plan buildings.

- monopitch roofs
- duopitch roofs
- hipped roofs

A typical chart is shown in Figure 2.4. and Table 2.8.

For buildings without internal partitions, internal pressure coefficients are related to the opening ratio, μ, which is defined as follows:

TABLE 2.8
External Pressure Coefficients for Vertical Walls of Rectangular Plan Buildings

Zone	A		B		C		D		E	
h/d	$C_{pe,10}$	$C_{pe,1}$	$C_{pe,10}$	$C_{pe,1}$	$C_{pe,10}$	$C_{pe,1}$	$C_{pe,10}$	$C_{pe,1}$	$C_{pe,10}$	$C_{pe,1}$
5	−1.2	−1.4	−0.8	−1.1	−0.5		+0.8	+1.0	−0.7	
1	−1.2	−1.4	−0.8	−1.1	−0.5		+0.8	+1.0	−0.5	
≤0.25	−1.2	−1.4	−0.8	−1.1	−0.5		+0.7	+1.0	−0.3	

μ = (Total area of openings at leeward and wind parallel sides)/(Total area of openings at windward, leeward, and wind parallel sides)

For a homogeneous distribution of openings in an approximately square building, c_{pi} may be taken as −0.25.

For closed buildings with internal partitions, the worst case may be assumed:

$$c_{pi} = 0.8, \text{ or } c_{pi} = -0.5$$

Special consideration is given to canopy roofs, buildings with more than one skin, solid boundary walls, fences and signboards, and special structures.

The process for determining wind loads on a building is summarized in Figure 2.5.

2.2.4 Determination of Snow Loads

Snow loads, s, are calculated based on a characteristic load, s_k, corresponding to a uniform depth of snow accumulated under calm conditions on flat ground. This value is modified to account for the shape of the roof and the effect of wind on the distribution of snow. (Note that the value of, s, may be reduced by exposure and thermal coefficients, which account for more severe wind conditions and reduced thermal insulation in the roof, but these are generally taken as 1.0). The characteristic values are based on snow maps for individual member states and relate s_k directly to geographical location.

The influence of roof geometry is accounted for by a snow load shape coefficient, μ, which is related to the roof pitch. Conventional pitched roofs, monopitch, multi-pitch, and cylindrical roofs are treated separately.

A typical chart providing values of the snow load shape coefficient in relation to roof slope is shown in Figure 2.6 and Table 2.9. Note that for monopitch roofs, two load cases should be considered: one in which the full snow load is applied across the entire roof, and a second in which half the snow load intensity is applied over the most unfavourable half of the roof. The second load case will rarely be critical.

Roofs with abrupt changes in height must account for the possibility of snow sliding from an upper level, and this is managed by a further shape coefficient.

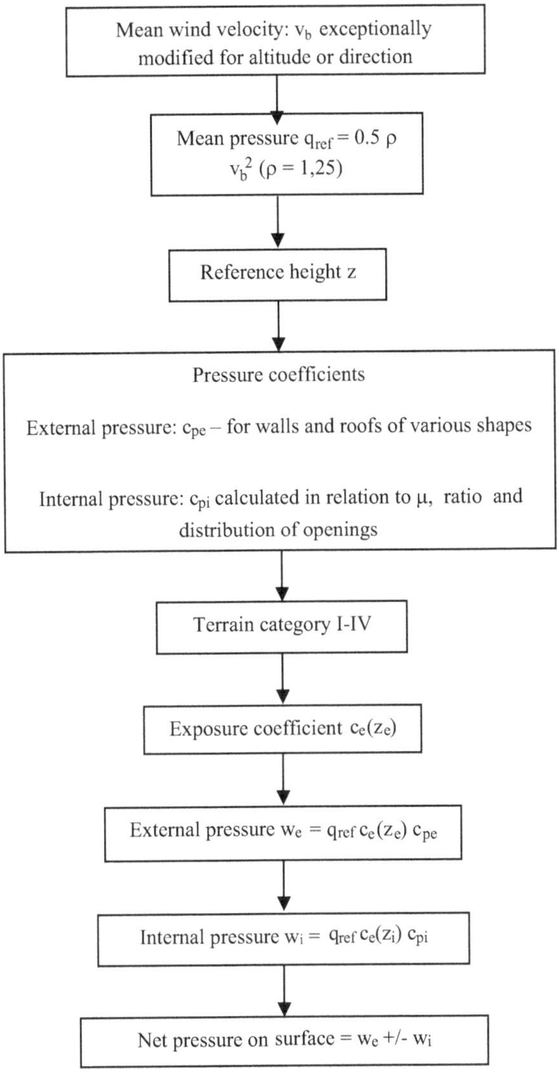

FIGURE 2.5 Flow chart for wind load calculation.

Rules are established to account for the effect of snow overhanging the edge of a roof. These apply to those parts of a roof that are cantilevered out beyond the walls, but the magnitude of the allowance is independent of the length of the overhang.

A flow chart for determining snow loads on roofs is shown in Figure 2.7.

2.2.5 Load Combination

The different loads discussed in the preceding subsections do not occur in isolation, but rather in combinations. Therefore, the designer must determine which

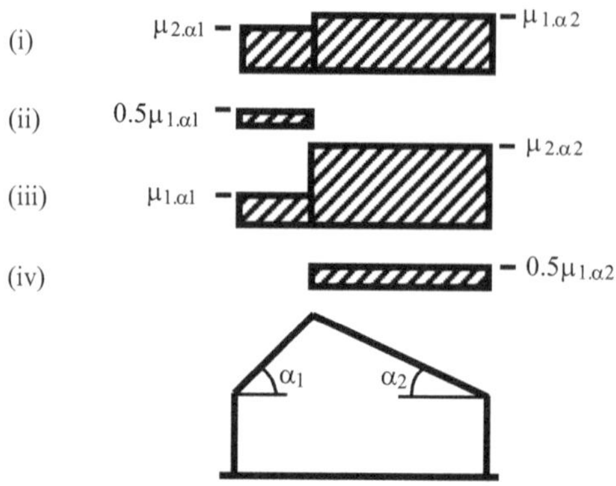

FIGURE 2.6 Snow load shape coefficient in relation to slope for duopitch roofs.

TABLE 2.9
Snow Load Shape Coefficients in Relation to Slope for Duo Pitch Roofs

Angle of pitch of roof	$0° \leq \alpha \leq 15°$	$15° \leq \alpha \leq 30°$	$30° \leq \alpha \leq 60°$	$\alpha \geq 60°$
Shape coefficient μ_1	0.8	0.8	$0.8(60-\alpha)/30$	0.0
Shape coefficient μ_2	0.8	$0.8+0.6(\alpha-15)/30$	$1.1(60-\alpha)/30$	0.0

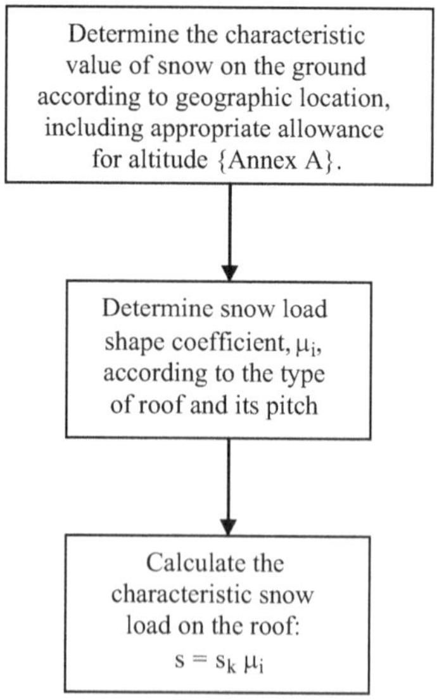

FIGURE 2.7 Flow chart for determining snow loads on roofs.

combination is most critical for the structure. If the individual loads, which have varying probabilities of occurrence and degrees of variability, were combined directly, the resulting load combination would exhibit a significantly reduced probability. Consequently, it is logical to reduce the magnitudes of the various components of a combination in accordance with their probabilities of occurrence. This is similar to the procedure used to reduce the imposed load intensities applied over large areas.

Past design practice involved using the worst combination of dead load with imposed load and/or wind load, allowing for increased stresses when the wind load was included (which is equivalent to reducing the load magnitudes).

These increases seem to be logical when imposed loads and wind loads act together because the probability that both will attain their maximum values simultaneously is greatly reduced. However, they are unjustified in the case of dead load and wind load, where the probability of occurrence remains virtually unchanged from that of the wind load.

A different and more rational approach to combining loads is employed in the EC limit states design method, which is based on statistical analyses of the loads and the structural capacities. Strength design is typically carried out for the most severe combination of actions under normal (termed persistent) or temporary (termed transient) conditions using the following equation, which is based on Equation 6.10 in EN1990

$$\sum_{j\geq 1} \gamma_{G,j} G_{k,j} + \gamma_{Q,l} Q_{k,l} + \sum_{i>1} \gamma_{Q,i} \psi_{0,i} Q_{k,i}$$

As per EN1990, ultimate limit states (other than fatigue) may be derived either from Equation 6.10 of EN 1990 or from Equations 6.10a and 6.10b. As stated by Gardner and Grubb (2008), the UK National Annex has permitted the use of either approach, although it should be noted that Equations 6.10a and 6.10b will yield more favourable combinations of actions (i.e. lower load factors). Furthermore, unless there is an unusually high ratio of dead load G_k to imposed load Q_k (e.g., $G_k > 4.5Q_k$ for the case of office floor loading, where $\psi 0 = 0.7$), only Equation 6.10b needs to be considered for strength (STR) verifications. For verifying equilibrium (e.g. assessing sliding or overturning as a rigid body), only Equation 6.10 may be applied. The load combination expressions, as they appear in the Eurocode, are provided below based on 6.10a and 6.10b:

$$\sum_{j\geq 1} \gamma_{G,j} G_{k,j} + \gamma_{Q,l} \psi_{0,i} Q_{k,l} + \sum_{i>1} \gamma_{Q,i} \psi_{0,i} Q_{k,i}$$

$$\sum_{j\geq 1} \gamma_{G,j} G_{k,j} + \gamma_{Q,l} Q_{k,l} + \sum_{i>1} \gamma_{Q,i} \psi_{0,i} Q_{k,i}$$

where ξ is the reduction factor for unfavourable permanent action G and is equal to 0.925 as per the UK annex.

As per the EC, the ultimate limit states categorize modes of failure of the structure into one or more categories, which are as follows:

a. EQU: Loss of static equilibrium of the structure or any part of it considered as a rigid body, where minor variations in the value or spatial distribution of actions from a single source are significant, and the strengths of construction materials or ground are generally not governing;
b. STR: Internal failure or excessive deformation of the structure or structural members, including footings, piles, basement walls, etc., where the strength of construction materials governs;
c. GEO: Failure or excessive deformation of the ground where the strengths of soil or rock are significant in providing resistance;
d. FAT: Fatigue failure of the structure or structural members.

The partial safety factors for design are presented in Tables 2.10 and 2.11.

Using the combination factors of $\psi_0=0.7$ and 0.5 for most variable actions and wind actions, as mentioned by Gardner and Grubb (2008), respectively, leads to the following common STR load combinations for buildings:

$1.35\ G+1.5\ Q_I+0.75\ Q_W$ (where the imposed load is leading in this case)
$1.35\ G+1.05\ Q_I+1.5\ Q_W$ (where the wind load is leading in this case)
$1.1\ G+1.5\ Q_I$, and
$1.1\ G+1.5\ Q_W$

TABLE 2.10
Design Values of Action (EQU and STR)

	Permanent Actions γ_G		Variable Actions γ_Q	
Ultimate Limit State	Unfavourable	Favourable	Leading Variable Action	Accompanying Variable Actions
EQU	1.1	0.9	1.5	$1.5\ \Psi_0$
STR/GEO	1.35	1.0	1.5	$0\ \Psi_0$

TABLE 2.11
Design Values of Action (STR) as per Equations 6.10a and b

	Permanent Actions γ_G		Variable Actions γ_Q	
Ultimate Limit State	Unfavourable	Favourable	Leading Variable Action	Accompanying Variable Actions
Equation 6.10a	1.35	1.0	1.5	$1.5\ \Psi_0$
Equation 6.10b	$\xi 1.35$	1.0	1.5	$1.5\ \Psi_0$

The last two equations check the structural equilibrium and its stability for overturning, sliding, or uplift.

For serviceability limit states, the partial factors for actions should be taken as 1.0 unless otherwise specified in EN 1991 to EN 1999.

Serviceability limit states in buildings should consider criteria related, for example, to floor stiffness, differential floor levels, storey sway, and building sway, as well as roof stiffness. Stiffness criteria may be expressed in terms of limits for vertical deflections and vibrations. Sway criteria may be expressed in terms of limits for horizontal displacements.

In summary, the serviceability criteria should be specified for each project and agreed upon with the client.

2.2.6 Solve the Example by EC

2.2.6.1 Examples

Calculate the load on the beam and column for four-storey office building. Consider a main floor beam spanning 10 m and supporting secondary beams that span 8 m.

2.2.6.1.1 I Dead Loads

The floor construction of a commercial office building consists of

Sand/cement screed thickness = 50 mm
Reinforced concrete slab thickness = 150 mm
Plaster from gypsum mortar for ceiling = 12 mm

The densities for each of these materials can be found in Table 2.1. These can be used to determine the dead load intensity (kN/m^2).

Screed weight = $0.05 \times 20 = 1.0$ kN/m^2
Slab weight = $0.15 \times 24 = 3.6$ kN/m^2
Plaster weight = $0.012 \times 15 = 0.18$ kN/m^2

Including allowances for partitions and services, the calculations may be set out as follows:

Partitions = 1.0 KN/m^2
Services = 0.3 KN/m^2
Total characteristic load = 6.08 kN/m^2

2.2.6.1.2 II-Imposed Loads

Consider the case of a commercial office building. The occupancy relates to category B (offices), and from Table 1.2, the corresponding imposed floor load intensity is 3.0 kN/m^2.

For beams and columns attracting load from wide areas, this value can be reduced.

$$A = 10.0 \times 8.0 = 80 \text{ m}^2$$

The corresponding reduction factor α is then given as follows:

$$a_A = 5y_o/7 + 10/A$$

For imposed floor loads (excluding storage areas), $\psi_o = 0.7$.

$$a_A = 5 \times 0.7/7 + 10/80 = 0.5 + 0.125 = 0.625$$

The imposed load on the beam can be reduced by this factor.

If the beam is subject to the dead load intensity specified in 7.1 above, the total beam load is as follows:

$$\text{Dead load} = 6.08 \times 8.0 \times 10.0 = 486.4 \text{ kN}$$

$$\text{Imposed load} = 0.625 \times 3.0 \times 8.0 \times 10.0 = 150 \text{ kN}$$

For typical ultimate limit state calculations, these would be combined using appropriate partial safety factors (1.35 for dead loads and 1.5 for imposed loads).

$$\text{Design load to beam (ULS)} = 1.35 \times 486.4 + 1.5 \times 150 = 881.64 \text{ kN}$$

For typical serviceability limit state calculations, these would be combined using appropriate partial safety factors (1.0 for dead loads and 1.0 for imposed loads).

$$\text{Design load to beam (SLS)} = 486.4 + 150 = 636.4 \text{ kN}$$

For columns

$$a_n = \{2 + (n-2)\psi_0\}/n$$

where n is the number of storeys supported.

Thus, for a column supporting four storeys,

$$a_n = \{2 + (4-2) \times 0.7\}/4 = 0.85$$

If the column supports two of the beams detailed in the previous example at each storey, the total dead and imposed column loads are as follows:

$$\text{Dead load} = 4 \times 486.4 = 1{,}945.6 \text{ kN}$$

$$\text{Imposed load} = 4 \times 0.85 \times 3.0 \times 8.0 \times 10.0 = 816 \text{ kN}$$

For typical ultimate limit state calculations, these would be combined using appropriate partial safety factors (1.35 for dead loads and 1.5 for imposed loads).

$$\begin{aligned}\text{Design load to column (ULS)} &= 1.35 \times 1{,}945.6 + 1.5 \times 816 \\ &= 2{,}626.56 + 1{,}224 \\ &= 3{,}850.56 \text{ kN}\end{aligned}$$

2.3 EXAMPLE FOR WIND LOADS

The wind affects a building located in a suburban area, 30 m high and with plan dimensions of 20 m × 20 m. The building is enclosed with internal partitions.

Mean wind velocity $v_b = 36$ m/s
Mean wind pressure $q_b = 0.5 \times 1.25 \times (36)^2/1{,}000 = 0.81$ kN/m^2

Reference height: consider the building in two zones, $z_e = 20$ m and $z_e = 30$ m.
Terrain category III ($k_t = 0.22$; $z_e(m) = 0.3$; $z_{\min}(m) = 8$; $\varepsilon = 0.37$).
Exposure coefficients:

$c_e(20) = 2.25$;
$c_e(30) = 2.55$

Pressure coefficients:

Internal pressure coefficient $c_{pi} = 0.8$ or -0.5

External pressure coefficients:

$h/d = 30/20 = 1.5$
e = smaller of b (= 20) and $2h$ (= 60), so $e = 20$
d = e, so consider wind pressure on the side walls of the building as uniform (i.e. pressures in zones B and C are all equal), except for the leading corner (zone A).

	A	B	D	E
C_{pe10}	−1.4	−0.8	+0.8	−0.6
C_{pe1}	−1.2	−1.1	+1.0	−0.6

External wind pressure:
Up to 20 m:

$$A: w_{e(20)} = 0.81 \times 2.25 \times (-1.4) = -2.55 \text{ kN/m}^2$$
$$B: w_{e(20)} = 0.81 \times 2.25 \times (-0.8) = -1.46 \text{ kN/m}^2$$
$$D: w_{e(20)} = 0.81 \times 2.25 \times (+0.8) = +1.46 \text{ kN/m}^2$$
$$E: w_{e(20)} = 0.81 \times 2.25 \times (-0.6) = -1.09 \text{ kN/m}^2$$

From 20 to 30 m:

$$A: w_{e(30)} = 0.81 \times 2.55 \times (-1.2) = -2.48 \text{ kN/m}^2$$
$$B: w_{e(30)} = 0.81 \times 2.55 \times (-0.8) = -1.65 \text{ kN/m}^2$$
$$D: w_{e(30)} = 0.81 \times 2.55 \times (+0.8) = +1.65 \text{ kN/m}^2$$
$$E: w_{e(30)} = 0.81 \times 2.55 \times (-0.6) = -1.24 \text{ kN/m}^2$$

Internal wind pressure:

Up to 20 m

Positive (pressure):

$$w_{i(20)} = 0.81 \times 2.25 \times (+0.8) = +1.46 \text{ kN/m}^2$$

$$\text{Negative(suction)}: w_{i(20)} = 0.81 \times 2.25 \times (-0.3) = -0.55 \text{ kN/m}^2$$

From 20 to 30 m:

$$\text{Positive(pressure)}: w_{i(30)} = 0.81 \times 2.55 \times (+0.8) = +1.65 \text{ kN/m}^2$$

$$\text{Negative(suction)}: w_{i(30)} = 0.81 \times 2.55 \times (-0.3) = -0.62 \text{ kN/m}^2$$

Net wind loads are calculated by combining internal and external wind pressures algebraically. It is important to maintain consistency regarding wind direction and the direction of internal pressure.

Note that for rectangular (not square) buildings, it is necessary to consider wind in two orthogonal directions.

2.4 LOAD COMBINATIONS

2.4.1 Span Continuous Beam (ULS for Failure of Structure)

For the consideration of the failure of structural elements, partial safety factors are derived from Table 2.12. In the case of a three-span continuous beam, several conditions must be considered (Figure 2.8). These correspond to different combinations of loading across the three spans. If these are designated W_1, W_2, and W_3, and the beams

TABLE 2.12
Live Load Member Factor (K_{LL})

Interior column	4
Exterior columns without cantilever slabs	4
Edge columns with cantilever slabs	3
Corner columns with cantilever slabs	2
Edge beams without cantilever slabs	2
Interior beams	2
All other members not identified, including the following:	1
Edge beam with cantilever slabs	
Cantilever beams	
One-way slabs	
Two-way slabs	
Members without provisions for continuous shear transfer normal to their span	

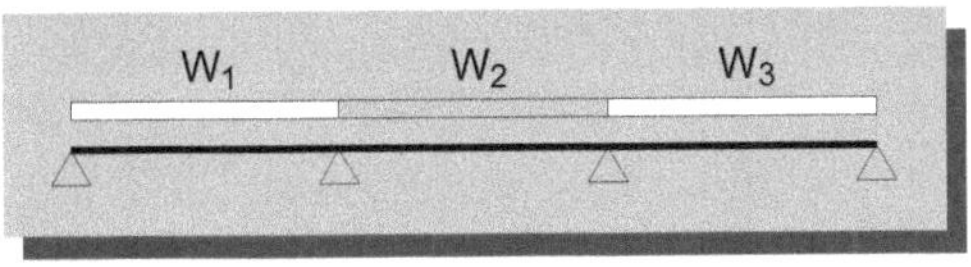

FIGURE 2.8 Example 2.

each span 10 m, are at 8 m centres, and are subject to the loadings calculated in sections 7.1-2 above, the load combinations to be considered are as follows:

1. Maximum bending in span 1:
   $W_1 = 1.35\ G_k + 1.5\ Q_k$;
   $W_2 = 1.35\ G_k$;
   $W_3 = 1.35\ G_k + 1.5\ Q_k$
   $W_1 = W_3 = 884$ kN
   $W_2 = 1.35 \times 488 = 659$ kN
2. Maximum bending in span 2:
   $W_1 = 1.35\ G_k$;
   $W_2 = 1.35\ G_k + 1.5\ Q_k$;
   $W_3 = 1.35\ G_k$
   $W_1 = W_3 = 659$ kN
   $W_2 = 884$ kN

For maximum loads in the supporting internal columns, the maximum load would be considered for all spans, i.e. $W_1 = W_2 = W_3 = 884$ kN.

2.5 EXAMPLE FOR STOREY BUILDING SINGLE-BAY WITH BALCONY (ULS – LOSS OF EQUILIBRIUM)

For the consideration of loss of equilibrium, partial safety factors are derived from Table 2.10 (Figure 2.9).

For a two-storey single-bay building with a cantilever, several conditions must be examined. These correspond to different combinations of dead, imposed, and wind loading, including varying patterns of vertical load on the span and cantilever at the first floor and roof. If the loads are designated W_{r1} and Wr_2 (for the roof span and cantilever), W_{f1} and W_{f2} (for the floor span and cantilever), and W_h is the horizontal load due to wind, the following load combinations should be considered:

$\psi_0 = 0.6$ (snow load); $\psi_0 = 0.6$ (wind loads); $\psi_0 = 0.7$ (floor imposed loads)
$W_{r2} = 1.35\ G_{kr} + 1.5\ Q_{kr}$
$W_{f1} = 1.0\ G_{kf}$
$W_{f2} = 1.35\ G_{kf}$

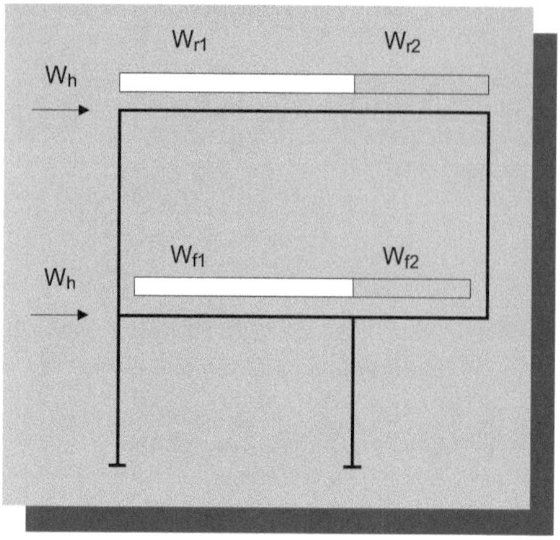

FIGURE 2.9 Example 3.

Considering imposed roof loads dominant,

$$
\begin{aligned}
W_{r1} &= 1.0\ G_{kr} \\
W_{r2} &= 1.35\ G_{kr} + 1.5\ Q_{kr} \\
W_{f1} &= 1.0\ G_{kf} \\
W_{f2} &= 1.35\ G_{kf} + \psi_0\ \text{(floor)}\ (1.50\ Q_{kf}) \\
&= 1.35\ G_{kf} + 0.7 \times 1.50\ Q_{kf} \\
&= 1.35\ G_{kf} + 1.05\ Q_{kf} \\
W_h &= \psi_0\ \text{(wind)}\ (1.5\ W_k) \\
&= 0.6 \times 1.5\ W_k \\
&= 0.9\ W_k
\end{aligned}
$$

Considering imposed floor loads dominant,

$$
\begin{aligned}
W_{r1} &= 1.0\ G_{kr} \\
W_{r2} &= 1.35\ G_{kr} + \psi_0\ \text{(snow)}\ (1.50\ Q_{kr}) \\
&= 1.35\ G_{kf} + 0.6 \times 1.50\ Q_{kr} \\
&= 1.35\ G_{kf} + 0.9\ Q_{kr} \\
W_{f1} &= 1.0\ G_{kf} \\
W_{f2} &= 1.35\ G_{kf} + 1.5\ Q_{kf} \\
W_h &= \psi_0\ \text{(wind)}\ (1.5\ W_k) \\
&= 0.6 \times 1.5\ W_k \\
&= 0.9\ W_k
\end{aligned}
$$

Considering wind loads dominant,

$$
\begin{aligned}
W_{r1} &= 1.0\ G_{kr} \\
W_{r2} &= 1.35\ G_{kr} + \psi_0\ \text{(snow)}\ (1.50\ Q_{kr})
\end{aligned}
$$

$$= 1.35\ G_{kf} + 0.6 \times 1.50\ Q_{kr}$$
$$= 1.35\ G_{kf} + 0.9\ Q_{kr}$$
$$W_{f1} = 1.0\ G_{kf}$$
$$W_{f2} = 1.35\ G_{kf} + \psi_0\ (\text{floor})\ (1.50\ Q_{kf})$$
$$= 1.35\ G_{kf} + 0.7 \times 1.50\ Q_{kf}$$
$$= 1.35\ G_{kf} + 1.05\ Q_{kf}$$
$$W_h = 1.5\ W_k$$

2.6 EXAMPLE: ULS COMBINATION OF ACTION ON A CANOPY

Example ULS combinations of actions for a shelter

The shelter is subjected to the following loads:

Self-weight	G_{k1}
Permanent imposed load	G_{k2}
Snow imposed load	Q_{k1}

1. Check for EQU – Static Equilibrium

   Factors to be taken for the verification of overturning are presented in Figure 2.10.
2. Check (STR) – Verification of resistance of a column

   The partial factor to be used for permanent loads in the verification of maximum compression stresses and of bending with axial force in the column is the same ($\gamma_G = 1.35$) for all spans.

   The variable imposed load is distributed over the full length of the canopy in the first case, and only on half of it for the verification of bending with axial force. The combinations are presented in Figures 2.10–2.12.

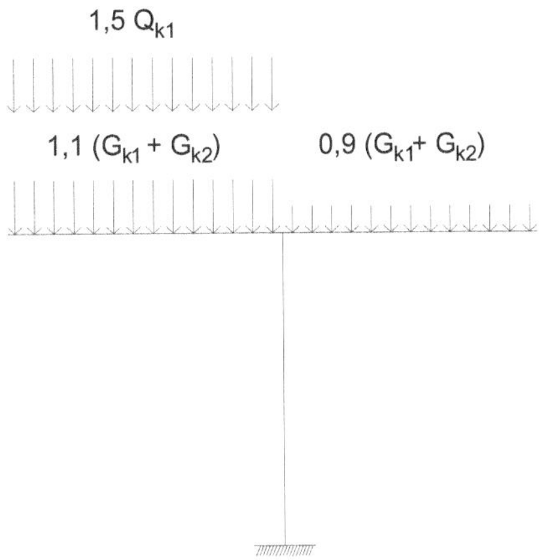

FIGURE 2.10 Load combination for verification of static equilibrium.

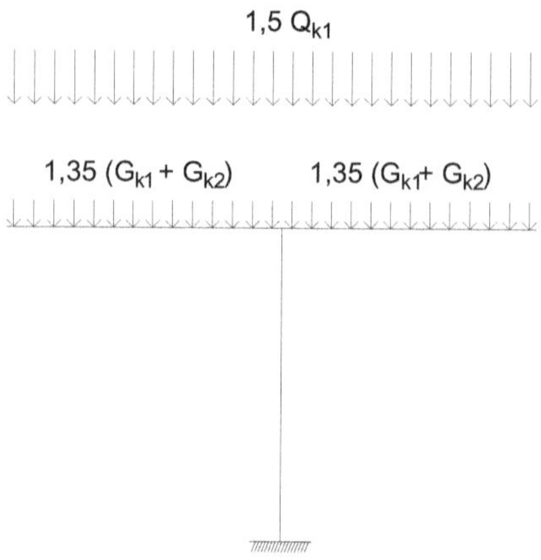

FIGURE 2.11 Load combination for the verification of compression stresses in the column.

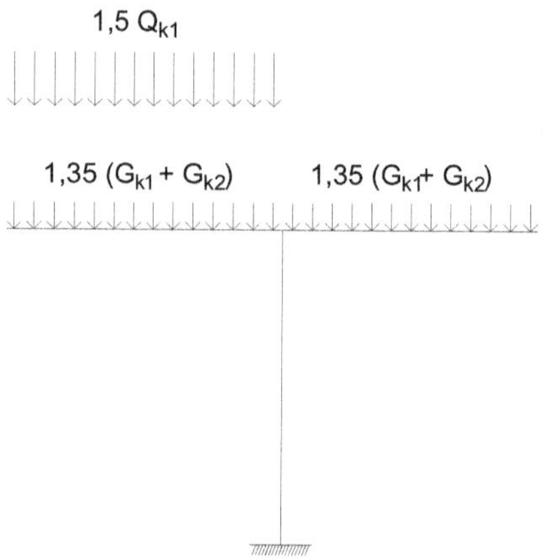

FIGURE 2.12 Load combination for the verification of bending with axial force in the column.

2.7 EXAMPLE FOR CONTINUOUS BEAM

Apply the load combination in the continuous beam from three spans as presented in Figures 2.13 and 2.14. The case of a frame structure in two storeys with wind action is presented in Figure 2.15.

1. Check (EQU) to verify stability
2. Check (STR) for the straining action

Basic combinations for the verification of the superstructure (STR) as per equation 6.10 in EN 1990

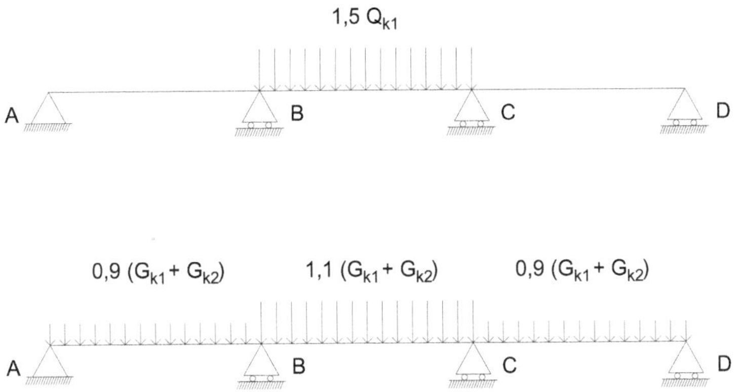

FIGURE 2.13 Load combination for the verification of holding down devices at the end bearings.

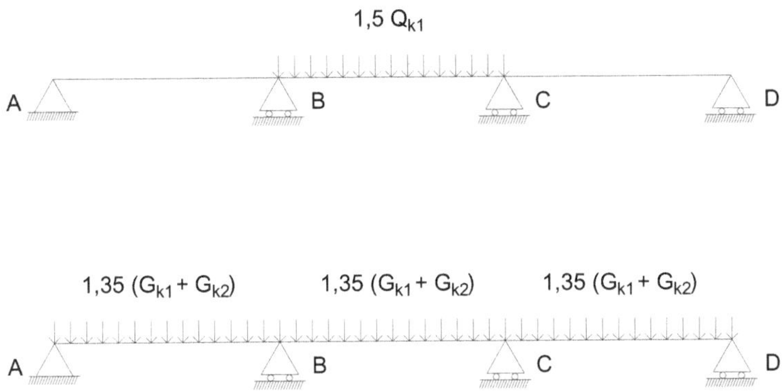

FIGURE 2.14 Load combination for the verification of bending moment in the BC span.

1. Favourable vertical loads:

$$G_k + 1.5\ F_{K,W}$$

2. Unfavourable vertical loads:

$$1.35G_k + 1.5\left(F_{K,W} + 0.5Q_{K,n} + 0.7Q_{k,\text{es}}\right) = 1.35G_k + 1.5F_{k,w} + 0.75Q_{k,n} + 1.05\ Q_{k,\text{es}}$$

Snow load is leading:

$$1.35Gk + 1.5\left(Q_{k,n} + 0.7Q_{K,es} + 0.6\ F_{K,W}\right) = 1.35G_k + 0.9F_{k,w} + 1.5Q_{k,n} + 1.05\ Q_{k,\text{es}}$$

Service load is leading:

$$1.35G_k + 1.5\left(Q_{k,\text{es}} + 0.5Q_{k,n} + 0.6F_{k,w}\right) = 1.35G_k + 1.5Q_{k,\text{es}} + 0.75Q_{k,n} + 0.9F_{k,w}$$

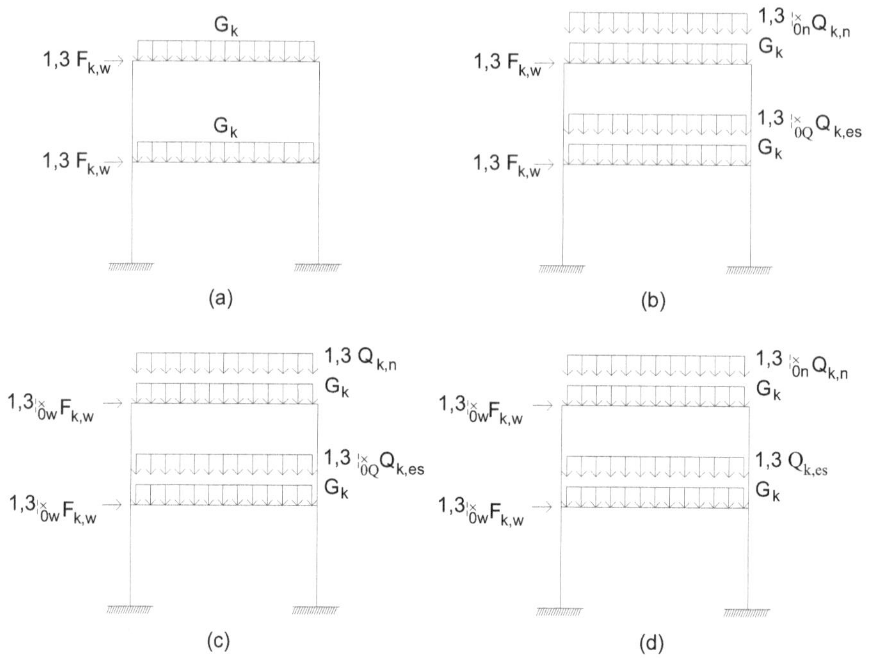

FIGURE 2.15 Basic combinations for the verification of the foundation: (a) wind predominant, favourable vertical loads; (b) wind predominant, unfavourable vertical loads; (c) snow load predominant; (d) service load predominant.

2.8 LOADS BASED ON AISC

The term "loads" has the same meaning in different codes and standards; however, it changes to "action" in European codes to be more generic.

2.8.1 Dead Load

The dead loads are the same as in the European code, as they depend on the geometry and the material density of the members.

2.8.2 Live Load

For live loads, the calculation is different. The following table presents the live loads as per ASCE7-16.

The reduction in uniform live loads as per ASCE is a function of the member types, in addition to the area. For which a value of $K_{LL}A_T$ of 37.16 m² (400 ft²) or more is permitted to be designed for a reduced live load in accordance with the following formula:

$$L = L_o\left(0.25 + \frac{4.57}{\sqrt{K_{LL}A_T}}\right)$$

where

L = reduced design live load per m^2 (ft^2) of area supported by the member
L_o = unreduced design live load per m^2 (ft^2) of area supported by the member as in Table 2.13
K_{LL} = live load element factor as per Table 2.12
A_T = tributary area in m^2 (ft^2).

TABLE 2.13
Minimum Uniformly Distributed Loads, L_o and Min. Concentrated Live Loads

Occupancy	L_o, kN/m^2 (psf)	Live Load Reduction Permitted	Multiple Storey Live Load Reduction Permitted	Concentrated kN (lb)
Apartments				
Access floor system				
• Office use	2.4(50)	√	√	8.9 (2)
• Computer use	4.79(100)	√	√	8.9(2)
Armories and drill rooms	7.18 (150)			
Assembly areas				
• Fixed seats	2.87(60)			
• Lobbies	4.79(100)			
• Movables seats	4.79(100)			
• Platform	4.79(100)			
• Stage floors	7.18(150)			
Corridor				
• First floor	4.79(100)			
Hospital				
• Operating rooms, laboratories	2.87(60)	√	√	4.45(1)
• Patient rooms	1.92(40)	√	√	4.45(1)
• Corridor above first floor	3.83 (80)	√	√	4.45(1)
Libraries				
• Reading rooms	2.87 (60)	√	√	4.45(1)
• Stack room	7.18 (150)	√	√	4.45(1)
• Corridor above first floor	3.83 (80)		√	4.45(1)
Office buildings				
• File and computers rooms shall be designed for heavier loads based on anticipated occupancy	4.79 (100)	√	√	8.90(2.0)
• Lobbied and first floor corridors	2.40 (50)	√	√	8.90(2.0)
• Offices	3.83 (80)	√	√	8.90(2.0)
• Corridors above first floor				

L shall not be less than 0.50 L_o for members supporting one floor, and L shall not be less than 0.40 L_o for members supporting two or more floors.

In the case of heavy live loads, live loads that exceed 4.79 kN/m^2 (100 lb/ft^2) shall not be reduced.

Reduction in roof live loads: The minimum uniformly distributed roof live loads, L_{oE}, in Table 4.3-1 as per ASCE7-16, for all types of structures, are detailed and summaries in Table 2.13.

Where uniform roof live loads are reduced to less than 0.96 kN/m^2 (20 lb/ft^2) and are applied to the design of structural members arranged to create continuity, the reduced roof live load shall be applied to adjacent spans or to alternate spans, whichever produces the greatest unfavourable load effect.

Structures of fabric construction supported by a skeleton structure are permitted to be designed for a reduced roof live load, as above or under other controlling combinations of loads, whichever produces the greater load effect. In structures such as greenhouses, where special scaffolding is used as a work surface for workers and materials during maintenance and repair operations, a lower roof load than that in Table 2.13 shall not be used unless approved by the Authority Having Jurisdiction. In such structures, the minimum roof live load shall be 0.58 kN/m^2 (12 psf).

$$L_r = L_o R_1 R_2 \text{ where } 0.58 \le L_r \le 0.96$$

where

- L_r = reduced roof live load per m^2 (ft^2) of horizontal projection supported by the member and
- L_o = unreduced design roof live load per m^2 (ft^2) of horizontal projection supported by the member (see Table 4.3-1).

The reduction factors R_1 and R_2 shall be determined as follows:

$$\begin{aligned} R_1 &= 1\left(A_T \le 18.58 \text{ m}^2\right) \\ &= 1.2 - 0.011\, A_T (18.58 \text{ m}^2) < A_T < 55.74 \text{ m}^2 \\ &= 0.6\left(A_T \ge 55.74 \text{ m}^2\right) \end{aligned}$$

where A_T is the tributary area in ft^2 (m^2) supported by the member and

$$\begin{aligned} R_2 &= 1 \ (F \le 4) \\ &= 1.2 - 0.05F \ (4 < F < 12) \\ &= 0.6 \ (F \ge 12) \end{aligned}$$

where for a pitched roof, F is the number of inches of rise per foot (in SI: $F = 0.12 \times$ slope, with slope expressed in percentage points) and, for an arch or dome, F is the rise-to-span ratio multiplied by 32.

2.8.3 Wind Load

The concept of wind load in ASCE is the same as in Ec-1, serving as the theoretical background.

$$q_z = 0.613 K_z K_{zt} K_d K e V^2 \text{kN/m}^2;\ V \text{ in m/s as per ASCE } (26.10\text{-}1.\text{si})$$

where

K_z = velocity pressure exposure coefficient, as in Table 2.14.
K_{zt} = topographic factor, clearly classified in ASCE, see Section 26.8.2. This factor depends on the topography around the building; if there are no hills, ridges, or escarpments nearby, this factor equal to 1.0.
K_d = wind directionality factor, as in Table 2.15.
K_e = ground elevation factor, as in Table 2.16.
V = basic wind speed, corresponds to a 3-second gust speed at 10 m (33 ft) above the ground in Exposure Category C.
q_z = velocity pressure at height z.

TABLE 2.14
Velocity Pressure Exposure Coefficients, K_h and K_z

Height above Ground Level, z_m (ft)	Exposure		
	B	C	D
0–4.6(0–15)	0.57 (0.70)	0.85	1.03
6.1 (20)	0.62 (0.70)	0.90	1.08
7.6 (25)	0.66 (0.70)	0.94	1.12
9.1 (30)	0.70	0.98	1.16
12.2(40)	0.76	1.04	1.22
15.2(50)	0.81	1.09	1.27
18.0 (60)	0.85	1.13	1.31
21.3 (70)	0.89	1.17	1.34
24.4 (80)	0.93	1.21	1.38
27.4 (90)	0.96	1.24	1.40
30.5 (100)	0.99	1.26	1.43
36.6 (120)	1.04	1.31	1.48
42.7 (140)	1.09	1.36	1.52
48.8 (160)	1.13	1.39	1.55
54.9 (180)	1.17	1.43	1.58
61 (200)	1.20	1.46	1.61
76.2 (250)	1.28	1.53	1.68
91.4 (300)	1.35	1.59	1.73
106.7 (350)	1.41	1.64	1.78
121.9 (400)	1.47	1.69	1.82
137.2 (450)	1.52	1.73	1.86
152.4 (500)	1.56	1.77	1.89

TABLE 2.15
Factor, K_d for Wind Directionally

Structure Type	K_d
• Buildings for main wind force resisting system	0.85
• Building component and cladding	
• Arched roof	
• Solid free standing walls, roof top	
• Equipment and solid free standing and attached signs	
Open signs and single-plane open frame	
Circular domes	1.0
Chimneys, tanks, and similar structures	
Square	0.90
Hexagonal	0.95
Octagonal	1.0
Round	1.0
Octagonal with non-axisymmetric structure system	0.95
Round with non-axisymmetric structure system	0.95
Trussed tower	
• Triangular, square, or rectangular	0.85
• All other cross sections	0.95

TABLE 2.16
Ground Elevation Factor, K_e

Ground Elevation above Sea Level, *m* (ft)	Ground Elevation Factor
0	1
305 (1,000)	0.96
610 (2,000)	0.93
914 (3,000)	0.90
1,219 (4,000)	0.86
1,524 (5,000)	0.83
1,829 (6,000)	0.80

The exposure categories presented in ASCE as follows:

Exposure Categories

Exposure B: For buildings or other structures with a mean roof height less than or equal to 9.1 m (30 ft), Exposure B shall apply where the ground surface roughness, as defined by Surface Roughness B prevails in the upwind direction for a distance greater than 457 m (1,500 ft). For buildings or other structures with a mean roof height greater than 9.1 m (30 ft) Exposure B shall apply where Surface Roughness, B prevails in the upwind direction for a distance greater than 2,600 ft (792 m) or 20 times the height of the building or structure, whichever distance is greater.

Exposure C: This shall apply in all cases where Exposure B or D does not apply.

Exposure D: This shall apply where the ground surface roughness, as defined by Surface Roughness D, prevails in the upwind direction for a distance greater than 1,524 m (5,000 ft) or 20 times the building or structure height, whichever is greater. Exposure D shall also apply where the ground surface roughness immediately upwind of the site is B or C, and the site is within a distance of 183 m (600 ft) or 20 times the building or structure height, whichever is greater, from an Exposure D condition as defined in the previous sentence.

As per ASCE recommendations, in cases where a site is located in the transition zone between exposure categories, the category resulting in the largest wind forces shall be used.

Enclosed and Partially Enclosed Rigid and Flexible Buildings. Design wind pressures for the MWFRS of buildings of all heights in (lb/ft²) kN/m², shall be determined by the following equation:

$$p = q\mathrm{GC}_p - q_i\left(\mathrm{GC}_{pi}\right)$$

The value of K_z can be taken as 0.7 for low-rise buildings with a height less than 9.1 m (30 ft).

As a conservative approach, K_e can be taken as 1.0.

$q = q_z$ for windward walls evaluated at height z above the ground.

$q = q_h$ for leeward walls, side walls, and roofs evaluated at height h.

$q_i = q_h$ for windward walls, side walls, leeward walls, and roofs of enclosed buildings, and for negative internal pressure evaluation in partially enclosed buildings.

$q_i = q_z$ for positive internal pressure evaluation in partially enclosed buildings, where height z is defined as the level of the highest opening in the building that could affect the positive internal pressure. For buildings sited in wind-borne debris regions, glazing that is not impact-resistant or protected with an impact-resistant covering shall be treated as an opening in accordance with ASCE Section 26.12.3. For positive internal pressure evaluation, q_i may conservatively be evaluated at height.

G = gust effect factor; for a rigid building or other structure, this is permitted to be taken as 0.85.

G_f determined in accordance with ASCE Section 26.11.5 shall be substituted for G.

C_p = external pressure coefficient

The value of C_p for walls and roofs is given in Table 2.17.

TABLE 2.17
C_p Coefficient for Walls

Surface	L/B	C_p
Windward	All values	0.8
	0–1	−0.5
Leeward wall	2	−0.3
	≥4	−0.2
Side wall	All values	−0.7

TABLE 2.18
GCpi Values as per Building Opening Classification

1-Enclosed buildings	+0.18
	−0.18
2-Partially enclosed buildings	+0.55
	−0.55
3-Partially open buildings	+0.18
	−0.18
4-Open buildings	0.0

C_P for the roof depends on the roof inclination angle. In the case of a horizontal roof, most of the wind load is suction. Different C_p values for varying angles are presented in Figure 27.3-1 in ASCE7-16.

The internal pressure coefficient (GCpi) is determined by certain enclosed criteria as follows:

1. Enclosed building $A_o < 0.01\ A_g$ or 0.37 m² (4ft²) and $A_{oi}/A_{gi} \leq 0.2$
2. Partially enclosed building: if $A_o > 1.1\ A_{oi}$ and A_o >the lesser of 0.01 A_g or 0.37 m² (4ft²) and $A_{oi}/A_{gi} \leq 0.2$
3. Partially open building: any building that does not qualify for criteria 1 or 24
4. Open buildings: if each wall is at least 80% open

where A_o is the total area of the openings and A_g is the gross area of the wall. A_{oi} and A_{gi} refer to the respective sums of the areas of the building envelope (walls and roof) for openings and gross values (Table 2.18).

2.8.4 Load Combination as per ASCE

The load combination as per ASCE7-16 in case of LRFD design will be as follows:

1. 1.4D
2. 1.2D + 1.6L + 0.5 (L_r or S or R)

3. $1.2D + 1.6\,(L_r \text{ or } S \text{ or } R) + (1.0L \text{ or } 0.5W)$
4. $1.2D + 1.0W + 1.0L + 0.5\,(L_r \text{ or } S \text{ or } R)$
5. $0.9D + 1.0W$
6. $1.2D + E_v + E_h + L + 0.2S$
7. $0.9D - E_v + E_h$
8. $1.2D + 1.2T + 0.5L$
9. $1.2D + 1.6L + 1.0T$

where

D is the dead load, *L* is the live load, L_r is the roof live load, *S* is the snow load, *R* is the rain load, *W* is the wind load, and E_v and E_h indicate the effects of earthquakes in vertical and horizontal directions, respectively. *T* generally denotes the cumulative effect of self-straining forces and effects arising from contraction or expansion due to environmental or operational temperature changes, shrinkage, moisture changes, creep in component materials, movement caused by differential settlement, or combinations thereof.

2.9 PROBABILITY-BASED LIMIT STATES DESIGN

European codes are based on limit state design. The AISC adopts two approaches: allowable strength design and load resistance factor design (LRFD), both of which align with limit state design principles.

Working stress design, or allowable stress design, has been used since before the mid-1980s for concrete and steel structures, although work began on concrete structures earlier than on steel structures. To this day, some codes still use allowable stress design for steel structures.

The working stress methods defined in previous codes and specifications required that the stresses calculated from the most adverse combination of loads must not exceed the specified permissible stresses. These specified stresses were derived after accounting for non-linear stability and material effects on the strength of isolated members. In effect, they were expressions of ultimate strengths divided by safety factors (SF). Thus, it can be expressed as Working stress $\leq$ Permissible stress $\approx$ Ultimate stress/SF.

As per Trahair et al. (2008), it was traditional to use safety factors of approximately 1.7.

In fact, loads calculated in specifications are typically represented as a single value; however, the actual calculation is a probabilistic distribution. When the member resistance capacity is calculated using code equations, it provides one value. Thus, our calculation is classified as deterministic. Conversely, if every parameter is treated as a variable represented by a probability distribution, the member resistance capacity will be a probabilistic distribution, and the calculation is termed a probabilistic calculation.

From Figure 2.16, it can be observed that the probability of structural failure occurs in the intersection zone between the two curves: load and resistance. Structural failure may manifest as partial collapse, total collapse, or member failure. This failure

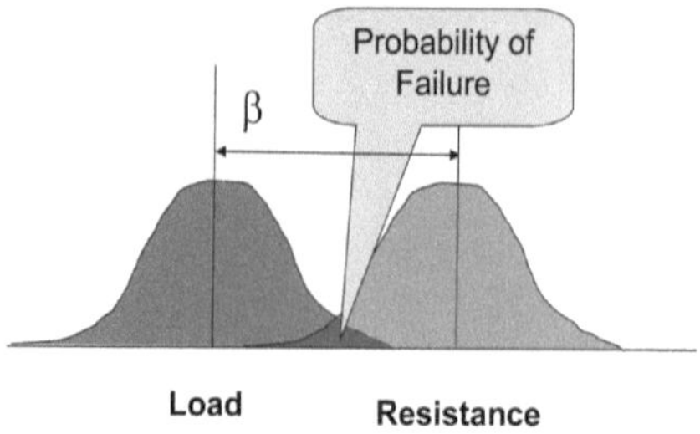

FIGURE 2.16 Probability of failure diagram.

occurs when the load value is at its highest while the strength is at its lowest, which can be attributed to deterioration, poor quality control, or incorrect design.

From a practical perspective, the load may increase due to a change in operation mode, or multiple loads can accumulate simultaneously, such as when all furniture is gathered in one room for painting, or during parties.

Strength is often compromised due to inadequate quality control during construction or concrete deterioration from steel bar corrosion, spalling of the concrete cover, or cracks in structural concrete members of older buildings.

The general limit states design philosophy (has been adopted in the development of probability-based design criteria. A limit state is a condition in which a structure or structural element becomes unfit for its intended purpose. For most structures, limit states can be categorized into two groups:

1. Ultimate limit states, which relate to structural failure or collapse of part or all of the structure, potentially leading to loss of life or significant financial loss, and should occur with a low probability.
2. Serviceability limit states, which pertain to disruptions in the functional use of the structure or to damage or deterioration of the structure.

Limit states design is a process that involves: (a) identifying all potential ways in which the structure might fail to fulfil its intended purpose; (b) determining acceptable levels of safety for each limit state; and (c) the designer's consideration of each significant limit state.

Probability-based limit states design customarily employs loads or load effects that are multiplied by load factors and resistances, which are further multiplied by resistance factors in a set of checking equations of the general form:

$$\text{Factored resistance} \geq \text{effects of factored load}$$

The conceptual framework for probability-based design in this study is supported by classical reliability theory (Ang and Cornell, 1974).

As per Elreedy (2013), in the development of limit state codes, the probability of failure pF is usually related to a parameter β, called the safety index, by the transformation:

$$\Phi(-\beta) = pF,$$

where the function Φ is the cumulative frequency distribution of a standard normal variate. The relationship between β and pF, shown in Figure 2.16, indicates that an increase in β of 0.5 corresponds to a decrease in the probability of failure by approximately an order of magnitude. The concept of the safety index was employed to derive the partial factors for EC3, as presented in the sections on strength and actions discussed in this chapter.

REFERENCES

Allen, D.E. (1976). Limit States Design: A Probabilistic Study, *Canadian Journal of Civil Engineering*, Vol. 2, No. 1, pp. 36–49.

Ang, A.H. and Cornell, C.A. (1974). Reliability Based Structural Safety and Design, *Journal of the Structural Division*, Vol. 100, No. 9, pp. 1755–1769.

Elreedy, M.A. (2013). *Reinforced Concrete Structural Reliability*, CRC Press, Boca Raton, FL.

El-Reedy, M.A., Ahmed, M.A., and Khalil, A.B. (2000). *Reliability Analysis of Reinforced Concrete Columns*. Ph.D. thesis, Cairo University, Giza, Egypt.

Gardner, L. and Grubb, P.J. (2008). *Guide to Eurocode Load Combinations for Steel Structures*, British Constructional Steelwork Association.,UK

Trahair, N.S., Bradford, M.A., Nethercot, D.A., and Gardner, L. (2008). *The Behaviour and Desig of Steel Structures to EC3*, 4th edition, Taylor Francis, London.

FURTHER READING

ASCE (2016). *Minimum Design Loads for Buildings and Other Structures*, ASCE/SEI 7-16, American Society of Civil Engineers, Reston, VA.

BSI (2005). *Eurocode 1: Actions on Structures: Part 1.4* General Actions – *Wind Actions*, BS EN 1991-1-4, British Standards Institution, London.

BSI (2005). General Actions – *Densities, Self-Weight, Imposed Loads for Buildings*, BS EN 1991-1-1, British Standards Institution, London.

Lui, E.M. (1999). Structural Steel Design, in *Structural Engineering Handbook*, Chen Wai-Fah, ed., CRC Press LLC, Boca Raton, FL.

Process Industry Practices (n.d.). *Recommended Practices for Machinery Installation and Installation Design*, PIP REIE 686.

3 Tension Member

3.1 INTRODUCTION

Tension members are structural elements subjected to pure tensile forces. The selection of their cross section is one of the simplest and most straightforward problems encountered in steel design. Since stability is of minor concern with tension members, the problem is reduced to selecting a section with sufficient area to carry the design load without exceeding the allowable tensile stress as determined by the factor of safety.

Once the magnitude of the force to be resisted has been established and the strength of the material ascertained, the required cross-sectional area can be readily calculated. However, the connection of tension members is, as with other member types, a very important consideration that, in many cases, may govern the design of the member and is therefore a major criterion in the design and selection of a section.

Generally, tension members are designed using rolled sections (angles, I, H, channels), bars, or flats.

A structural member is considered a compression member if it is under a compressive load, either alone or with other loadings. This category includes columns, beam columns, plates, component parts of frames, and, for instance, the compression flange of beams or plate girders.

The two main differences between tension and compression members are as follows:

- Tension members are held straight by means of tensile loads, whereas, in the case of compression members, the compressive loads tend to bend the member out of the plane of loading.
- For riveted and bolted connections, the net area governs the strength of a tension member, while for compression members, the rivets and holes are assumed to fill the holes.

For tension members, it is well known that stability phenomena are not a criterion in the design; however, it is necessary to prevent sagging. If a tension member is too long or used to support vibrating equipment, the slenderness ratio ($\lambda = L/i$), where L is the tension member length and i is the radius of gyration, should not be higher than 300.

It is worth mentioning that the radius of gyration is denoted as r in AISC and as i in European codes; in this book, we will present it as (i).

3.2 SLENDERNESS RATIO

For tension members, it is well known that stability phenomena do not govern the design. However, to prevent sagging in tension members that are too long and are used to support vibrating equipment such as fans or compressors, it is necessary to

DOI: 10.1201/9781003307303-3

establish a stiffness criterion. This criterion is based on the slenderness ratio of a member, which is obtained from the following equation:

$$\lambda = L_b / i$$

where

L_b = the buckling length
i = the radius of gyration corresponding to the member's effective buckling length (L_b)

The approximate values for the radius of gyration for sections composed of angles are provided in Table 3.1.

TABLE 3.1
Approximate Value for *i*, for Section Composed of Angles

Case	Section of Members	i_z	i_y
1	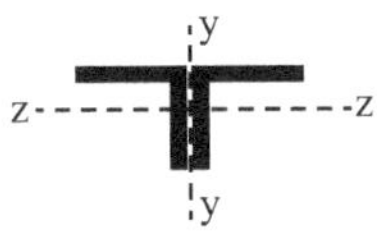	0.3a	
2	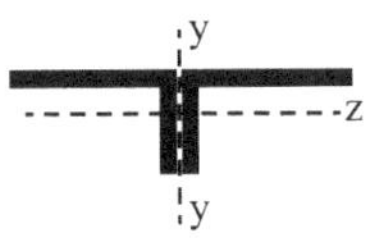 *a:b* = *1.5:1*	0.28a	0.48a
3	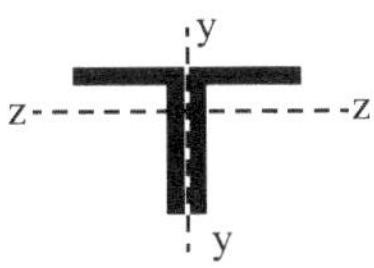 *a:b* = *1.5:1*		0.3a
4	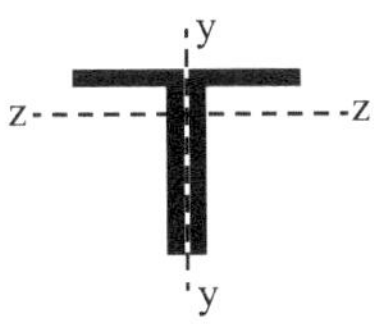 *a:b* = *2:1*		0.3a
5		0.2a	

(*Continued*)

TABLE 3.1 (*Continued*)
Approximate Value for *i*, for Section Composed of Angles

Case	Section of Members	i_z	i_y
6	 *a:b* = *1.5:1*	0.14a	
7	 *a:b* = *2:1*	0.1a	
8	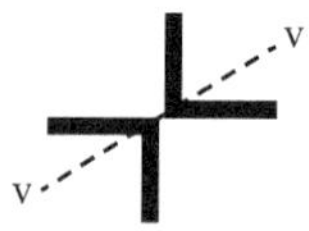	0.385	

Here, *a* is the angle length and is the bigger length for unequal angle and *b* is the smallest length for the unequal angle.

3.3 CONNECTIONS IN TENSION MEMBER

It is usual to assume that the distribution of tensile stresses in a tension member is uniform. Connection details may affect the validity of this assumption in two ways. Firstly, if bolting is used, the cross-sectional area is reduced at bolt holes, and the stresses around the holes are increased locally, as shown in Figure 3.1. Secondly, some eccentricity in connections is often inevitable, resulting in induced secondary moments. These problems may be accounted for by using an effective net area rather than the gross area in the calculation of the design plastic resistance.

3.3.1 Plastic Stress Distribution

The plastic stress can be achieved, as shown in Figure 3.2 As long as the peak stress is smaller than the yield stress F_y, the stress will follow curve 1. As the load increases, f_{max} becomes equal to F_y, as shown in curve 2. When the load is increased again, the metal immediately adjacent to the hole yields at a constant stress, similar to that of a tension test specimen.

However, the stress in the portion further removed from the hole will continue to increase, causing the distribution of stress to change successively to that shown in curves 3 and 4, as presented in Figure 3.2, with an increasingly larger portion of the metal yielding.

As long as some of the material remains stressed below the yield point, the member will not exhibit any sudden elongation at a constant load. In the end, with a further increase in *P*, yielding spreads over the entire net section, as indicated by curve 5, which presents uniform stresses equal to the yield for both sides around the hole.

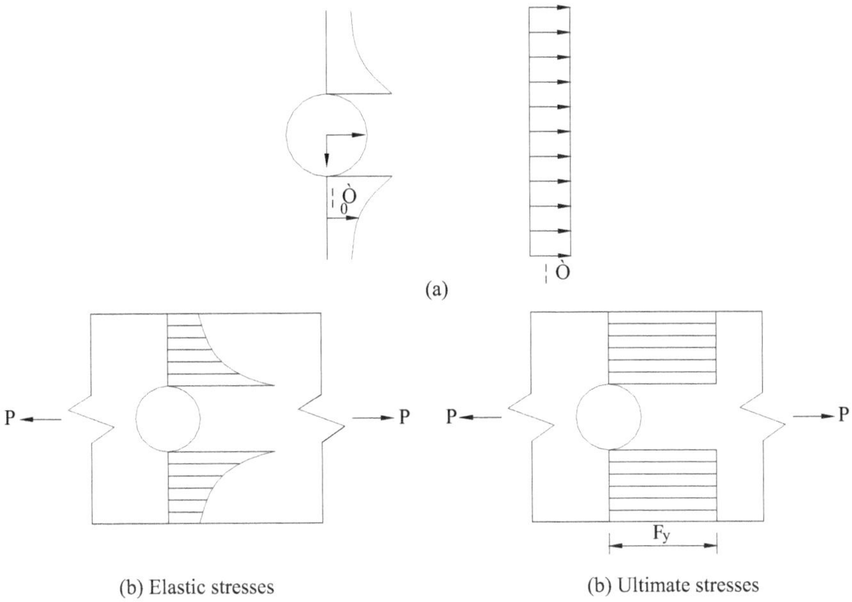

FIGURE 3.1 Distribution of stress across a section with holes.

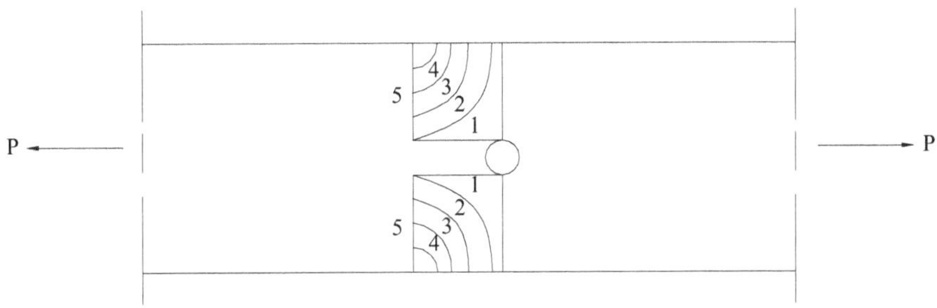

FIGURE 3.2 Plastic stress distribution.

In this case, permanent damage to the member occurs through excessive deformation, and the stress concentration disappears altogether. The stress becomes uniform over the net section, and the load at which general yielding occurs is given by $P_y = F_y.A_{net}$.

This is the plastic stress redistribution that accounts for the fact that a hole weakens a steel tension member only in the ratio A_{net}/A_{gross}, regardless of the degree of initial elastic stress concentration. This serves as one example of the fundamental importance of the ductility of steel structures.

If the member was made from brittle materials such as cast iron or glass, it would have failed when the peak stress f_{max} reached the tensile stress, which for such material is practically equal to the proportion limit, without the benefit of plastic redistribution.

3.3.2 Effect of Dynamic Loading

The above is true only for static loading, where the loads are applied gradually and remain substantially constant throughout the structure's lifetime. In contrast, for structures such as bridges, cranes, transmission, and antenna towers, the stresses change greatly and frequently due to the nature of the loads, such as moving loads on bridges and cranes or wind loads on towers. Should such structures fail, they may do so due to fatigue caused by repetitive loading rather than yield under excessive static loading.

In this case, plastic stress redistribution contributes little, and the strength of a structure is significantly reduced by the presence of stress raisers. It follows that when loads are such that the possibility of fatigue exists, structures should be designed to reduce stress concentrations where possible; in addition, allowable stresses must be reduced to accommodate the detrimental influence of the remaining stress raisers.

While structural steel is ductile in most normal situations, it can become brittle under certain conditions, such as low temperatures, high loading speeds, unfavourable chemical compositions, and unusual states of stress. It fails in the same sudden and complete manner as cast iron or glass. This very undesirable eventuality is known as brittle facture. Under such circumstances, plastic stress redistribution does not develop, and stress concentrations significantly contribute to the weakness of the member.

3.4 RESISTANCE OF THE CROSS SECTION

The European code is more sophisticated than the AISC; thus, an attempt is made to present EC3.

For members not connected by bolting, the design tensile resistance of the cross section is defined as the design plastic resistance of the gross cross section.

$$N_{pl.Rd} = \frac{Af_y}{\gamma_{M0}} \tag{3.1}$$

where

- A is the gross area of the cross section
- f_y is the yield strength of the steel
- γ_{M0} is the partial safety factor for steel.

For members connected by bolting, the section resistance is weakened by the reduction of the cross-sectional area due to the presence of holes, and an additional check is required for the stress concentrations induced by the holes (as seen in Figure 3.1). The ductility of steel is necessary. Although the assumption is that at the ultimate limit state, the stress distribution across the net section is uniform, the design ultimate resistance of the net section is calculated as follows:

$$N_{u.Rd} = 0.9\frac{A_{net}f_u}{\gamma_{M2}} \tag{3.2}$$

where

A_{net} is the net area of the cross section
f_u is the ultimate tensile strength of steel
γ_{M2} is the partial safety factor for the resistance of the net section, which will likely be a different value from that for γ_{M0}

In equation 3.2, the 0.9 factor is a reduction coefficient that accounts for unavoidable eccentricities, stress concentrations, and other stresses that cannot be calculated. The design tension resistance ($N_{t.Rd}$) is thus taken as the lesser value of equations 3.1 and 3.2 and compared with the design value of the applied tensile force (N_{sd}).

The main question is "Why does the code allow the yield strength to be exceeded at the net section by using f_u rather than f_y in equation 3.2?"

It is implicitly assumed that the failure of a tension member can be described by its deformation. Consider, for example, a tension member in which the connection accounts for 5% of the total length. If the strain in the connection region at the ultimate load – which causes yielding in the gross section remote from the connection – is, say, ten times the yield strain, then the increase in length along the member would be in the connection:

$$\Delta l_c = \frac{5}{100} 10\varepsilon_y L_{tot} = 0.5\varepsilon_y L_{tot}$$

In the remainders of the member,

$$\Delta l_m = \frac{95}{100} \varepsilon_y L_{tot} = 0.95\varepsilon_y L_{tot}$$

Hence, the elongation within the connection zone is much smaller than the elongation of the whole member. Thus, "failure" will have occurred due to excessive deformation of the complete member, provided, of course, that the resistance of the net section to fracture ($N_{u.Rd}$) is not less than the plastic resistance of the gross cross section ($N_{pl.Rd}$).

However, there are two cases where it is deemed necessary to limit the stress at the net section. The first concerns connections designed to be slip resistant at the ultimate limit state. For these connections, the design plastic resistance of the net section ($N_{net.Rd}$) is limited to the yield strength at the net section:

$$N_{net.Rd} = \frac{A_{net} f_y}{\gamma_{M0}}$$

$$N_{net.Rd} = \frac{A_{net} f_y}{\gamma_{M0}}$$

The second case concerns members in which ductile behaviour is required (for seismic design, for example). In such cases, it is necessary to ensure that the limiting condition is the yield of the gross section, not failure at the net section. Thus,

$$N_{uRd} \geq Npl.Rd \tag{3.3}$$

This will be satisfied if,

$$\frac{A_{net}}{A} \geq \frac{\left[f_y/f_u\right]\left[\gamma_{M2}/\gamma_{MO}\right]}{0,9} \tag{3.4}$$

For S275 and S355 (less than 40 mm thick) and using the EC3 boxed values for γ_{M2} and γ_{MO}, the minimum values of A_{net}/A for ductile behaviour are 0.81 and 0.88, respectively.

If the actual ratio falls below this value, the details of the connection or the size of the cross section will need to be amended accordingly.

For members where ductile behaviour is not specifically required, no allowance for bolt holes is necessary unless the A_{net}/A ratio falls below 0.81 for S275 or 0.88 for S355, since above these values the design tension resistance will always be controlled by the plastic resistance of the gross cross section, as in equation 3.1.

3.5 DETERMINATION OF THE NET AREA

The net area of the cross section is the gross area less deductions for bolt holes and other openings.

For each fastener hole, the deduction is the gross cross-sectional area of the hole(Figure 3.3). When fasteners are not staggered, the total area to be deducted from any cross section perpendicular to the member axis is the maximum sum of the sectional areas of the holes. When fasteners are staggered, the total area to be deducted is the greater of the area for holes crossing a perpendicular cross section or the sum of the areas of all holes in any diagonal or zig-zag line across the member, less $s^2t/4p$ for each gauge space in the chain of holes, as shown in Figure 3.3.

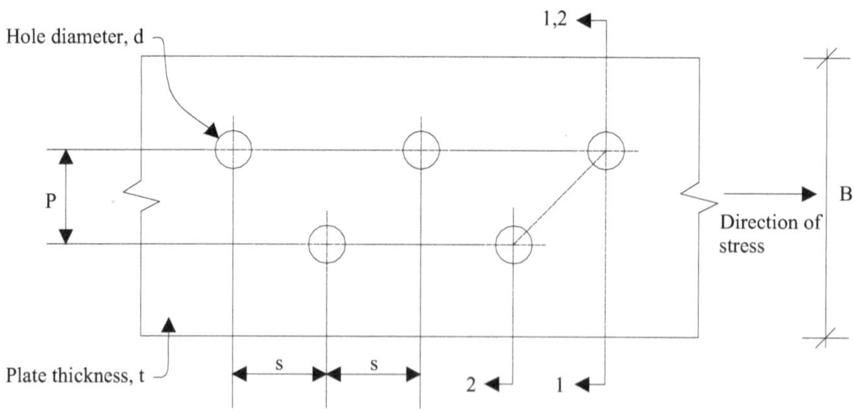

FIGURE 3.3 Bolts distribution.

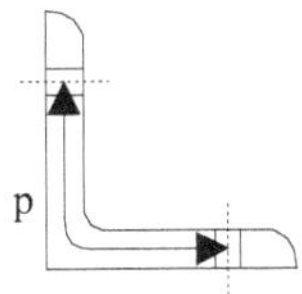

FIGURE 3.4 Net area.

On Section 1.1,

$$\text{Net area} = Bt - dt - \text{The minimum area is taken as } A_{net}$$

On Section 2.2,

$$\text{Net area} = Bt - 2dt + \frac{s^2 t}{4p} - \text{The minimum area is taken as } A_{net}$$

where

t is the plate thickness
d is the hole diameter
s is the staggered pitch parallel to the member axis
p is the hole spacing perpendicular to the member axis

In the case of the angle, where the holes are in more than one plane, p is measured along the centre thickness of the section (Figure 3.4).

3.5.1 Angles Connected by One Leg

Where members are connected unsymmetrically, or the member itself is unsymmetric, such as angles, tees, and channels, the eccentricity of the connection should be taken into consideration. In the particular case of an angle connected by a single row of bolts in one leg, the member may be treated as concentrically loaded, and the design ultimate resistance is based on a modified net section determined as shown in Figure 3.5.

where

$\beta_2 = 0.4$ if $p_1 \leq 2.5d$
$\beta_2 = 0.7$ if $p_1 \geq 5.0d$
$\beta_3 = 0.5$ if $p_1 \leq 2.5d$
$\beta_3 = 0.7$ if $p_1 \geq 5.0d$
A_{net} is the net area of the angle.

For an unequal angle connected by its smaller leg, A_{net} is taken as the net section area of an equivalent equal angle.

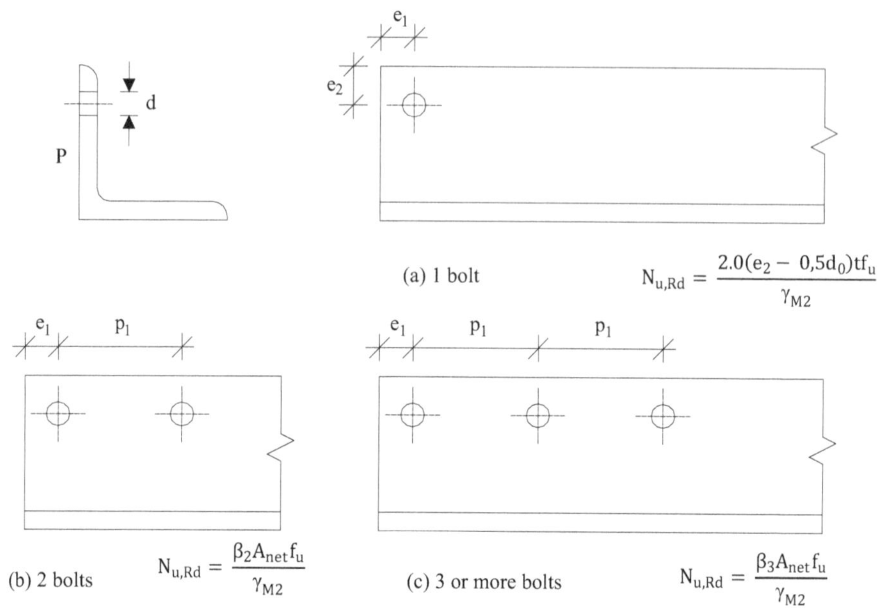

FIGURE 3.5 Bolts distribution in one angle.

For welded angles connected by one leg through welding rather than bolting, the eccentricity inherent in the connection may be accounted for by using an "effective" cross-sectional area, treating the member as concentrically loaded. Examples of the effective cross-sectional area to be adopted are shown in Figure 3.6. The welding process in detail will be covered in Chapter 9.

EC3 suggests that sections connected through outstands, T sections, and channels may be treated in a similar way, but it does not provide details. One approach might be to calculate the net area of the cross section (A_{net}) for a single T section connected through its flange or a single channel section connected through its web, as the net area of the connected part of the cross section plus half the area of the outstand element(s).

This net area is then used to calculate the design ultimate resistance of the net section as in equation 3.2, which, provided it is less than the design plastic resistance of the gross section as presented in equation 3.1, would be the design tension resistance.

3.6 SERVICEABILITY, CORROSION AND FATIGUE

As tension members transfer load very efficiently, they tend to have relatively small cross-sectional areas. This makes ties susceptible to excessive elongation under axial load, which may permit large drift of a structure if the tie forms part of a bracing system, as well as lateral deflection under self-weight. Light rolled sections are also easily damaged during transport. For these reasons, good practice limits the slenderness of tension members to approximately 300 for principal members and 400 for

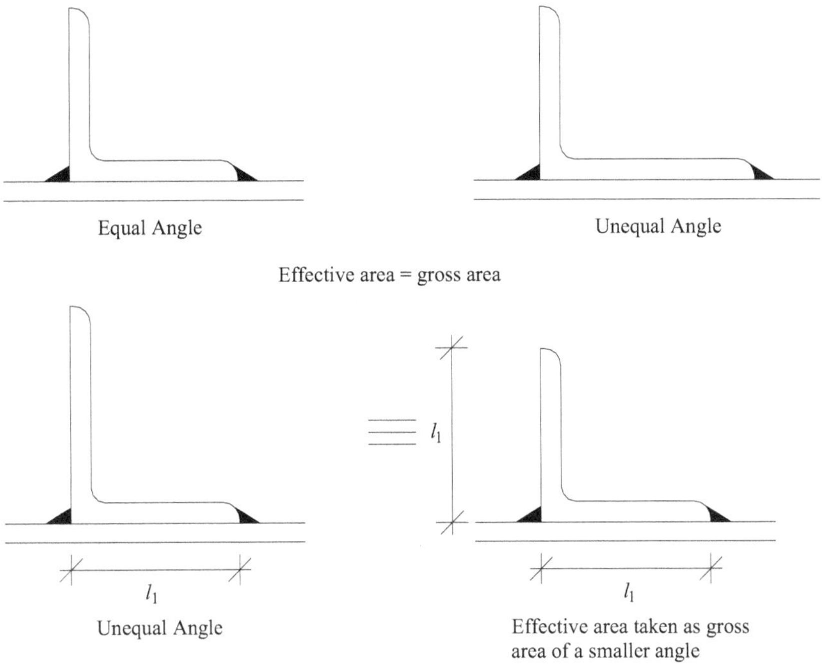

FIGURE 3.6 Effective areas of angles connected by one leg by welding.

secondary members. A further consideration is the consequences of material loss due to corrosion, which are greater for tension members since they are relatively highly stressed.

3.7 EFFECTIVE NET AREA OF EYEBAR MEMBER

Eyebars are usually used in lifting equipment, and they are welded onto the skid of the equipment or any steel structure. The shape of the padeye, along with the dimensions and parameters that affect the design, is shown in Figure 3.7.

Some restrictions concerning the general geometry of eyebars connected links are necessary. The relationships between the width of the body, the diameter of the pin, and the diameter of the head or net section adjacent to the pin hole are such that "dishing" of the head or the region adjacent to the pin hole will be avoided. If dishing of the head occurs, it will lead to an associated loss of strength of the member. The restriction on the width-to-thickness ratio of the body is included to provide a compact section and to avoid large diameter pins.

Timoshenko and Gooider have summarized the current state of knowledge regarding eyebars in the elastic range. With respect to safe design, the following guides should be followed:

1. $wt \geq P/0.6F_y$
2. $2bt \geq P/0.45\ F_y$

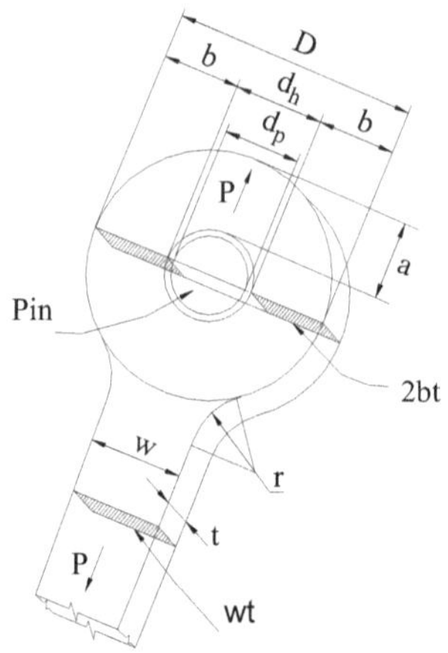

FIGURE 3.7 Eyebars dimensions.

3. $3-t \geq 12\,\text{mm}$
4. $w \leq 8t$
5. $1.33w \leq 2\text{b} \leq 1.5w$
6. $d_p \geq 7/8w$
7. $a=b$
8. $r \geq D$
9. $d_h - d_p \leq 1\,\text{mm}$

3.8 SAG RODS FOR STEEL FRAME BUILDINGS

For roofs of steel frame industrial buildings with sleeper slopes than 1:4, sag rods are used to provide lateral support, especially for purlins consisting of a channel cross section. Channels have a low resistance against lateral bending, so sag rods can be used to enhance lateral support for purlins, thereby reducing lateral bending. This tie rod is shown in Figure 3.8.

Industrial buildings have grits running between the columns along the vertical walls, as shown in Figure 3.9.

According to Machaley (2001), the following recommendations can lead to economical design:

1. For light roofs with corrugated steel sheets as the covering material, sag rods will be needed at one-third points if the spacing between trusses is greater than 6.0 m
2. For spacing less than 6.0 m with the same covering materials as mentioned above, sag rods at midpoints will be sufficient.
3. For heavier covering materials, such as cement tiles or other similar, the one-third points will be necessary for spacing greater than 4.0 m.
4. For spacing less than 4.0 m with the same covering material as mentioned above, mid-point will be satisfactory.

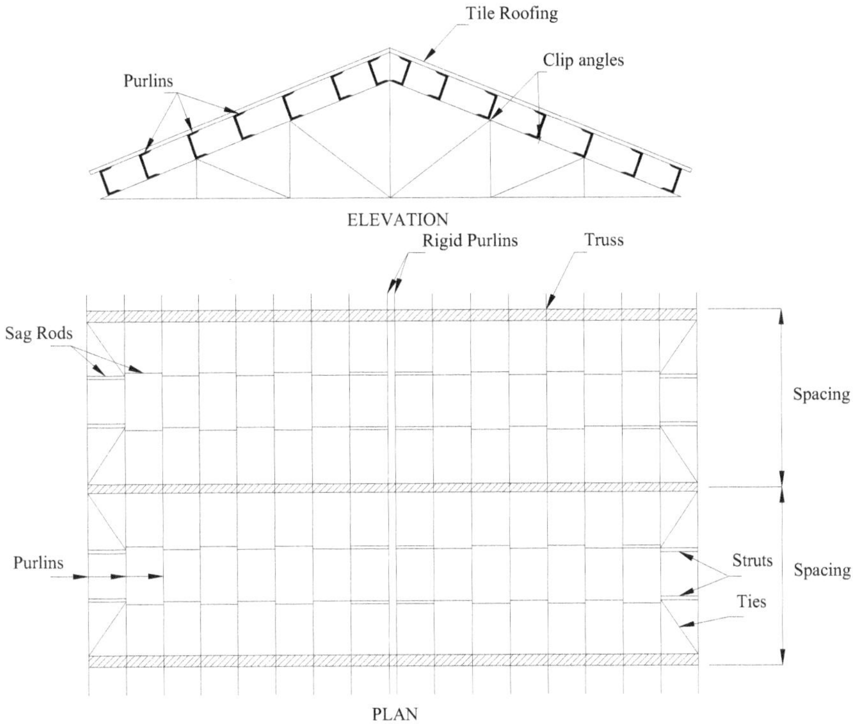

FIGURE 3.8 Tie roofing.

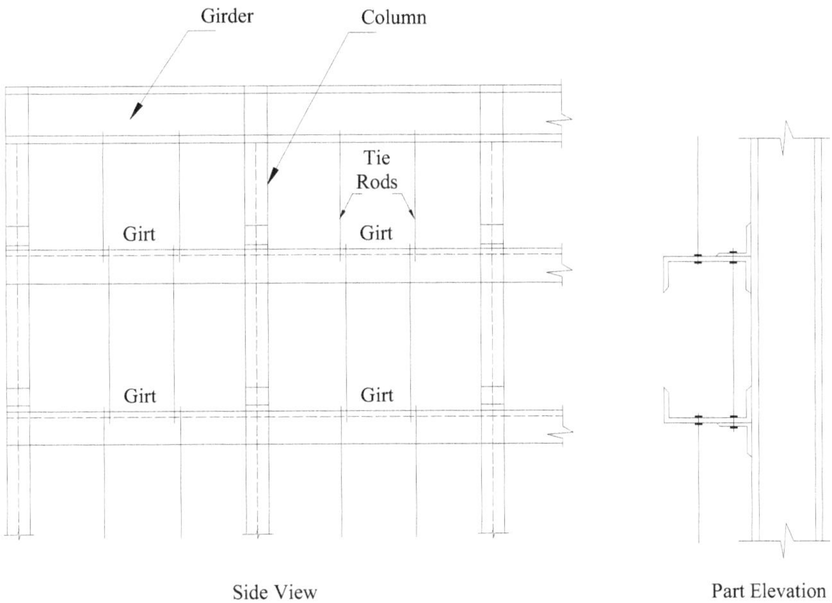

FIGURE 3.9 Tie rod for the girt.

Regarding stiffness conditions, as discussed previously, a common practice is to use rod diameters no less than 1/500 of their length, even though stress calculations may permit much smaller sizes. For practical construction considerations, it is usually desirable to limit the minimum diameter of sag rods to 15 mm. For sag rods used for purlins of roof trusses, the maximum force in a sag rod will occur in the top sag rod because it must support the sum of the forces in the lower sag rods. It is general practice to use the same diameter for both the top and lower sag rods.

Example 3.1: Tension Member

Design the lower chord tension member. Design force = 30 tons; geometric length of the member $L_x = 300$ cm (using bolts 16 mm for bolted connection) (Figure 3.10).

For 2 equal angle $i = 0.3\ a$

For one angle $i = 0.2\ a$

For star angles $i = 0.385\ a$

Solution:

$$1 - \lambda_{max} = KL/i = 300$$

$$i = 300/300 = 1$$

$$a = 1/0.3 = 3.33 \text{ cm}$$

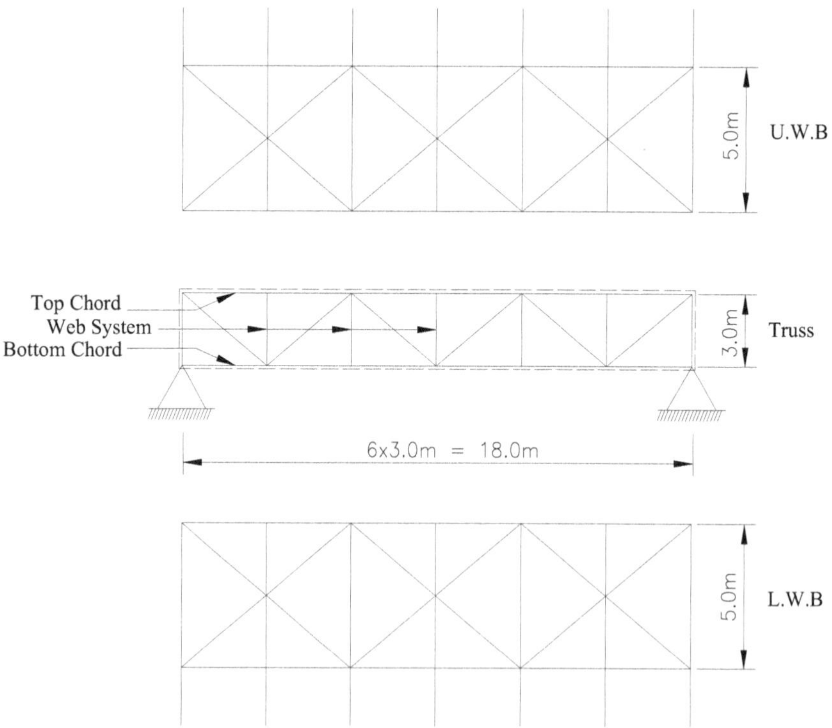

FIGURE 3.10 Upper and lower bracing system.

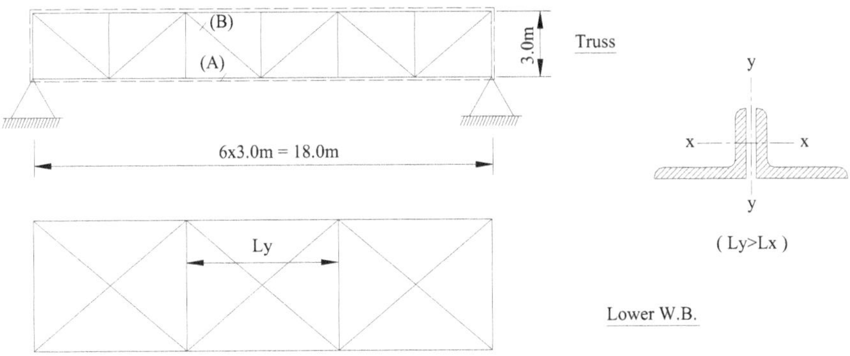

FIGURE 3.11 Buckling length to the truss.

Construction guideline:

$$D = 16 \text{ mm (bolts size)}$$
$$\phi = 16 + 2 \text{ mm} = 18 \text{ hole size}$$
$$a - t < 3\phi < 4.8 \text{ cm}$$

Example 3.2

Design the previous tension member (*A*) from the previous example, considering the shown wind bracing system (Figure 3.11).

The member being designed is a bottom chord member, so choose two angles back to back. The buckling lengths are $L_x = 3$ m and $L_y = 6.0$ m.

Example 3.3

Design a tension member with the following parameters: Force = 6.0 tons, $L_x = L_y = 5.8$ m; the connections are welded.

Example 3.4

Design a tension member given that the applied force is 6.0 tons. This member is a diagonal member with $L = 3.6$ m and a bolted connection (ϕ = 16 mm). In this case, it will consist of one equal angle as the length is less than 5.50 m

Example 3.5

Design an eyebar with a force of 80 tons, as presented in Figure 3.12. Use steel 37 for both plates and pins, with $F_y = 240$ Mpa. The pin is subjected to double shear stresses, as presented in Figure 3.13.

Net section req.:

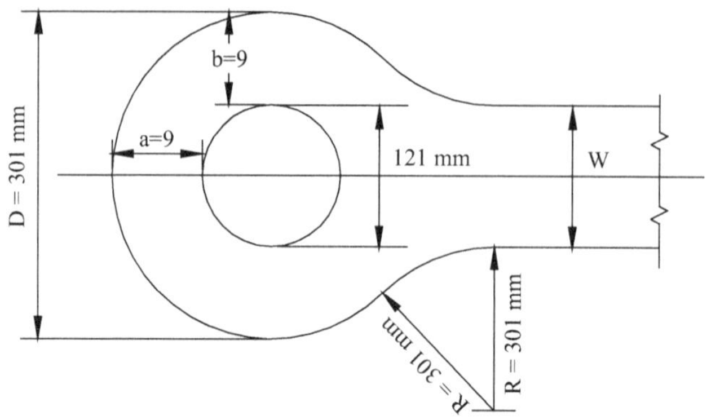

FIGURE 3.12 Stresses on the padeye.

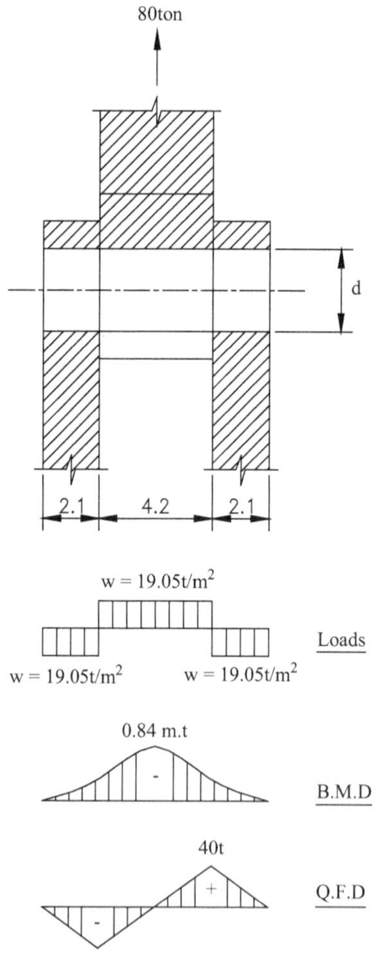

FIGURE 3.13 Dimensions of the padeye.

3.9 TENSION MEMBER DESIGN BY AISC

The AISC guidelines are similar to those of the EC, but the symbols will differ. As per AISC22, there is no maximum slenderness limit for members under tension forces. However, there are recommendations suggesting that the limit should be equal to or less than 300, although this does not apply to rods.

The tensile strength will be calculated using the following equation:

- Check yielding failure

$$P_n = F_y A_g \tag{3.5}$$

The strength reduction factor ϕ_t is equal to 0.9.

- Check rupture failure

$$P_n = F_u A_e \tag{3.6}$$

The rupture strength reduction factor ϕ_r is equal to 0.75.

where

A_e = effective net area,
A_g = gross area of the member,
F_y = specified minimum yield stress,
F_u = specified minimum tensile strength,

In calculating the net area, An, for tension and shear, the hole diameter will be considered as z in (2 mm) greater than the nominal dimension of the hole.

In the case of multiple bolt holes extending across a part in any diagonal or zigzag line, the net width of the part shall be determined by deducting from the gross width the sum of the diameters or slot dimensions provided in this section for all holes in the chain, and adding, for each gage space in the chain, the quantity $s^2/4g$,

where g is the transverse centre-to-centre spacing (gage) between fastener gage lines, and s is the longitudinal centre-to-centre spacing (pitch) of any two consecutive bolt holes.

The effective net area will be calculated from the following equation:

$$A_e = U \cdot A_n \tag{3.7}$$

where U is the shear lag factor, which will be calculated as follows based on the following cases:

Case 1: All tension members where the tension load is transmitted directly to each of the cross-sectional elements by fasteners or welds (except as in Cases 4, 5, and 6).

$$U = 1.0$$

Case 2: All tension members, except HSS, where the tension load is transmitted to some but not all of the cross-sectional elements by fasteners or by longitudinal welds in combination with transverse welds. Alternatively, Case 7 is permitted for *W*, *M*, *S*, and *HP* shapes and Case 8 is permitted for angles.

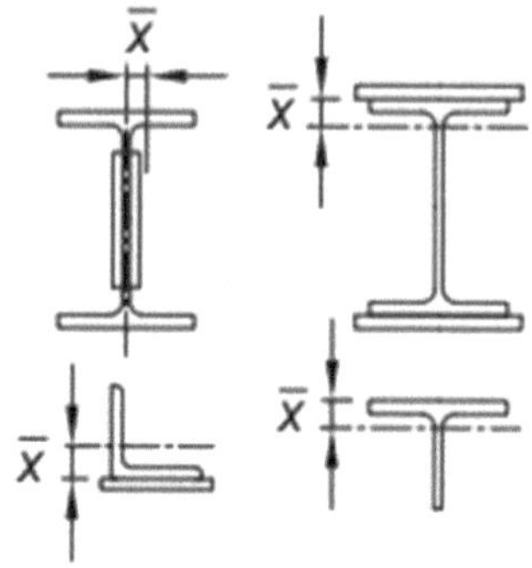

$$U = 1 - x'/l \tag{3.8}$$

Case 3: All tension members where the tension load is transmitted only by transverse welds to some but not all of the cross-sectional elements.
$U = 1.0$ and An = area of the directly connected elements

Case 4: Plates, angles, channels with welds at heels, tees, and W shapes with connected elements, where the tension load is transmitted by longitudinal welds only. The definition of x' is the connection eccentricity.

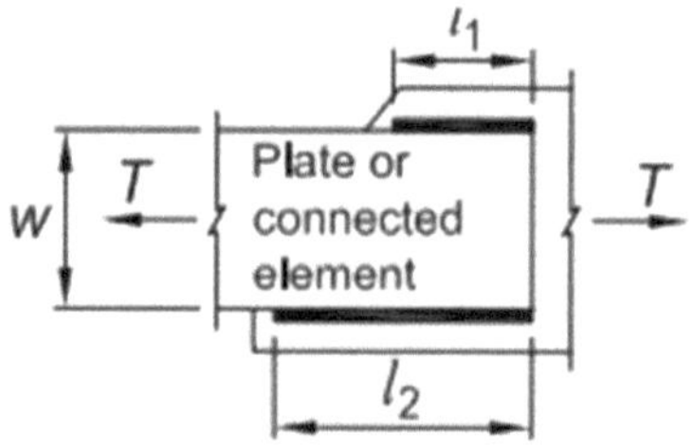

$$l = (l_1 + l_2)/2 \tag{3.9}$$

$$U = \frac{3l^2}{3l^2 + W^2}\left(1 - \frac{x'}{l}\right) \tag{3.10}$$

Case 5: Round and rectangular HSS with a single concentric gusset through slots in the HSS

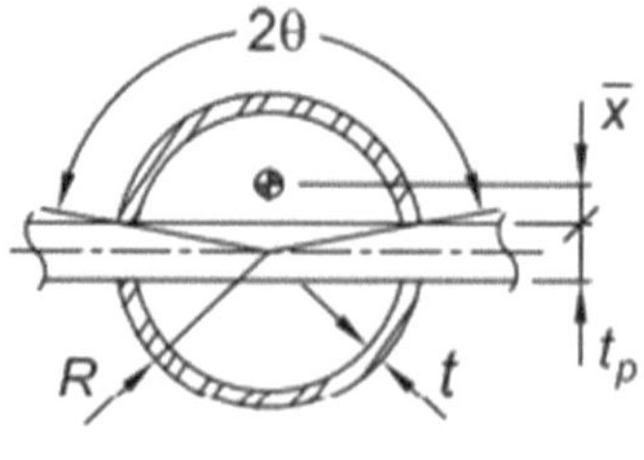

$$x' = \frac{R\sin\theta}{\theta} - 0.5t_p \tag{3.11}$$

$$U = \left[1 + \left(\frac{x'}{l}\right)^{3.2}\right]^{-10} \tag{3.12}$$

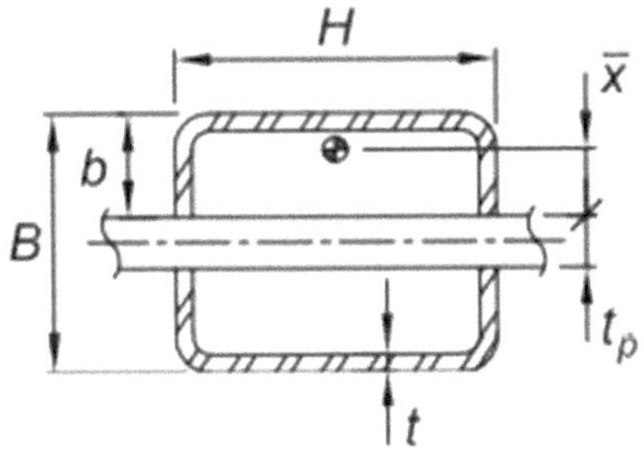

$$x' = b - \frac{2b^2 + tH - 2t^2}{2H + 4b - 4t} \tag{3.13}$$

$$U = 1 - x'/l \tag{3.14}$$

Case 6: Rectangular HSS with two-side gusset plates.

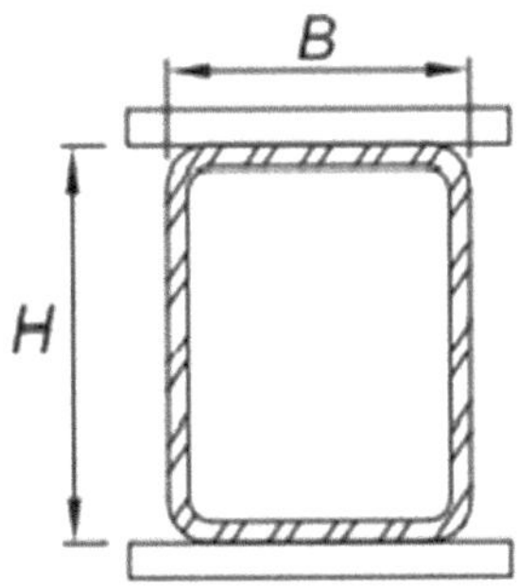

$$U_B = \frac{3l^2}{3l^2 + B^2} \tag{3.15}$$

$$U_H = \frac{3l^2}{3l^2 + H^2} \tag{3.16}$$

$$U = \frac{BU_B + HU_H}{H + B} \tag{3.17}$$

Case 7: *W*-, *M*-, *S*-, or *HP*-shapes, or tees cut from these shapes. (If U is calculated per Case 2, the larger value is permitted to be used.

a. with flange connected with three or more fasteners per line in the direction of loading
$bf \geq 0.67d$ $U=0.9$
$bf < 0.67d$ $U=0.85$

b. with web connected with four or more fasteners per line in the direction of loading $U=0.7$

Case 8: Single and double angles. (If U is calculated per Case 2, the larger value is permitted to be used.)

a. with four or more fasteners per line in the direction of loading; $U=0.8$

b. with three fasteners per line in the direction of loading (for fewer than three fasteners per line in the direction of loading, use Case 2); $U=0.60$

3.10 PIN-CONNECTED MEMBERS

The design tensile strength, $\phi_t Pn$, of pin-connected members shall be the lower value determined according to the limit states of tensile rupture, shear rupture, bearing, and yielding.

a. For tensile rupture

$$Pn = Fu(2tb_e) \tag{3.18}$$

$$\phi_t = 0.75$$

b. For shear rupture

$$P_n = 0.6C_r F_u A_{sf} \tag{3.19}$$

where A_{sf} is the area of the path of shear failure

$$A_{sf} = 2t(a + d/2) \tag{3.20}$$

Cr = reduction factor for shear rupture on pin-connected members = 1.0

If $d_h - d \leq 32$ in. (1 mm)

= 0.95 when 32 in. $< d_h - d \leq z$ in. (1 mm $< d_h - d \leq 2$ mm)

a = shortest distance from the edge of the pin hole to the edge of the member measured parallel to the direction of the force,

$b_e = 2t + 0.63$ in. (= $2t$ + 16, mm), but not more than the actual distance from the edge of the hole to the edge of the part measured in the direction normal to the applied force,

d = diameter of pin,

d_h = diameter of hole, and

t = thickness of plate.

It is required to check the bearing in the case of pin connection in addition to the tension; the strength reduction factor for bearing, ϕ, is equal to 0.75.

The nominal bearing strength Rn will be calculated as follows:

$$R_n = 1.8\ F_y A_{pb} \tag{3.21}$$

where A_{pb} is the projected area in bearing, in^2 or mm^2.

The limitations of pin connected members as per AISC are follows:

a. The pin hole shall be located midway between the edges of the member in the direction normal to the applied force.
b. When the pin is expected to provide for relative movement between connected parts while under full load, the diameter of the pin hole shall not be more than 32 in. (1 mm) greater than the diameter of the pin for pins less than 3 in. (75 mm) in diameter, and not more than z in. (2 mm) greater than the diameter of the pin for pins of 3 in. (75 mm) in diameter or greater.
c. The width of the plate at the pin hole shall not be less than $2b_e + d$, and the minimum extension, a, beyond the bearing end of the pin hole, parallel to the axis of the member, shall not be less than 1.33b_e.
d. The corners beyond the pin hole are permitted to be cut at 45° to the axis of the member, provided the net area beyond the pin hole, on a plane perpendicular to the cut, is not less than that required beyond the pin hole parallel to the axis of the member.

FURTHER READING

AISC. (2022). *An American National Standard Specification for Structural Steel Buildings*, ANSI/AISC 360-22, American Institute of Steel Construction, Chicago, IL.

British Standards Institute. (2006). *Eurocode 3: Design of Steel Structures– Part 1–8: Design of Joints*, BSI, London.

Machaley. (2001). Behaviour analysis and design for steel structure.

Timoshenko, S.P. and Goodier, J.N. (1970). *Theory of Elasticity*, 3rd edition, McGraw Hill, New York.

4 Compression Member

4.1 INTRODUCTION

Columns are structural elements subjected to a pure compression force; this can describe columns under special loading conditions but generally refers to compressed pin-ended struts. Their role is clearly evident in supporting roof trusses, system girders, or bracing members. The primary concern for compression members is buckling. If they are affected by significant bending moments in addition to axial loads, they are referred to as beam columns (Figure 4.1).

This chapter focuses on compression members only, but it is worth mentioning that it is rare to find a member subjected to a pure compression force, as in most cases there are eccentricities with the axial loads, and transverse forces are generally not negligible. The design of compression members represents an elementary case that leads to an understanding of compression effects in the study of beam–column design. Because most steel compression members are rather slender, buckling can occur.

This chapter describes the design of the different kinds of compression members and explains the behaviour of both stocky and slender columns; the buckling curves used to design slender columns are also provided. Figure 4.2 shows a steel column.

FIGURE 4.1 Steel column with attached beams.

DOI: 10.1201/9781003307303-4

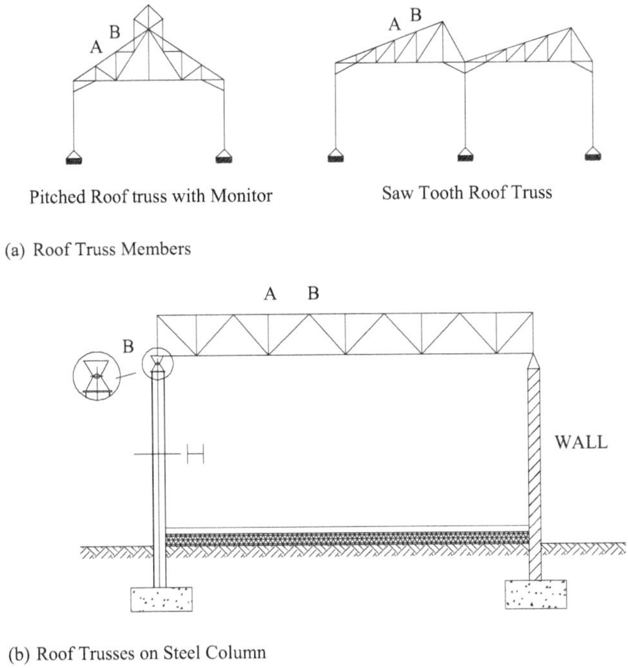

FIGURE 4.2 Types of columns and compression members.

4.2 STUB COLUMNS

The stub column refers to a very short column or a column with a higher section that provides a high value for the radius of gyration, resulting in a very low slenderness ratio that renders them unaffected by the overall buckling of the member. In such cases, buckling will not govern the design, and the compressive capacity of the member is dictated by the compressive resistance of the cross section, which is a function of section classification. The section classification is presented in Section 4.2.1. This aligns with AISC guidelines for compact and non-compact sections.

For sections in Class 1, 2, and 3, the applied load can increase until the cross section reaches plasticity without being affected by local buckling. Therefore, in this case, the compression member is designed for plastic capacity:

$$N_{c.Rd} = Af_y / \gamma_{M0} \tag{4.1}$$

In the case of cross sections in Class 4, local buckling in one or more elements of the cross section prevents the attainment of member failure at yield, and the design is governed by the resistance to local buckling:

$$N_{c.Rd} = A_{eff} f_y / \gamma_{M1} \tag{4.2}$$

where A_{eff} is the area of the effective cross section determined by considering that the effective area *Aeff* will be calculated for the cross section subject only to stresses due to uniform axial compression. For non-symmetrical cross sections, the possible

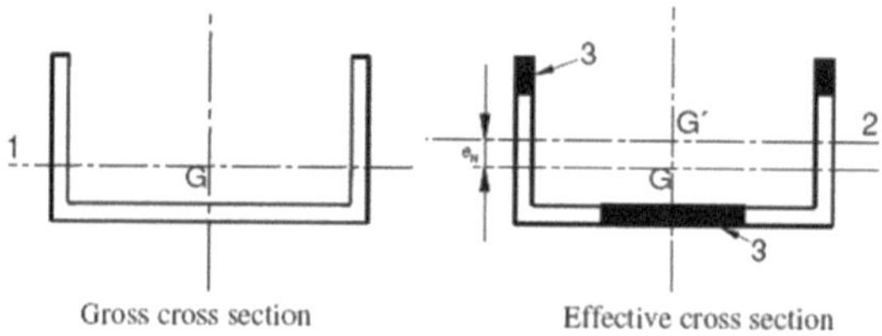

FIGURE 4.3 Class 4 cross sections under axial force.

shift *eN* of the centroid of the effective area *Aeff* relative to the centre of gravity of the gross cross section is shown in Figure 4.3, where (*G*) and (*G*′) are centroids of the gross and effective cross sections, respectively. Axis (1) is the centroid axis of the gross section, axis (2) is the centroidal axis of the effective cross section, and zone (3) represents the non-effective zone.

4.2.1 Section Classification

As per EC3, which also aligns with AISC principles, sections are classified based on the following criteria, and the four classes of cross sections are defined as follows:

1. Class 1 cross sections are those that can form a plastic hinge with the required rotation capacity from plastic analysis without a reduction in resistance.
2. Class 2 cross sections are those that can develop their plastic moment resistance but have limited rotation capacity due to local buckling.
3. Class 3 cross sections are those in which the stress in the extreme compression fibre of the steel member, assuming an elastic distribution of stresses, can reach the yield strength, but local buckling is likely to prevent the development of plastic moment resistance.
4. Class 4 cross sections are those in which local buckling will occur before the attainment of yield stress in one or more parts of the cross section.

In Class 4 cross sections, effective widths may be used to make the necessary allowances for reductions in resistance due to the effects of local buckling.

The classification of a cross section depends on the width-to-thickness ratio of the parts subject to compression as presented in Figure 4.4.

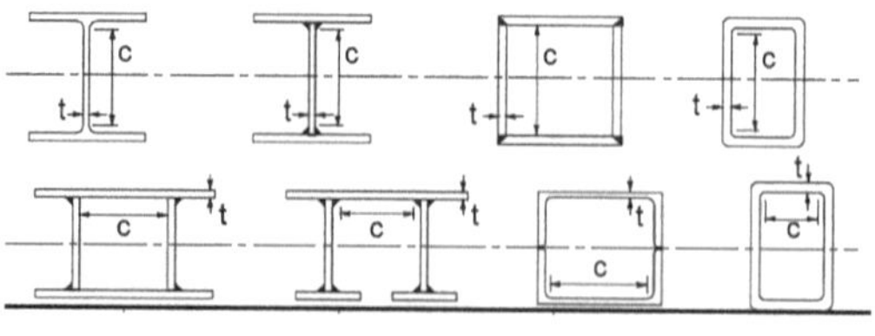

FIGURE 4.4 The effective web depth to its length for different section

Compression parts include every part of a cross section that is either totally or partially in compression under the load combination considered.

The various compression parts in a cross section (such as a web or flange) can, in general, be in different classes. In this case, the cross section is classified according to the highest (least favourable) class of its compression parts.

Alternatively, the classification of a cross section may be defined by quoting both the flange classification and the web classification. The limiting proportions for Class 1, 2, and 3 compression parts should be obtained from Table 4.1. A part that fails to satisfy the limits for Class 3 should be classified as Class 4.

However, when verifying the design buckling resistance of a member, the limiting proportions for Class 3 should always be obtained from Table 4.2.

Cross sections with a Class 3 web and Class 1 or 2 flanges may be classified as Class 2 cross sections with an effective web.

When the web is considered to resist shear forces only and is assumed not to contribute to the bending and normal force resistance of the cross section, the cross section may be designed as Class 2, 3, or 4 sections, depending only on the flange class.

In case of steel angle with long leg h and short leg, b, with thickness, t, it will be class 3 and to fulfill these requirements $h/t <= 15\,\varepsilon$ and $b + h/2t <= 11.5\,\varepsilon$.

In case of tubular section with outside diameter, d, and thickness, t. It will follow class 1 or 2 or 3 if d/t is less than or equal to $50\,\varepsilon^2$, $70\,\varepsilon^2$ and $90\,\varepsilon^2$ respectively

where ε calculate from this equation:

$$\varepsilon = \sqrt{\frac{235}{f_y}} \tag{4.3}$$

Table 4.3 and 4.4 presents the class with width to thickness ratio in case of compression with bending.

TABLE 4.1
Values of λ_1

Steel Grade	Thickness Range (mm)	Yield Strength, f_y (MPa)	λ_1
S275	$t \leq 16$	275	86.8
S275	$16 < t \leq 40$	265	88.4
S335	$t \leq 16$	355	76.4
S335	$16 < t \leq 40$	345	77.5

TABLE 4.2
Imperfection Factors

Buckling curve	a_o	a	b	C	d
Imperfection factor, α	0.13	0.21	0.34	0.49	0.76

Source: EC-3 Table 6.1.

TABLE 4.3
Define class based on web depth to thickness ratio

class	Under bending	Under compression	Under bending and compression
1	$c/t \leq 72\varepsilon$	$c/t \leq 72\varepsilon$	If $\alpha > 0.5$; $c/t \leq (396\varepsilon/13\,\alpha - 1)$ If $\alpha \leq 0.5$; $c/t \leq 36\varepsilon/\alpha$
2	$c/t \leq 83\varepsilon$	$c/t \leq 38\varepsilon$	If $\alpha > 0.5$; $c/t \leq (456\varepsilon/13\,\alpha - 1)$ If $\alpha \leq 0.5$; $c/t \leq 41.5\varepsilon/\alpha$
3	$c/t \leq 142\varepsilon$	$c/t \leq 42\varepsilon$	If $\psi > -1$; $c/t \leq (42\varepsilon/0.67+0.33\psi)$ If $\psi \leq -1$; $c/t \leq 62(1-\psi)(-\psi)^{0.5}$

Where, α is the percentage of compression to the total web length

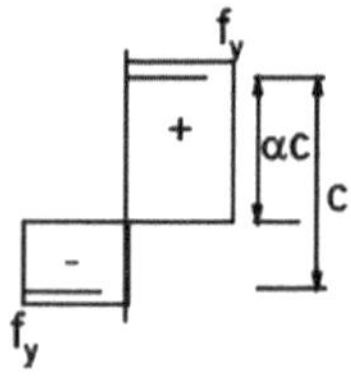

ψ is the partial percentage of stresses in case of bending and compression

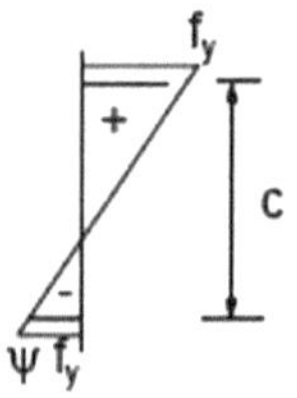

TABLE 4.4
Define class based on flange length to its thickness

Class	Under compression	Under bending and compression (Flange tip in compression)	Buckling Length, L_{cr}
1	$c/t \leq 9\varepsilon$	$c/t \leq 9\varepsilon/\alpha$	$c/t \leq 9\varepsilon/\alpha(\alpha)^{0.5}$
2	$c/t \leq 10\varepsilon$	$c/t \leq 10\varepsilon/\alpha$	$c/t \leq 10\varepsilon/\alpha(\alpha)^{0.5}$
3	$c/t \leq 14\varepsilon$	$c/t \leq 21\varepsilon(k_\sigma)^{0.5}$	

Where;

α is the percentage of the compression part

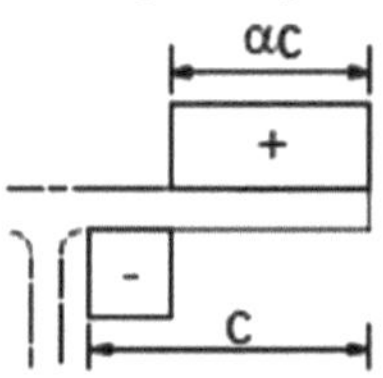

K_σ is discussed in EN1993-1-5

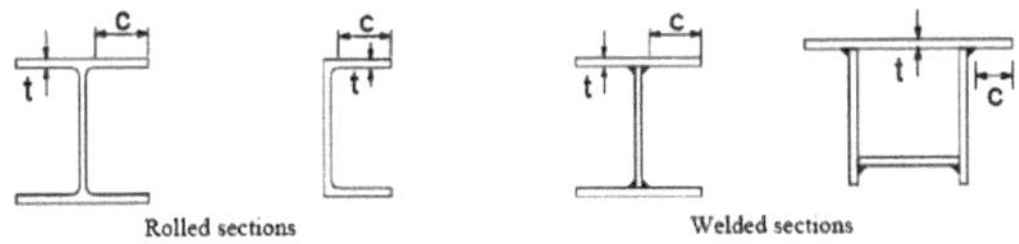

FIGURE 4.5 The effective flange length for different section

4.3 SLENDER STEEL COLUMNS

These columns depend on their slenderness. They exhibit two different types of behaviour: those with high slenderness present a quasi-elastic buckling behaviour, whereas those of medium slenderness are very sensitive to the effects of imperfections.

If ℓ_{cr} is the critical length, the Euler critical load N_{cr} is equal to

$$N_{cr} = \frac{\pi^2 EI}{\ell_{cr}^2} \tag{4.4}$$

and it is possible to define the Euler critical stress σ_{cr} as follows:

$$\sigma_{cr} = \frac{N_{cr}}{A} = \frac{\pi^2 EI}{\ell_{cr}^2 A} \tag{4.5}$$

By introducing the radius of gyration, $i = \sqrt{I/A}$, and the slenderness, $\lambda = \ell_{cr}/i$ for the relevant buckling mode, equation 4.5 becomes

$$\sigma_{cr} = \frac{N_{cr}}{A} = \frac{\pi^2 E}{\lambda^2} \tag{4.6}$$

Plotting the curve σ_{cr} as a function of λon a graph (Figure 4.6), with the horizontal line representing perfect plasticity, $\sigma = f_y$, it is interesting to note the idealized zones representing failure by buckling, failure by yielding, and safety.

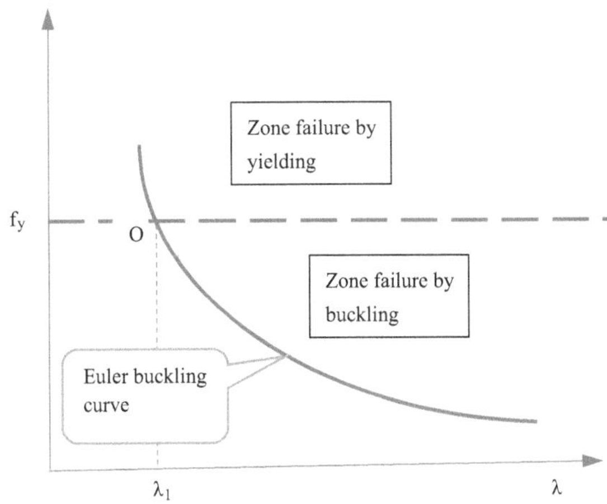

FIGURE 4.6 Euler buckling curve and modes of failure.

As shown in Figure 4.6, the intersection point O of the two curves represents the maximum theoretical value of slenderness for a column compressed to the yield strength. This limiting slenderness, when σ_{cr} is equal to the yield strength of the steel is given by

$$\lambda_1 = \pi\left[E/f_y\right]^{0.5} = 93.9\varepsilon \tag{4.7}$$

The differences between the idealized curve and the real behaviour of steel columns are significant. As shown in Figure 4.7, columns generally fail by inelastic buckling before reaching the Euler buckling load, due to various imperfections in the "real" element, resulting from initial out-of-straightness, residual stresses, eccentricity of applied axial loads, and strain hardening. These imperfections all affect buckling and, therefore, the ultimate strength of the column. Experimental studies of real columns yield results as shown in Figure 4.7 which is based on the results of experimental study on members under axial load as presented in ECCS (1976).

Compared to the theoretical curves, the real behaviour shows greater differences in the range of medium slenderness than in the range of large slenderness. In the zone of medium values of λ (representing most practical columns), the effect of structural imperfections is significant and must be carefully considered. The greatest reduction in the theoretical value occurs in the region of limiting slenderness λ_1. The lower bound curve is obtained from a statistical analysis of test results and represents the safe limit for loading.

A column will be considered slender if its slenderness is greater than that corresponding to the point of inflexion of the lower bound curve, as per ESDEP from experimental tests shown in Figure 4.7. The ultimate failure load for such slender columns is close to the Euler critical load (*Ncr*) and is thus independent of the yield stress.

Columns of medium slenderness are those whose behaviour deviates most from Euler's theory. When buckling occurs, some fibres have already reached the yield strength, and the ultimate load is not simply a function of slenderness; the more

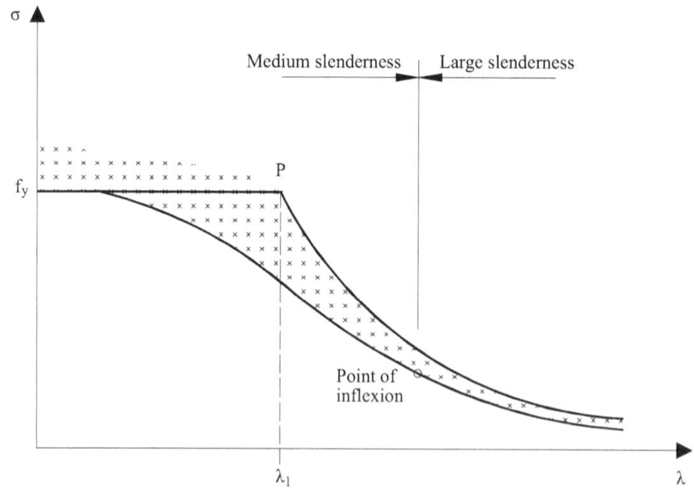

FIGURE 4.7 Real column test results and buckling curves.

numerous the imperfections, the larger the difference between the actual and theoretical behaviour. Out-of-straightness and residual stresses are the imperfections that have the most significant effect on the behaviour of this type of column.

The presence of residual stresses in an intermediate-length steel compression member may cause a significant reduction in its buckling resistance. Residual stresses are established during the cooling of a hot-rolled or welded steel member (and during plastic deformation, such as cold rolling). The contraction of the late cooling regions of the member induces residual compressive stresses in the early cooling regions, which are balanced by equilibrating tensile stresses in the late cooling regions. In hot-rolled I-section members, the flange–web junctions are the least exposed to cooling influences, and so these are the regions of residual tensile stress as discussed by Trahir et al (2008).

Residual stresses can be distributed in various ways across the section as shown in Figure 4.8. Residual stresses combined with axial stresses cause yielding to occur in the cross section, thereby reducing the effective area able to resist the axial load.

where

σ_R = residual stress and N/A = axial stress N/Area

When added, they are either less than the yield strength f_y or reach f_y, causing the part to yield, i.e. fail.

An initial out-of-straightness, *eo*, produces a bending moment that generates a maximum bending stress σ_B (see Figure 4.9a), which when added to the residual stress, σ_R gives the stress distribution shown in Figure 4.9b. If σ_{max} is greater than the yield stress the final distribution will be partly plastic, and sections of the member will have yielded in compression, as shown in Figure 4.9c.

4.4 NON-DIMENSIONAL SLENDERNESS

In EC3, non-dimensional slenderness $\bar{\lambda}$ is defined as follows:

$$\bar{\lambda} = \left[\frac{Af_y}{N_{cr}}\right]^{0.5} \tag{4.8}$$

which may be written by using equations 4.5 and 4.6 and can be conveniently represented as

$$\bar{\lambda} = \left(\frac{\lambda}{\lambda_1}\right) \tag{4.9}$$

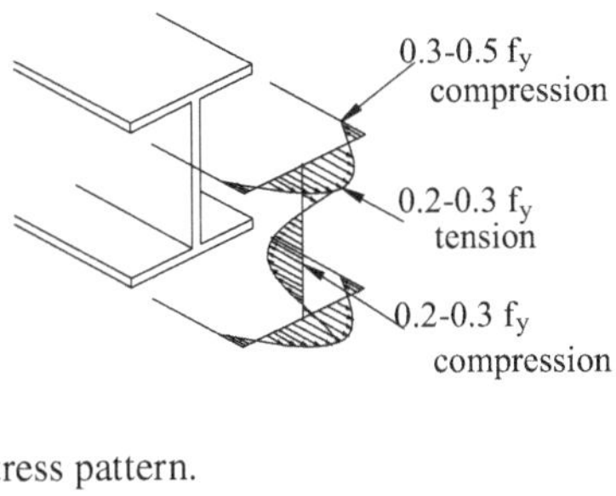

FIGURE 4.8 Residual stress pattern.

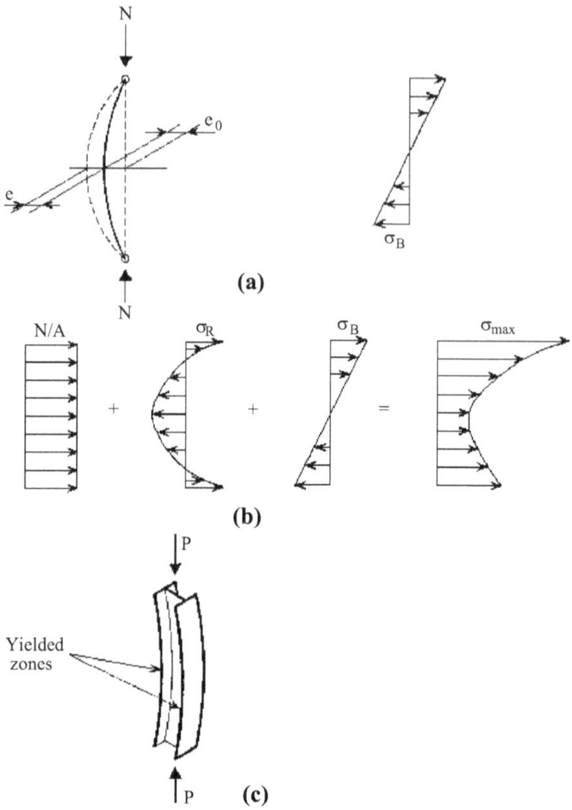

FIGURE 4.9 Applied and residual stresses in compression members.

4.5 BUCKLING CURVES

The buckling curves are based on the results of more than 1,000 tests on various types of members, including I, H, T, and angle sections, as well as pipes, with different values of slenderness (between 55 and 160). A probabilistic approach, using the experimental strength in conjunction with a theoretical analysis, allows curves to be drawn that describe column strength as a function of the reference slenderness. A half sine wave geometric imperfection with a magnitude equal to 1/1,000 of the length of the column, along with the effect of residual stresses specific to each type of cross section, is accounted for.

The buckling curves (a_0, a, b, c or d) are shown in Figure 4.10. These curves provide the value for the reduction factor χ of the column's resistance as a function of the reference slenderness for different kinds of cross sections, with reference to different values of the imperfection factorα).

EC3 expresses the buckling curves by the mathematical expression for χ:

$$\chi = \frac{1}{\phi + \left[\phi^2 - \bar{\lambda}^2\right]^{0.5}} \leq 1 \tag{4.10}$$

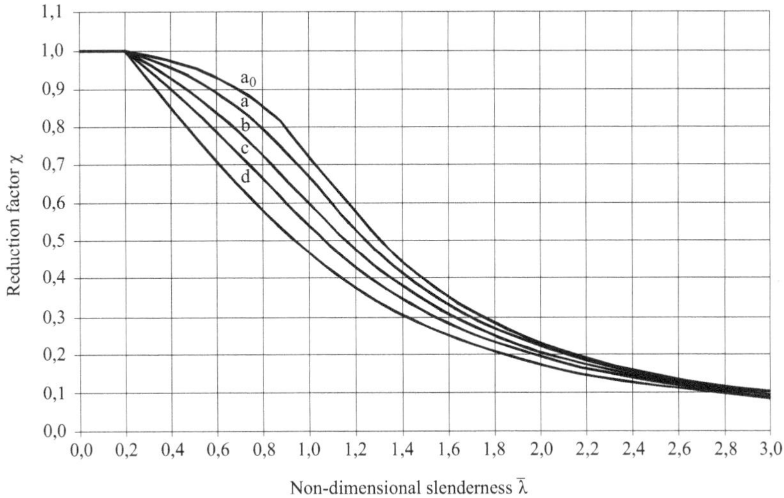

FIGURE 4.10 Buckling curves (Adapted from EC3).

where

$$\phi = 0.5\left[1+\alpha\left(\bar{\lambda}-0.2\right)+\bar{\lambda}^2\right] \quad (4.11)$$

The imperfection factor α depends on the shape of the column cross section considered, the direction in which buckling can occur (y-axis or z-axis), and the fabrication process used for the compression member (hot-rolled, welded, or cold-formed). Values for α, which increase with imperfections, are provided in Table 4.2.

Table 4.5 helps in selecting the appropriate buckling curve based on the type of cross section, its dimensional limits, and the axis about which buckling can occur.

For U-channel, T, or solid sections, follow curve c, and for angle steel sections, follow curve *b*.

4.6 COLUMN IN BRACED AND UNBRACED FRAME

As per the previous equation of buckling, the main element is to define the buckling length, which is very important and is a function of the structure type, whether it allows side sway or prevents side sway.

In the case of columns in braced multi-storey frames, buckling lengths are normally taken as equal to the column length (L_{cr} = 1.0L). The lower values of buckling lengths are presented in the following Table 4.6, which is a function of the rotational restraint that exists at the column end.

In the case of a single-storey building, the side rails and purlins are usually attached to the columns and beams, so the buckling lengths will be reduced in the direction of the minor axis (*z-z*). As these secondary members are considered bracing points, the member will buckle around the major axis (*y-y*). As per NCCI SN031, the

TABLE 4.5
Selection of Appropriate Buckling Curve for a Cross Section

	Cross section		Limits	Buckling about Axis	Buckling Curve	
					S235,275,355, 420	S460
1. Rolled Sections	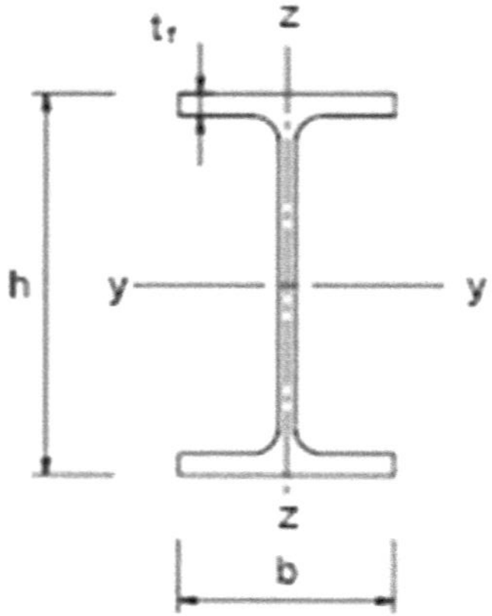	$h/b > 1.2$	$t_f \le 40\,mm$	*y-y*	*A*	a_o
				z-z	*B*	a_o
			$40 < t_f \le 100\,mm$	*y-y*	*b*	*a*
				z-z	*c*	*a*
		$h/b \le 1.2$	$t_f \le 100mm$	*y-y*	*b*	*a*
				z-z	*c*	*a*
			$t_f > 100\,mm$	*y-y*	*d*	*c*
				z-z	*d*	*c*

(Continued)

TABLE 4.5 (*Continued*)
Selection of Appropriate Buckling Curve for a Cross Section

	Cross section	Limits	Buckling about Axis	Buckling Curve S235,275,355, 420	Buckling Curve S460
2. Welded Sections	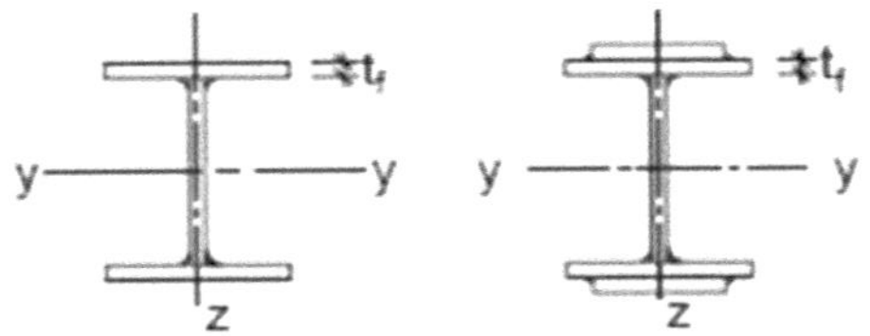	$t_f \leq 40$ m	y-y	b	b
			z-z	c	c
		$t_f > 40$ mm	y-y	C	C
			z-z	D	d
3. Hollow Sections		Hot finished	Any	A	a_o
		Vold finished	Any	C	c
4. Welded Box Sections	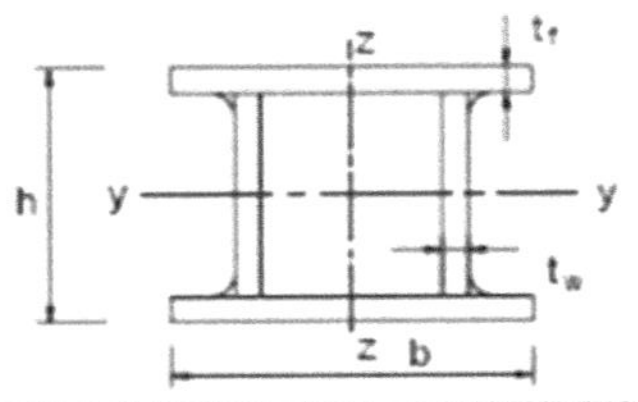	Generally otherwise below	Any	B	b
		Thick web $a > 0.5\ t_f$ $b/t_f < 30$ $h/t_w < 30$	Any	C	c

TABLE 4.6
Buckling Length for Different End Constrains

End Condition	L_{cr}
Effectively restrained against rotation at *b* ends	0.7L
Partially restrained against rotation at both ends	0.85L
Restrained against rotation at one end	0.85L
Not restrained against rotation at both ends	1.0L

TABLE 4.7
Buckling Length for One-storey Building Column

Connection to Base	Connection to Eave Level	Buckling Length, L_{cr}
Pinned	Fixed	1.5L
Fixed	Pinned	1.5L
Fixed	Fixed	1.2L

Table 4.7 presents the in-plane buckling length of a column in the major axis direction in a typical industrial building.

4.7 DESIGN STEPS FOR COMPRESSION MEMBERS

To design a member subjected to a compressive axial force, it is necessary to evaluate its two effective lengths in relation to the two principal axes, considering the expected connections at its ends. The verification procedure should then proceed as follows:

- geometric characteristics of the shape and its yield strength give the reference slenderness $\bar{\lambda}$.
- χ is calculated, taking into account the forming process and the shape thickness, using one of the buckling curves and $\bar{\lambda}$.

The design buckling resistance of a compression member is then taken as

$$N_{b,Rd} = \chi \frac{Af_y}{\gamma_{M1}} \tag{4.12}$$

for class 1, 2, and 3 cross sections, and

$$N_{b,Rd} = \chi \frac{A_{eff} f_y}{\gamma_{M1}} \tag{4.13}$$

for class 4 cross sections.

If this is higher than the design axial load, the column is acceptable; if not, another larger cross section must be chosen and checked.

Solved Examples

Example.4.1: For Column

The inner column in the building at base level is selected. This member has a length of 4.335 m and is composed of an HEB 340 section in steel S 355.

In this column, the bending moments (and the shear force) may be neglected; the **design axial force** (compression) obtained from the previous analysis is given by ***NEd*** **= 3,326.0 kN**.

Contact us to receive the Excel sheet at www.elreedyman.com

Buckling resistance

λy= 29.66
λz= 57.57
$\lambda 1$= 76.44
$\lambda y1$= $\lambda y/\lambda 1$= 0.39
$\lambda z1$= $\lambda z/\lambda 1$= 0.75

Calculate reduction factor χ_{min}

h/b= 1.13 <1.2
tf= 21.5 <= 100
αy= 0.34
αz= 0.49
Flexural buckling around y-curve b (α=0.34)
Flexural buckling around z-curve b (α=0.49)

$\lambda z1$ > $\lambda y1$

α curve c > α curve b

χmin is χz

φ_z= 0.92
χ_z 0.69

Section safety verification

$$N_{b,Rd} = \chi_z A f_y / \gamma_{M1}$$

$N_{b,RD}$= 4195.73 kN > N_{ED}= 3326.00 kN Ok Safe

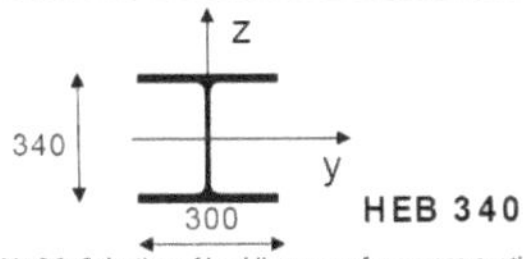

Table 6.2: Selection of buckling curve for a cross-section

Cross section		Limits	Buckling about axis	Buckling curve S 235 S 275 S 355 S 420	S 460
Rolled sections	h/b > 1.2	$t_f \le 40$ mm	y – y z – z	a b	a_0 a_0
		40 mm < $t_f \le 100$	y – y z – z	b c	a a
	h/b ≤ 1.2	$t_f \le 100$ mm	y – y z – z	b c	a a
		t_f > 100 mm	y – y z – z	d d	c c

Table 6.1: Imperfection factors for buckling curves

Buckling curve	a_0	a	b	c	d
Imperfection factor α	0.13	0.21	0.34	0.49	0.76

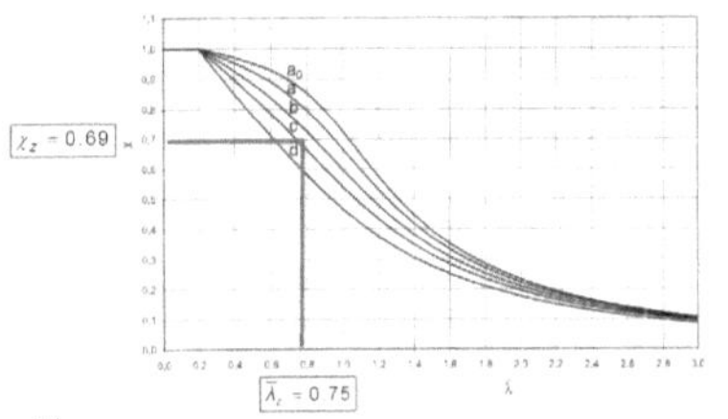

4.8 COMPRESSION MEMBER IN AISC

The same concept applies in AISC as it does for EC3, but the equations for buckling are different, though they should provide similar results; AISC 22 will be discussed here.

ELREEDYMAN www.elreedyman.com	Client :-		Page no.	
	Project :-		REV. No :-	
			Calcs By :-	
	Title :-		Check. By :-	
			Approve By :-	
EC3 Column Design Calculation	Calculation sheet		Date:	

Column Design Calculation

INPUT DATA

MEMBER SIZE:	HEB340	
YIELD STRESS OF STEEL (Fy):	355	Mpa
ELASTIC MODULUS (E):	210,000	MPa
EFFECTIVE UNBRACED LENGTH ABOUT X AXIS (KLy):	4.34	m
EFFECTIVE UNBRACED LENGTH ABOUT Y AXIS (KLz):	4.34	m
γ_{M0}	1.00	
γ_{M1}	1.00	

DESIGN LOADING

ALLOWABLE STRESS DESIGN

AXIAL LOAD (COMPRESSION) (N_{ED}):	3326	kN
BENDING MOMENT ABOUT X AXIS (MyED):	0.0	kN-m
BENDING MOMENT ABOUT Y AXIS (Mz):	0	kgf-m

MEMBER PROPERTIES

GENERAL

A =	170.90	cm2
d =	340.00	mm
tw =	12.00	mm
bf =	300.00	mm
tf =	21.50	mm
h = T =	340	mm
r=	27	mm
hc =	340.00	mm
kc=	0.75	

Y- AXIS

Iy =	36660	cm^4
Wpl,y =	13.18	cm^3
ry =	14.65	cm
Zy =	0.00	cm^3

Z - AXIS

Iz =	9690	cm^4
Wpl,z =	39.42	cm^3
rz =	7.53	cm

Web in compression (Table 5.2 of EC3-1-1)

e= 0.81

c/t= 20 < 33ε 26.84934

(class1)

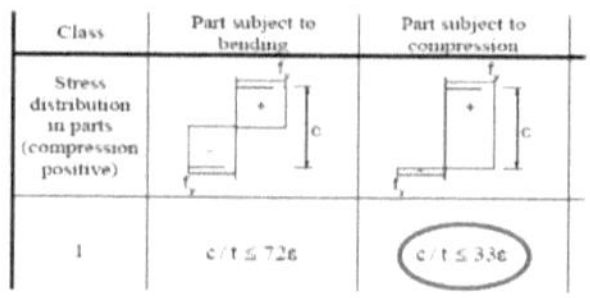

Flange in compression (Table 5.2 of EC3-1-1)

C/t 5.44 < 9ε 7.322549

(class1)

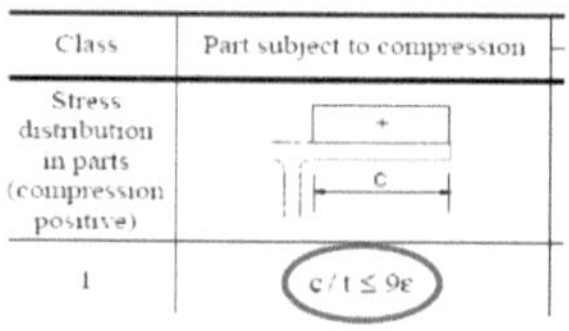

Cross section verification

NC,RD= 6066.95 kN > NED= 3326 kN OK safe

4.8.1 Compression Member

4.8.1.1 Strength Reduction Factor LRFD

The reduction factor is as follows:

$$\varphi_c = 0.90$$

4.8.1.2 Effective Column Length

The effective length, L_c, for the calculation of member slenderness, L, shall be determined in accordance with Chapter C or Appendix 7, where L_c = effective length of the member, in. (mm) = KL K = effective length factor; L = laterally unbraced length of the member, in. (mm); r = radius of gyration, in. (mm)

In general

$$L_c / r < 200$$

4.8.1.3 Flexural Buckling of Members without Slender Elements

The nominal compressive strength, P_n, shall be determined based on the limit state of flexural buckling:

$$P_n = F_n \cdot A_g$$

The nominal stress, F_n, is determined as follows:

a. If $L_c/r \leq 4.71\ (E/f_y)^{0.5}$
   Or

$$F_y / F_e \leq 2.25$$

$$F_n = \left(0.658^{\frac{F_y}{F_e}}\right) F_y$$

b. If $L_c/r > 4.71\ (E/f_y)^{0.5}$
   Or

$$F_y / F_e > 2.25$$

$$F_n = 0.877 F_e$$

A_g = gross area of the member, in.^{2} (mm^2)
E = modulus of elasticity of steel, ksi (MPa) = 29,000 ksi (200,000 MPa)
$F_e = (\pi)^2\ E/(L_c/r)^2$

4.8.1.4 Single-Angle Compression Members

The nominal compressive strength for single members will be calculated as the lowest value from the above equation in Section 4.8.1.3 or the flexural-torsional buckling.

And this is considered only if

$$b/t > 0.71\left(E/f_y\right)^{0.5}$$

where

$$F_e = \left(\frac{F_{ey} + F_{ez}}{2H}\right)\left[1 - \sqrt{1 - \frac{4F_{ey}F_{ez}H}{\left(F_{ey} + F_{ez}\right)^2}}\right]$$

$$F_{ez} = \left(\frac{\pi^2 EC_w}{L_{cz}^2} + GJ\right)\frac{1}{A_g\bar{r}_o^2}$$

$F_{ey} = (\pi)^2\, E/(L_{cy}/r_y)^2$
H = flexural constant

$$= 1 - \frac{x_o^2 + y_o^2}{\bar{r}_o^2}$$

r_o = polar radius of gyration about the shear centre, in. (mm)
x_o, y_o = coordinates of the shear centre with respect to the centroid, in. (mm)

In case of single angle AISC, the effects of eccentricity on single-angle members are permitted to be neglected, and the member evaluated as axially loaded using one of the effective slenderness ratios, provided it complies with the following s:

1. Members are loaded at the ends in compression through the same leg.
2. Members are attached by welding or by connections with a minimum of two bolts.
3. There are no intermediate transverse loads.
4. L_c/r, as determined in this section, does not exceed 200.
5. For unequal leg angles, the ratio of long leg width to short leg width is less than 1.7.
    a. For angles that are individual members or are web members of planar trusses with adjacent web members attached to the same side of the gusset plate or chord
        1. For equal-leg angles or unequal-leg angles connected through the longer leg
           If $L/r_a \le 80$
           $L_c/r = 72 + 0.75(L/r_a)$
           If $L/r_a > 80$
           $L_c/r = 32 + 1.5(L/r_a)$
    b. For angles that are web members of box or space trusses with adjacent web members attached to the same side of the gusset plate or chord
    c. If $L/r_a \le 75$
    d. $L_c/r = 60 + 0.8\ (L/r_a)$
    e. If $L/r_a > 75$
    f. $L_c/r = 45 + (L/r_a)$
       r_a = radius of gyration about the geometric axis parallel to the connected leg, in. (mm)

4.8.1.5 Round HSS

The effective area, A_e, is determined as follows:
If $D/t \leq 0.11E/F_y$

$$A_e = A_g$$

If $0.11\ (E/F_y) < D/t < 0.45\ E/F_y$,

$$A_e = \left(0.67 + 0.038E/F_y\left(D/t\right)\right)A_g$$

where D = outside diameter of round HSS, in. (mm), and t = thickness of wall, in. (mm).

FURTHER READING

CEN (2005). *Eurocode 3: Design of Steel Structures - Part 1-1: General Rules and Rules for Buildings*. 2005.

ECCS (1976). *Manual on stability of steel structures*, 2nd Edition, P022, ECCS Technical Committee 8 – Structural Stability, European Convention for Constructional Steelwork, Brussels.

Galambos, T.V., ed. (1988). *Guide to Stability Design Criteria for Metal Structure*, 4th edition, Structural Stability Research Council, John Wiley & Sons, Inc., New York.

SCI (2006). NCCI: Effective lengths of columns and truss elements in truss portal frame construction, https://www.steel-ncci.co.uk/.

Timoshenko, S.P. and Gere, J.M. (1961). *Theory of Elastic Stability*, McGraw-Hill, New York.

Trahair, N.S. and Bradford, M.A. (1994). *The Behaviour and Design of Steel Structures*, E&F Spon, London.

Trahair, N.S., Bradford, M.A., Nethercot, D.A., and Gardner, L. (2008). *The Behaviour and Design of Steel Structures to EC3*, 4th edition, Taylor Francis, USA.

5 Beam Design

5.1 INTRODUCTION

Beams are perhaps the most fundamental of structural components. They are structural members loaded transversely along their lengths and in the plane of their webs. Beams resist flexural, shear, and torsional stresses.

Regardless of the methods of design or the design code used, three aspects must be considered in beam design.

- Strength requirements necessitate that the cross section of a member be adequate in resisting the applied bending moments and shearing forces.
- Stability requirements ensure that the steel plate elements composing the cross section resist local buckling phenomena. The web should be designed to be safe against buckling caused by shear forces and bending moments. The cross section should possess adequate torsional strength to resist lateral–torsional buckling.
- Performance considerations require that the deformation of the member is not excessive under service loads. As such, limitations on deflections may sometimes govern beam design.

Beams are structural elements subjected solely to bending moments and shearing forces. The normal force is neglected (considered zero).

- Purlins
- Crane track girder
- Beams in flooring systems, as shown in Figure 5.1
- Girders of rigid beams (Figures 5.2 and 5.3)

A variety of section shapes and beam types may be used depending on the magnitude of loading and the span, as shown in Table 5.1 (Figure 5.4).

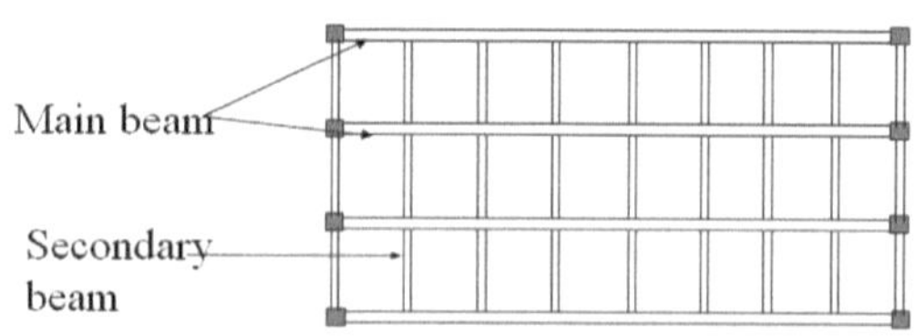

FIGURE 5.1 Floor structure system.

DOI: 10.1201/9781003307303-5

FIGURE 5.2 Photo for roof system.

FIGURE 5.3 Steel frame in Dubai metro station.

5.2 BEAM DESIGN BASED ON EC3

The next sections present the way of design the steel beams as per EC3, the concept approach is the same as AISC but there are some differences presented in next sections.

5.2.1 Restrain Beam

Steel beams can often be designed simply based on bending moment resistance. The design should ensure that the maximum applied moment is less than the moment resistance of the selected cross section. Furthermore, the beam should be stiff enough so

TABLE 5.1
Typical Beam Types for Various Applications

Beam Type	Span Range (m)	Notes
Angles	3–6	Used for roof purlins, sheeting rails, etc., where only light loads have to be carried.
Cold-formed sections	4–8	Used for roof purlins, sheeting rails, etc., where only light loads have to be carried.
Rolled sections: UB, IPE, UPN, HE	1–30	*Most frequently used type of section*; proportions selected to eliminate several possible types of failure.
Open web joists	4–40	Prefabricated using angles or tubes as chords and round bars for web diagonals; used in place of rolled sections.
Castellated beams	6–60	Used for long spans and/or light loads, depth of UB increased by 50%, web openings may be used for services, etc.
Compound sections e.g. IPE + UPN	5–15	Used when a single rolled section would not provide sufficient capacity; often arranged to provide enhanced horizontal bending strength as well.
Plate girders	10–100	Made by welding together three plates, sometimes automatically; web depth up to 3–4 m sometimes need stiffening.
Box girders	15–200	Fabricated from plate, usually stiffened; used for OHT cranes and bridges due to good torsional and transverse stiffness properties.

that it does not deflect excessively, which could affect serviceability. Beams that are unable to move laterally are termed "restrained", and are unaffected by out-of-plane buckling (lateral–torsional instability). Beams may be considered restrained if:

- Full lateral restraint is provided, for example, through the positive attachment of a floor system to the top flange of a simply supported beam, as illustrated in Figure 5.5. Full lateral restraint may be assumed to exist if the frictional or positive connection of a floor (or other) construction to the compression flange of the member is capable of resisting a lateral force of no less than 2.5% of the maximum force in the compression flange of the member.
- Adequate torsional restraint of the compression flange is provided, for example, by profiled roof sheeting.
- Closely spaced bracing elements are provided such that the minor axis slenderness is low (see unrestrained beams for details).

Additionally, sections bent about their minor axis cannot fail by lateral–torsional instability, and it is unlikely that sections with high torsional and lateral stiffness, as in the case of square or rectangular hollow sections, will fail in this way.

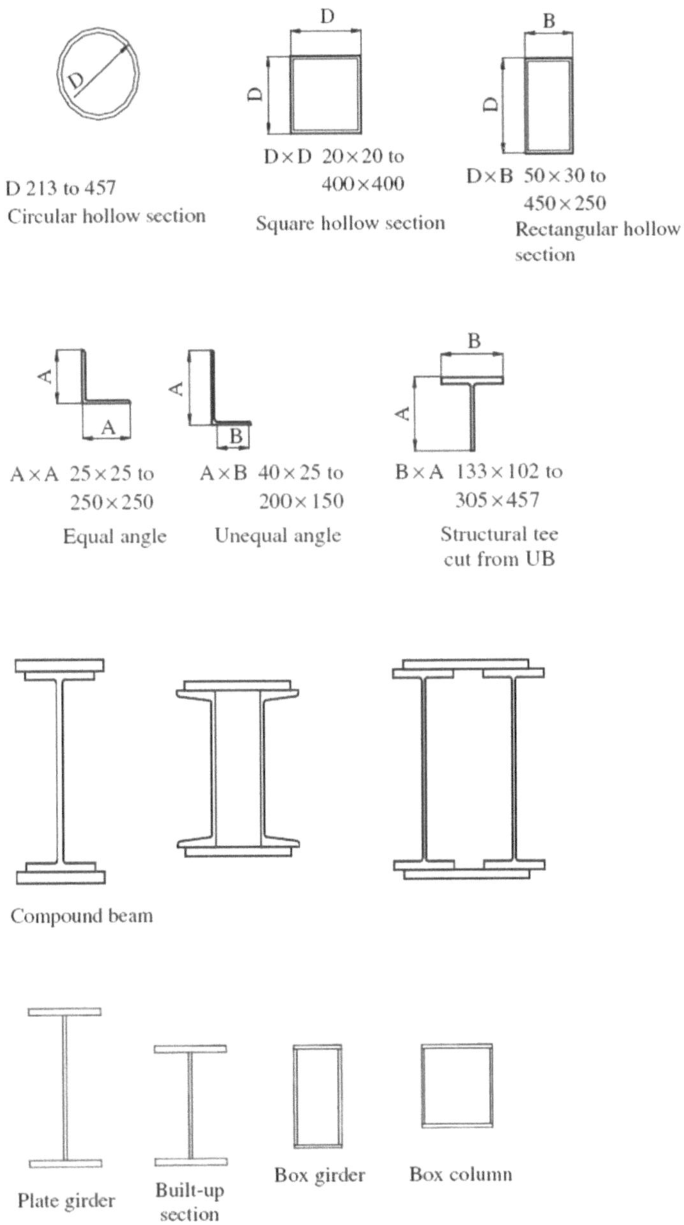

FIGURE 5.4 Types of steel sections.

It is assumed that there are adequate restraints on the beam to prevent lateral buckling. During the detailing phase, as noted in "Evil in Detail," the designer should verify this assumption and consider any restrictions.

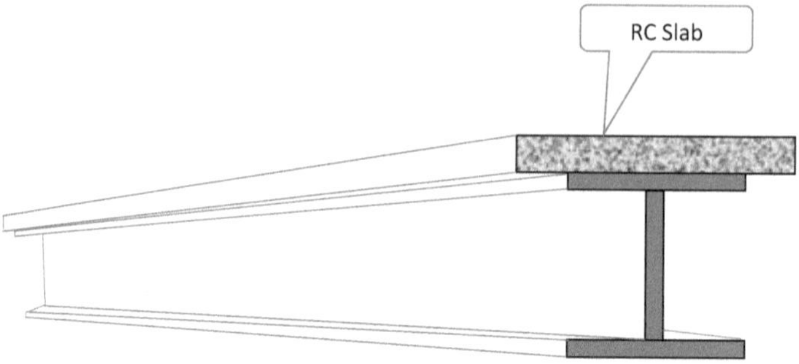

FIGURE 5.5 Fully laterally restrained beam.

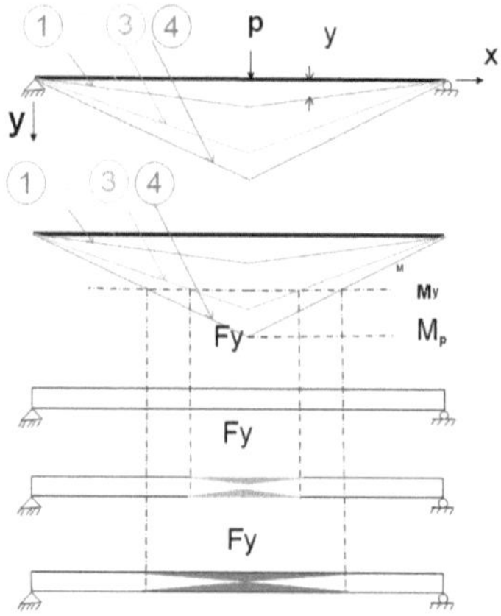

FIGURE 5.6 Simple beam failure stresses increasing steps.

5.2.2 Moment Resistance

The beam design is based on the failure steps shown in Figure 5.6. In this figure, the load, P, is applied to the centre of the beam, and the load is gradually increased, drawing the corresponding bending moment diagram. The beam failure passes through four steps:

1. The stress on the bottom and top fibre of the steel section is less than the yield strength
2. The stresses in the top and bottom fibres of the steel section are equal to the yield strength, and the moment is M_y
3. Part of the section has stresses equal to the yield strength
4. At this step, the entire cross section is affected by stresses equal to the yield strength, so we reach the plastic limit, and the moment is M_p.

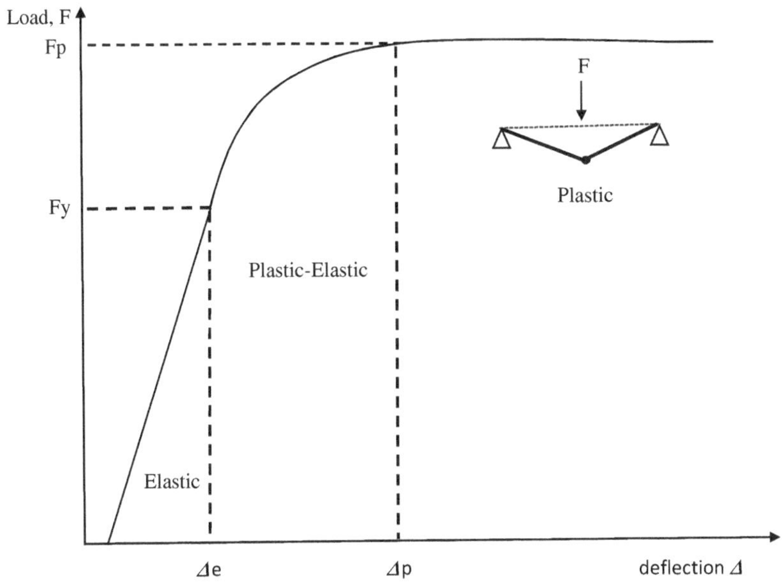

FIGURE 5.7 Behaviour of a simply supported beam.

Therefore, it is important to define whether the section is compact or non-compact. A compact section means the section can carry load until it reaches the plastic moment with local buckling occurring in the web or the flange. However, a non-compact section is one that can reach the yield only in top and bottom fibres without causing local buckling of the web or the flange.

In general, channels, equal angles, and unequal angles are always considered non-compact sections.

In a simple single-span beam, as shown in Figure 5.7, failure occurs when the design value of the bending moment, M_{Ed}, exceeds the design moment resistance of the cross section, $M_{c,Rd,b}$. The magnitude of this resistance depends on the section shape, material strength, and section classification.

For cases where the shear force, V_{Ed}, on the cross section can be considered small enough, its effect on the resistance moment may be neglected. This value of V_{Ed} is defined in clause 5.2.9 as not exceeding 50% of the design plastic shear resistance, $V_{pl,Rd}$. In such cases, no reduction in the resistance defined for bending and axial force will be applied.

The exception to this is when shear buckling reduces the section resistance.

For Class 1 and 2 cross sections, the design moment resistance, $M_{c,Rd}$, is determined as follows, as presented in Figure 5.7 for the section in plastic

$$M_{c,Rd} = M_{pl,Rd} = \frac{W_{pl} f_y}{\gamma_{M0}} \tag{5.1}$$

For Class 3 cross sections, the design moment resistance, $M_{c,Rd}$, is determined as follows:

$$M_{c,Rd} = M_{el,Rd} = \frac{Wy_{el,\min}}{\gamma_{M0}} \tag{5.2}$$

For a class 4 cross section, the design moment resistance, $M_{c,Rd}$, is determined as follows:

$$M_{c,Rd} = \frac{Wy_{eff,\min}}{\gamma_{M0}} \tag{5.3}$$

It is worth mentioning that in cases where bolt holes are located in the critical section of the flange subjected to tension, it is mandatory to ensure that the ratio of the net area to the gross area of the flange is not so small that the section would rupture at the net section before the gross section yields. This equation will apply to achieve that.

$$\frac{A_{f,net}0.9f_u}{\gamma_{M2}} \geq \frac{A_f f_y}{\gamma_{M0}} \tag{5.4}$$

Bolt holes in the tension zone of the web should be considered similarly; however, holes in the compression zone (both web and flange) may be ignored unless they are oversized or slotted.

In the case of a statically indeterminate structure such as a continuous beam, the attainment of the design moment resistance at the point of maximum moment, obtained from an elastic analysis, will not normally lead to collapse, as shown in Figure 5.8. In fact, the cross section at that point will behave as a hinge, providing the requisite rotational capacity. At this moment, the pattern of moments within the structure will alter from the original elastic distribution as successive hinges form. The redistribution of moments enables the structure to withstand loads beyond those that produce the first hinge until eventually, sufficient hinges have formed to turn the structure into another mechanism, leading to failure. This is referred to as plastic design, which requires a cross section to be class 1, capable of rotating while sustaining the plastic resistance moment. This analysis can be conducted using pushover analysis, which is available in most structural analysis software.

5.2.3 Shear Resistance

Bending governs the design of many steel beams; however, in the case of a short span with a heavy concentrated load, shear resistance can dictate the design. Figure 5.9 illustrates the pattern of shear stress in an I-section, assuming elastic behaviour. In general, all the shear force is carried by the web, and since the variation in shear stress across the web is quite small, it is sufficiently accurate for design to assume an average shear stress over the web.

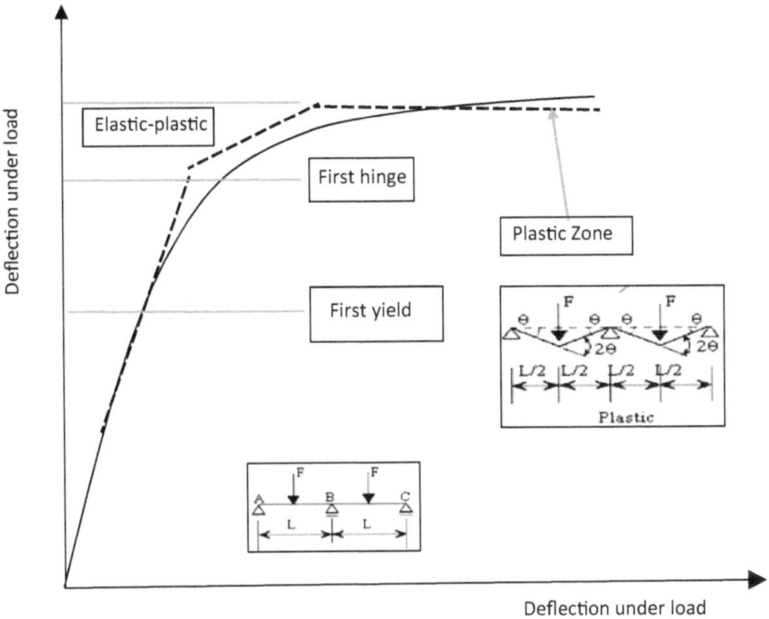

FIGURE 5.8 Load deflection curve for a statically indeterminate beam.

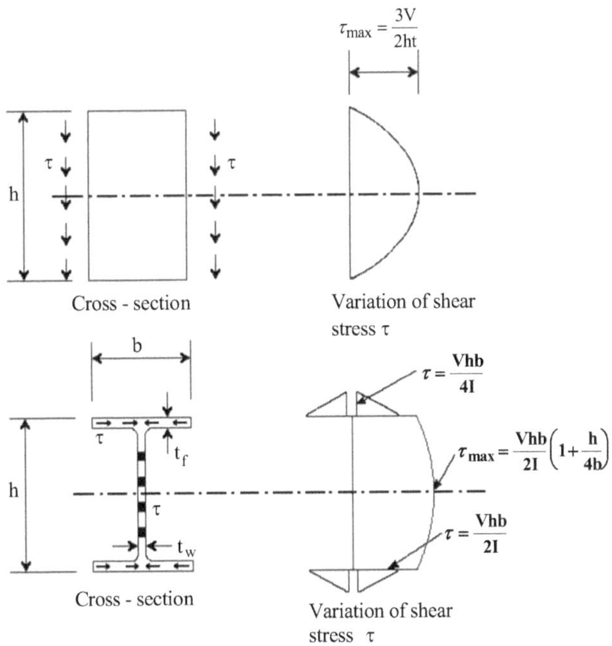

FIGURE 5.9 Distribution of shear force in beams.

Steel in shear yields at a stress approximately equal to $1/\sqrt{3}f_y$. Hence, the design value of the shear force, V_{Ed}, at each cross section is compared with the plastic shear resistance, $V_{pl,Rd}$, of the shear area, Av:

$$V_{pl,Rd} = \frac{A_v\left(f_y/\sqrt{3}\right)}{\gamma_{M0}} \tag{5.5}$$

The calculation of the shear area A_v may be taken as follows:

a. rolled I- and H-sections, load parallel to the web equal to $A-2bt_f+(t_w+2r)$ t_f but not less than $\eta\ h_w\ t_w$
b. rolled channel sections, load parallel to the web, equal to $A-2bt_f+(t_w+r)\ t_f$
c. rolled T-section, load parallel to the web equal to $0.9\ (A-bt_f)$
d. welded I, H and box sections, load parallel to the web equal to $\eta\ \Sigma(h_w\ t_w)$
e. welded I, H, channel and box sections, load parallel to flanges equal to $A-\Sigma(h_w\ t_w)$
f. rolled rectangular hollow sections of uniform thickness:
   load parallel to depth equal to $Ah/(b+h)$
   load parallel to width equal to $Ab/(b+h)$
g. circular hollow sections and tubes of uniform thickness equal to $2A/\pi$

where

A is the cross-sectional area
b is the overall breadth
h is the overall depth
h_w is the depth of the web
r is the root radius
t_f is the flange thickness
t_w is the web thickness (if the web thickness is not constant, t_w should be taken as the minimum thickness)
η is conservatively taken as 1.0.

For verifying the design elastic shear resistance, for I- or H-sections, the shear stress in the web may be taken as follows:

$$t_{ED} = V_{ED}/A_w \quad \text{in case of } A_f/A_w \geq 0.6$$

where A_f is the area of one flange

$A_w = h_w\ t_w$, which is the area of the web

5.2.4 Moment Resistance with High Shear

When the design shear force, V_{Ed}, exceeds 50% of the plastic shear resistance, $V_{pl,Rd}$, the cross-combinations of moment and shear force should be calculated using a reduced yield strength, f_y, given by

$$(1-p)f_y \tag{5.6}$$

The value of ρ is dependent on the ratio of the shear load to shear capacity and is given by

$$\rho = \left(\frac{2V_{Ed}}{V_{pl,Rd}} - 1\right)^2 \tag{5.7}$$

In the case of I cross sections with equal flanges and a bending moment applied about the major axis, the reduced design plastic resistance moment, $M_{yv.Rd}$, allowing for the shear force, may alternatively be obtained from this equation:

$$M_{y,v.Rd} = \frac{\left[W_{pl,y} - \frac{\rho A_w{}^2}{4t_w}\right]f_y}{\gamma_{M0}}, \text{but } M_{y,v.Rd} \le M_{y,c.Rd} \tag{5.8}$$

5.2.5 Biaxial Bending

In the case of a beam bent about both axes of the cross section, there is a plastic neutral axis inclined to the rectangular axes by an amount that depends on the ratio of the applied moments and the precise shape of the section. The interaction equation will be as follows:

$$\left[\frac{M_{y,Ed}}{M_{N,y,Rd}}\right]^{\alpha} + \left[\frac{M_{z,Ed}}{M_{N,z,Rd.}}\right]^{\beta} \le 1 \tag{5.9}$$

In the above expression, α and β are constants, which may conservatively be taken as unity; otherwise, they can be taken as shown in Table 5.2:

TABLE 5.2
Values for α and β

Sections	α	B
I and H	2	$5n \ge 1$
Circular hollow	2	2
Rectangular hollow	$\alpha = \beta = \frac{1.66}{1-1.13n^2} \le 6$	

$n = N_{Ed}/N_{pl,Rd}$.

5.2.6 Unrestrained Beam

As a rule of thumb, when a slender structural member is subjected to a load on its stiff plane, there exists a tendency for failure due to buckling in another, more flexible plane. In the case of a beam bent about its major axis, failure may occur in a manner involving both lateral deflection and twisting, a phenomenon known as lateral–torsional buckling.

The end of the cantilever would deflect in the vertical plane only, with no out-of-plane deflection, if the cantilever is perfectly straight, with a cross section that is initially stress-free and perfectly elastic. After increasing the load until the moment reaches a critical value, the tip of the cantilever will buckle, deflect laterally, and twist as presented in Figure 5.10.

A design approach for beams prone to failure by lateral–torsional buckling must necessarily account for a large number of factors, including section shape, the degree of lateral restraint, type of loading, residual stress pattern, and initial imperfections, and is therefore relatively complex. It is instructive to first consider a simple basic model, which can then be developed to include more general cases.

5.2.7 Elastic Buckling of a Simply Supported Beam

Figure 5.11 shows a perfectly elastic, initially straight, I, beam loaded by equal and opposite end moments about its major axis in the plane of the web. Assume the beam is unrestrained along its length, except at each end where the sections are prevented from twisting and lateral deflection but are free to rotate both in the

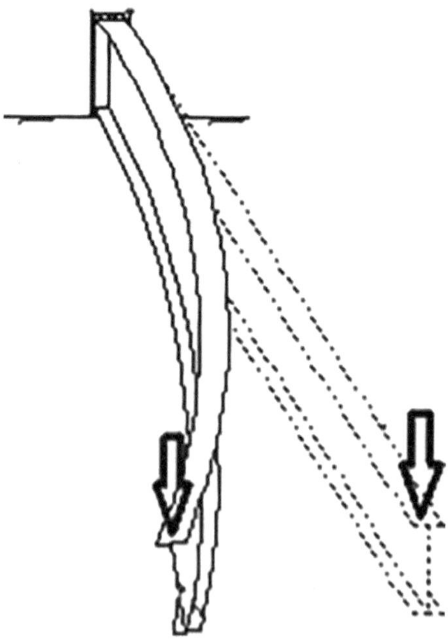

FIGURE 5.10 Lateral–torsional buckling of a slender cantilever beam.

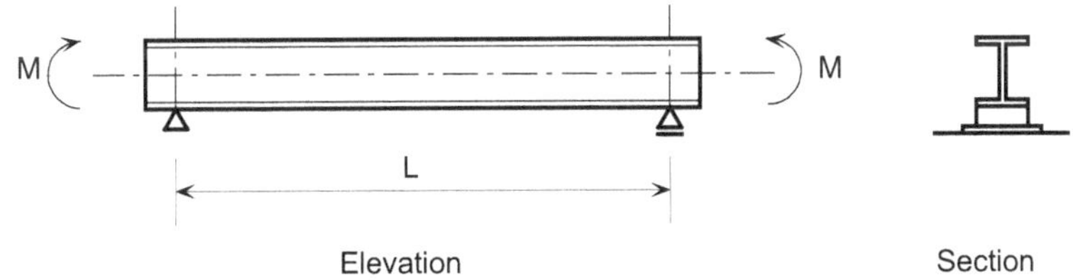

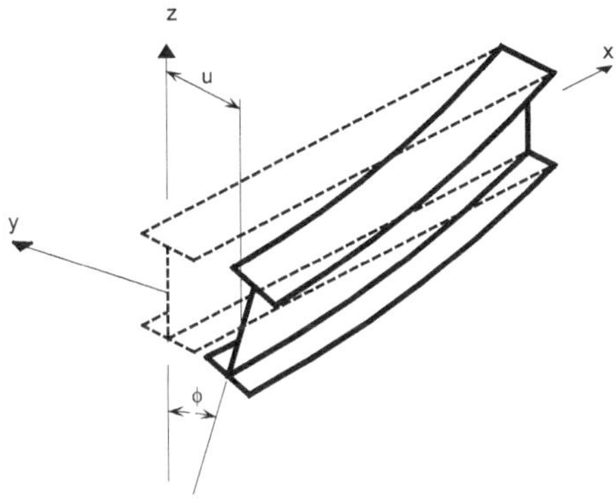

FIGURE 5.11 Lateral–torsional buckling of a simple I beam under uniform moment.

plane of the web and in plan. The buckled shape and resultant deformations are also shown in the figure, which presents half of the beam and the deformation at the mid-span.

The moment necessary to cause buckling may be determined by equating the disturbing effect of the applied end moments, acting through the buckling deformations, to the internal bending and torsional resistance of the section. The critical value of applied end moments, *the elastic critical moment* (M_{cr}), is found to be

The elastic critical moment may be calculated from the following general formula:

$$M_{cr} = C_1 \frac{\pi^2 EI_z}{(k_z L)^2} \left\{ \begin{array}{l} \left[\left(\dfrac{k_z}{k_w} \right)^2 \dfrac{I_W}{I_z} + \dfrac{(k_z L)^2 GI_T}{\pi^2 EI_z} + (C_2 z_g - C_3 z_j)^2 \right]^{0.5} \\ -(C_2 z_g - C_3 z_j) \end{array} \right\},$$

where

$z_g = z_a - z_s$, where z_a and z_s are the coordinates of the point of application of the load and of the shear centre, relative to the centroid of the cross section; these quantities are positive if located in the compressed section and negative if located in the tension section.

In case of c_2 and c_3 equal to zero,

$$M_{cr} = \frac{\pi^2 EI_z}{L^2}\left[\frac{I_w}{I_z} + \frac{L^2 GI_t}{\pi^2 EI_z}\right]^{0,5} \quad (5.10)$$

where

I_t is the torsion constant;
I_w is the warping constant;
I_z is the second moment of area about the minor axis;
L is the unrestrained length of the beam.

The critical bending moment in the equation mainly depends on flexural stiffness (EI_z) and torsional stiffness (GI_t and EI_w), which are direct consequences of the lateral and torsional components of buckling deformations. The relative importance of these factors will reflect the type of cross section considered.

As per equation 5.10, parameters compares values of the elastic critical moment (M_{cr}) for an I beam and an H-section with similar in-plane plastic moment capacities. Lateral–torsional buckling is a potentially more significant design consideration for a beam section that is much less stiff laterally and torsionally.

5.2.8 Development of a Design Approach

Real beams are not perfectly straight, nor is the material elastic. It is important to mention that there is an effect of residual stresses and strain hardening on lateral buckling strength. At high slenderness values, the behaviour is well represented by elastic buckling theory; however, for stocky beams, there is a complex interplay because inelastic behaviour causes a reduction in capacity. For very stocky beams, the capacity is limited by the plastic resistance of the section.

Applying a theoretical treatment to the problem would be too complex for routine design; thus, a combination of theory and test results is required to produce a reliable safe design approach.

Figure 5.12 compares a typical set of lateral torsional buckling test data with the theoretical elastic critical moments given by equation 5.1. A non-dimensionalized form of the plot has been used, which allows results from different test series with varying cross sections and material strengths to be compared directly via a non-dimensional slenderness, $\bar{\lambda}_{LT}$. For stocky beams $\left(\bar{\lambda}_{LT} < 0.4\right)$, the capacity is unaffected by lateral–torsional buckling and is governed by the plastic resistance moment of the cross section. Conversely, slender beams $\left(\bar{\lambda}_{LT} < 1.2\right)$ have capacities close to the theoretical elastic critical moment, M_{cr}. However, beams of intermediate slenderness, which cover many practical beams, are significantly adversely affected by inelasticity and geometrical imperfections; therefore, elastic theory provides an upper-bound solution. A design expression linking the plastic capacity of stocky beams with the elastic behaviour of slender beams is required. EC3 achieves this by using a reduction factor for lateral–torsional buckling, χ_{LT}.

The design buckling resistance moment ($M_{b.Rd}$) of a laterally unrestrained beam is thus taken as

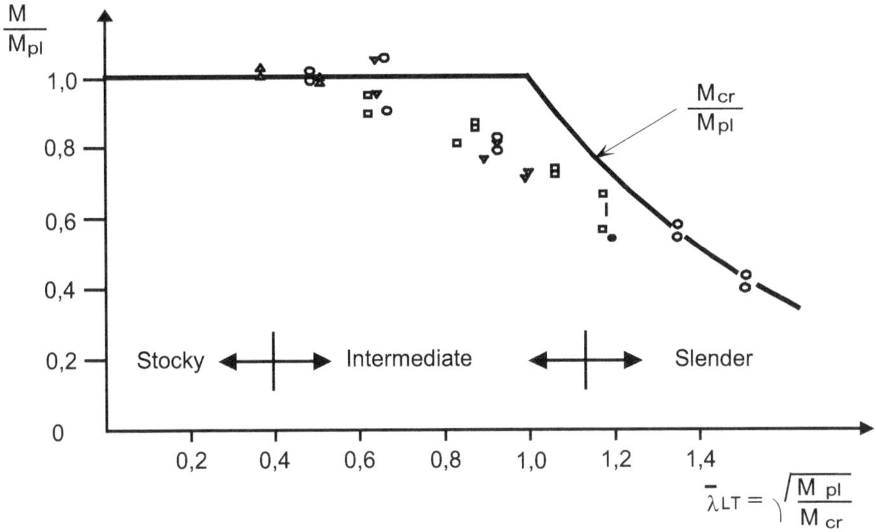

FIGURE 5.12 Comparison of test data with theoretical elastic critical moments (Adapted from EC3).

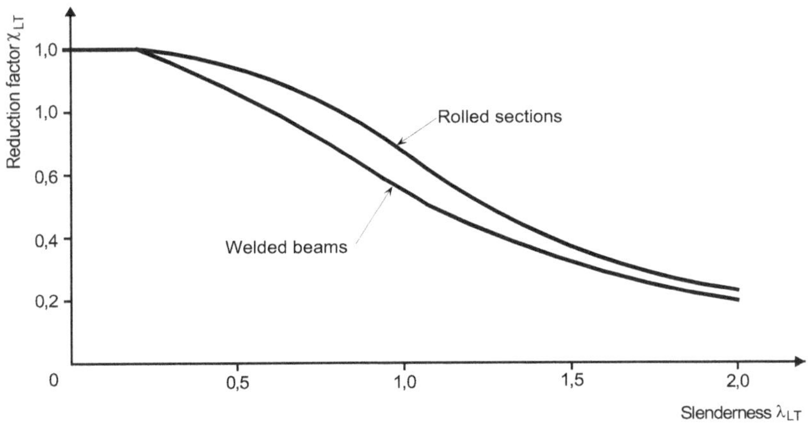

FIGURE 5.13 Lateral–torsional buckling reduction factor.

$$M_{b.Rd} = \chi_{LT}\beta_{wWpl.y}f_y/\gamma_{m1}$$

which is effectively the plastic resistance of the section multiplied by the reduction factor (χ_{LT}). Figure 5.13 shows the relationship between χ_{LT} and the non-dimensional slenderness, $\bar{\lambda}_{LT}$.

The curves shown are expressed by the following equation:

$$\chi_{LT} = \frac{1}{\varphi_{LT} + \left[\varphi_{LT}^2 - \bar{\lambda}_{LT}^2\right]^{0,5}} \tag{5.11}$$

where

$$\varphi_{LT} = 0.5\left[1 + a_{LT}\left(\bar{\lambda}_{LT} - 0.2\right) + \bar{\lambda}_{LT}{}^{2}\right] \tag{5.12}$$

in which a_{LT} is an imperfection factor. It can be taken as 0.21 for rolled sections and 0.49 for welded sections, which have more severe residual stresses. On the other hand, it can be taken from Table 5.3 after obtaining the buckling curve from Table 5.4:

$\bar{\lambda}_{LT}$, the non-dimensional slenderness, defined as $\sqrt{M_{plRd}/M_{cr}}$, may be calculated either by determining the plastic resistance moment and elastic critical moment from first principles or, more conveniently, by the relationship:

$$\bar{\lambda}_{LT} = \left[\frac{\lambda_{LT}}{\lambda_1}\right] \beta_w{}^{0,5} \tag{5.13}$$

where

$$\lambda_1 = \pi\left[\frac{E}{f_y}\right]^{0,5} \tag{5.14}$$

And

λ_{LT} may be calculated using appropriate expressions for a variety of section shapes. For example, for any section I- or H-section with equal flanges, and subjected to a uniform moment with simple end restraints:

$$\lambda_{LT} = \frac{L/i_z}{\left[1 + \frac{1}{20}\left[\frac{L/i_z}{h/t_f}\right]^2\right]^{0.25}} \tag{5.15}$$

TABLE 5.3
Recommended Value for Imperfection Factors for Lateral–torsional Buckling Curves

Buckling Curve	***a***	***b***	***C***	***d***
Imperfection factor α_{LT}	0.21	0.34	0.49	0.76

TABLE 5.4
Recommended Values for Lateral–torsional Buckling Curves for Cross Sections

Rolled I-section	$h/b \leq 2$	*A*
	$h/b > 2$	*B*
Welded I-section	$h/b \leq 2$	*C*
	$h/b > 2$	*D*
Other cross sections	–	*D*

5.2.9 Extension to Other Cases

5.2.9.1 Load Pattern

A uniform moment applied to an unrestrained beam is the most severe case for consideration of lateral–torsional buckling. An elastic analysis of alternative load cases results in higher values of elastic critical moments.

For example, in the case of an IPE 400 and the load applied on the top, the value of z_g will be as shown in this Figure 1:

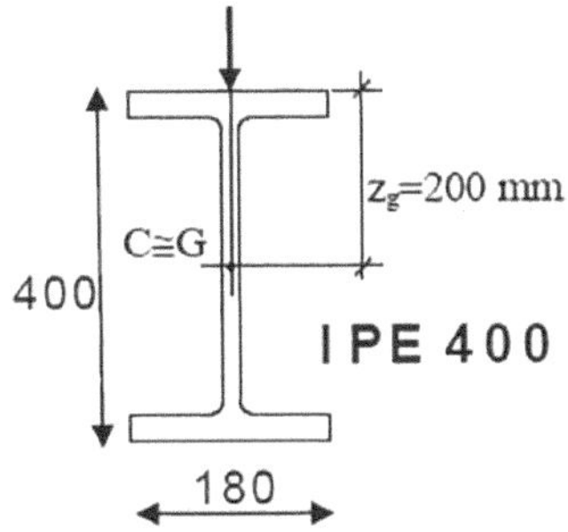

Z_j is calculated from the following equation:

$$z_j = z_s - \left(0.5\int_A \left(y^2+z^2\right)\left(z/I_y\right)dA\right)$$

z_j is a parameter that reflects the degree of asymmetry of the cross section in relation to the y-axis. It is zero for beams with a doubly symmetric cross section (such as I or H cross sections with equal flanges) and takes positive values when the flange with the largest second moment of area about z is the compressed flange at the cross section with the maximum bending moment.

Thus, in case of symmetrical section, the previous equation will be as follows:

$$M_{cr} = C_1\frac{\pi^2 EI_z}{(kL)^2}\left\{\sqrt{\left(\frac{k}{k_w}\right)^2\frac{I_w}{I_z}+\frac{(kL)^2 GI_t}{\pi^2 EI_z}+\left(C_2 z_g\right)^2}-C_2 z_g\right\}$$

The factor, k, refers to end rotation in the plane. K should be taken as no less than 1.0 unless justified. However, k_w refers to end wrapping; thus, special provision of wrapping fixity is made and k_w should be 1.0. Therefore, in the most common case, k and k_w are taken equal to 1.0.

C_1, C_2, and C_3 are coefficients depending on the shape of the bending moment diagram and on support conditions. Their values are provided in Tables 5.5 and 5.6 for some common situations (Boissonnade et al., 2007) In Tables 5.5 and 5.6, the support conditions are representative of the "standard case"; however, lateral–bending restraints and warping restraints may be taken into account through the parameters k_z and k_w described below:

TABLE 5.5
C1 Coefficient for Different Loadings with Two Hinge Supports

Bending Moment Diagram	ψ	C_1	C_3	
			$\psi_f \leq 0$	$\psi_f > 0$
	1.0	1.0	1.0	
		1.05	1.019	
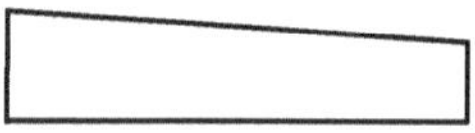	+3/4	1.14	1.0	
		1.19	1.017	
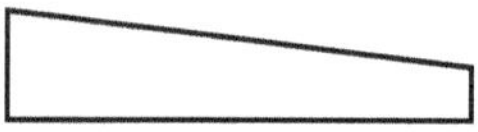	+1/2	1.31	1	
		1.37	1	
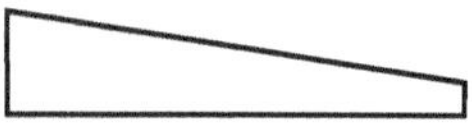	+1/4	1.52	1	
		1.60	1	
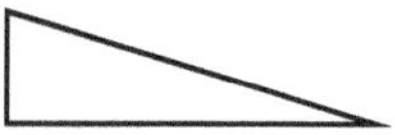	0	1.77	1	
		1.86	1	
	−1/4	2.06	1.0	0.85
		2.15	1.0	0.65
	−1/2	2.35	1.0	$1.3\text{-}1.2\psi_f$
		2.42	0.95	$0.77\text{-}\psi_f$
	−3/4	2.60	1.0	$0.55\text{-}\psi_f$
		2.45	0.85	$0.35\text{-}\psi_f$
	−1	2.60	$-\psi_f$	
		2.45	$-0.125\text{–}0.7\psi_f$	

K_z *is* equal to 1.0 and 0.5

In case of beams with end moments $C_2\, z_g = 0$.

$\psi_f = I_{fc} - I_{ft}/(I_{fc} + I_{ft})$ where I_{ft} and I_{ft} are the second moment of area for compressions and tension flanges, respectively, with respect to the weak axis of the section.

C_1 must be divided by 1.05 when $(\pi/K_{w.}L)\ (E\ I_w/G.I_t)^{0.5} \leq 1.0$ but $C_1 \geq 1.0$.

TABLE 5.6
Coefficients C1,C2, and C3 for Beams

Bending Moment Diagram	K_z	C_1	C_2	C_3
	1.0	1.12	0.45	0.525
	0.5	0.97	0.36	0.478
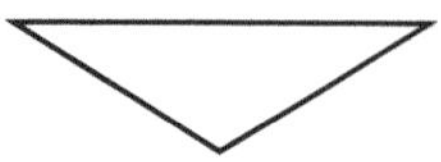	1.0	1.35	0.59	0.411
	0.5	1.05	0.48	0.338
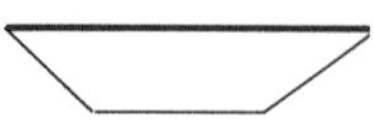	1.0	1.04	0.42	0.562
	0.5	0.95	0.31	0.539

By considering z_g equal to 0 in the case where the transverse load is applied at the shear centre or the bending moment is linear along a segment of a member subjected to lateral restraint.

For example, the elastic critical moment for a uniform moment is as follows:

$$M_{cr} = \frac{\pi}{L}\sqrt{EL_zGI_r}\sqrt{1+\frac{\pi^2 EI_w}{L^2 GI_t}} \tag{5.16}$$

But for a beam with a central concentrated point load, the maximum moment at the centre, at the point of buckling, is

$$M_{cr} = \frac{4.24}{L}\sqrt{EL_zGI_r}\sqrt{1+\frac{\pi^2 EL_w}{L^2 GI_t}} \tag{5.17}$$

which is 4.24/π higher than the base case. EC3 uses this ratio expressed as a factor, C_1, to allow for the loading arrangement, which is the bending moment diagram shape for a variety of loading cases, as shown in Table 5.6. C_1 appears as a simple multiplier in expressions for M_{cr}, as per EC3 eqn F.2, or as $1/\sqrt{C_1}$ in expressions for λ_{LT} (Table 5.7).

5.2.9.2 Level of Application of Load

The lateral stability of a beam is dependent not only on the arrangement of loads within the span but also on the level of application of the load relative to the centroid. Figure 5.14 illustrates the effect of placing the load above and below the centroid for a simple span with a concentrated point load. There are three cases, the worst case if the load applied on the top flange as it will increase the twisting moment on the member. The best case if the load applied in the bottom flange and the natural case if the load at centroid of the section.

TABLE 5.7
C1 Coefficient for Different Loadings

Bending Moment Shape	M_{max}	C_1
	M	1.0
	M	1.879
	M	2.752
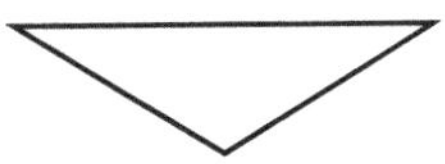	FL/4	1.365
	FL/8	1,132

Note: The above values correspond to an effective length factor *k* of 1,0.

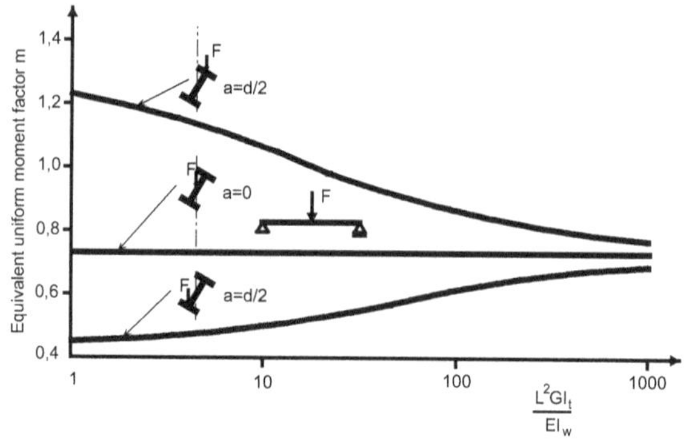

FIGURE 5.14 Effect of level of load application on beam stability (Adapted from EC3).

The influence of this behaviour becomes more significant as the depth of the section increases and/or the span reduces, i.e. as L^2GI_t/EI_w becomes smaller.

5.2.9.3 End Support Conditions

All of the foregoing has assumed end conditions which prevent lateral movement and twist but permit rotation in plan. End conditions that prevent rotation in plan enhance the elastic buckling resistance (in much the same way that column capacities are enhanced by rotational end restraints). A convenient way of including the effect of different support conditions is to redefine the unrestrained length as an effective length, or more precisely, with two effective length factors, k and k_w. The two factors reflect the two possible types of end fixity: lateral bending restraint and warping restraint. However, it should be noted that it is recommended that k_w be taken as 1.0 unless special provision for warping fixing is made. EC3 recommends k values of 0.5 for fully fixed ends, 0.7 for one free and one fixed end, and, of course, 1.0 for two free ends. The value of k is based on designer requirements for member behaviour.

5.2.9.4 Beams with Intermediate Lateral Support

Where beams have lateral restraints at intervals along the span, the segments of the beam between restraints may be treated in isolation, with the design of the beam based on the most critical segment. Lengths of beams between restraints should use an effective length factor k of 1.0 and not 0.7, as in the buckled shape the adjacent unrestrained length will buckle in sympathy.

5.2.9.5 Continuous Beams

Beams that are continuous over a number of spans may be treated as individual spans, taking into account the shape of the bending moment diagram within each span as a result of continuity using the C_1 factor.

5.2.10 Serviceability Checks

In addition to the strength checks outlined above, it is also necessary to assess the performance of beams at the serviceability limit states. Deflections and vibrations of beams must be limited to avoid adverse effects on the appearance or effective use of the structure, discomfort to the occupants, or damage to the finishes and contents of a building. Acceptable limits for deflections should be agreed upon between the client, designer, and competent authorities. For guidance, Table 5.8 provides recommended limiting values for vertical deflections of certain members under the characteristic load combination due to variable loads. Permanent loads should not be included. Circumstances may arise where greater or lesser values would be more appropriate. Other members may also need deflection limits.

For structures open to the public, it is important to ensure that oscillation and vibrations are not so severe that they cause discomfort to users. Verification of the suitability of a design may be achieved through dynamic analysis; however, in many cases, limiting the deflection is sufficient. For example, floors in dwellings and offices should have a lowest natural frequency of not less than 3 Hz (cycles/second). This condition can be satisfied if the instantaneous total deflection, as shown in Table 5.9 is less than 28 mm. For floors in gymnasia or dance halls, the lowest natural frequency should not be less than 5 cycles/second, considering that a deflection limit of 10 mm would satisfy this limit.

In case of flat roofs with slopes of less than 5°, there is a vulnerability to ponding if the roof deflects in such a way that pools of water may form. It is therefore necessary to check roof deflections carefully, including allowances for construction inaccuracies, settlements of foundations, deflection of roof materials, etc. (Figure 5.15).

The allowable **deflection** always depends on the project specifications, but generally, according to the AISC, the following limits apply:

For crane track girders = span/800
For floors and roofs with plastering = span/360

Example 5.1

Contact us to request the Excel sheet at www.elreedyman.com

610 × 229 UB125
$b_f = 229$ mm

TABLE 5.8
Recommended Limiting Values for Vertical Deflections

Type of Beam	Deflection Limit
Cantilevers	Length/180
Beams carrying plaster or other brittle finish	Span/360
Other beams (except purlins and sheeting rails)	Span/200
Purlins and sheeting rails	To suit the characteristics of particular cladding

TABLE 5.9
Suggested Limits for Calculated Deflection Based on BS5950

1. Vertical deflection of beam due to imposed load	
Cantilever	Length/180
Beams is not carrying plaster or brittle finished	Length/200
2. Horizontal deflection of columns due to imposed load and wind load	
Top of columns in single-storey building, except portal frames.	Height/300
In each storey building with more than one storey.	Height of that storey/300
3. Crane girders	
Vertical deflection due to static vertical wheel loads from overhead travelling crane	Span/600
Horizontal deflection(calculate)	Span/500

$h = 612.2\,\text{mm}$
$t_f = 19.6\,\text{mm}$
$t_w = 11.9$
$r = 12.7$
$A = 159\,\text{cm}^2$
$I_z = 3{,}932\,\text{cm}^4$
$i_z = 49.7$
$W_{ply} = 3{,}676\,\text{cm}^3$

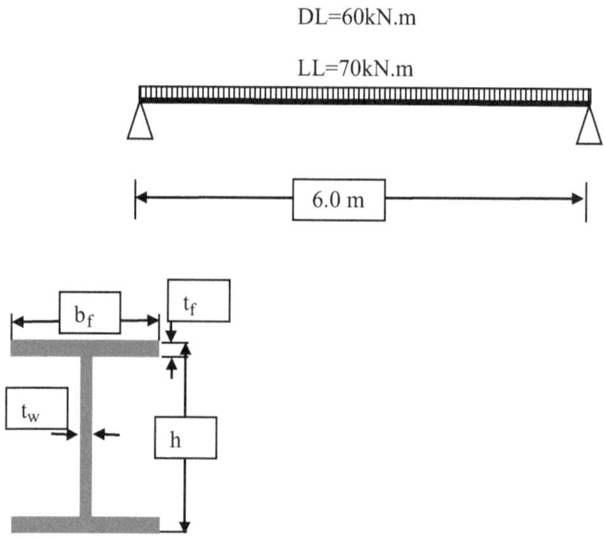

FIGURE 5.15 Example 1.

ELREEDYMAN www.elreedyman.com	Client :-		Page no.	
	Project :-		REV. No :- Calcs By :-	
	Title :-		Check. By :-	
			Approve By :-	
EC3 Beam column	Calculation sheet		Date:	

Beam Calculation

INPUT DATA

MEMBER SIZE:	UB125	
YIELD STRESS OF STEEL (Fy):	265	Mpa
ELASTIC MODULUS (E):	210,000	Mpa
UNBRACED LENGTH IN BENDING (l):	6.00	m
EFFECTIVE UNBRACED LENGTH ABOUT X AXIS (KLy):	6.00	m
EFFECTIVE UNBRACED LENGTH ABOUT Y AXIS (KLz):	6.00	m
γM0	1.00	
G	80770.00	MPa

DESIGN LOADING

ALLOWABLE STRESS DESIGN

BENDING MOMENT ABOUT X AXIS (MyED):	837.0	kN-m
BENDING MOMENT ABOUT Y AXIS (Mz):	0	kN-m
VED	558	kN

MEMBER PROPERTIES

GENERAL

A =	159.00	cm2	
d =	612.20	mm	
tw =	11.90	mm	
bf =	229.00	mm	
tf =	19.60	mm	
h = T =	612.2	mm	
I_t =	154	cm4	torsion constant
hc =	612.20	mm	
I_w =	3450000.00	cm6	Warping constant

Y - AXIS

Iy =	98610	cm^4
Sy =	3481.00	cm^3
ry =	25.00	cm
Wpl,y =	3676.00	cm^3

Z - AXIS

Iz =	3932	cm^4
Sz =	346.00	cm^3
rz =	5	cm
Wpl,z=	535.00	cm^3

e=	0.94				
cf/(tf ε)	5.19	<9	The flange is class 1		
cw/(twε)	48.87	<62	The web is class 1		
Mc,RD=	922.465	kN.m	>	837.00	
Av=	7654.28	mm2			
Vc_rd=	1171.09	kN	>	558.00	ok

N_{pl},Rd=AFy/γM0	4213.50	kN
N_{ED}<Npl,Rd		
N_{ED}<0.25 Npl,Rd		
Mpl,y,Rd=Wpl,y fy/M0	974.14	kN.m
a=A-2bt/A=	0.44	
n=N_{ED}/Npl,R	0.00	
Mn,y,Rd=Mpl,y,Rd (1-n)/(1-0.5a)		
Mn,y,Rd=Mpl,y,Rd (1-n)/(1-0.5a)	1245.24	kN.m
My,ED= 837.00 <	Mn,y,Rd	oK

$$M_{cr} = C_1 \frac{\pi^2 EI_z}{L^2}\sqrt{\frac{I_w}{I_z} + \frac{L^2 GI_t}{\pi^2 EI_z}}$$

Cross-section	Limits	Buckling curve
Rolled I-sections	h/b ≤ 2	a
	h/b > 2	b
Welded I-sections	h/b ≤ 2	c
	h/b > 2	d
Other cross-sections	-	d

Mcr= 964.34 kN.m

$$\bar{\lambda}_{LT} = \sqrt{\frac{W_y f_y}{M_{cr}}}$$

λLT= 0.98

h/b=	2.67	Curve b
α LT=	0.34	
λLT,0=	0.40	maximum value
β=	0.75	minimum value
φ_{LT}=	0.96	
χ_{LT}	0.71	

$$\Phi_{LT} = 0,5\left[1 + \alpha_{LT}\left(\bar{\lambda}_{LT} - \bar{\lambda}_{LT,0}\right) + \beta\bar{\lambda}_{LT}^2\right]$$

$$\chi_{LT} = \frac{1}{\Phi_{LT} + \sqrt{\Phi_{LT}^2 - \beta\bar{\lambda}_{LT}^2}}$$

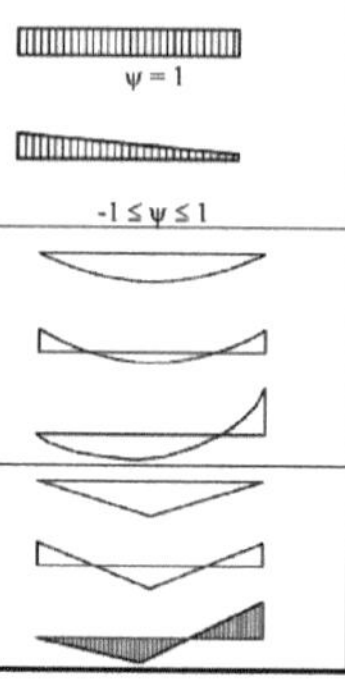

Moment distribution	k_c
ψ = 1	1,0
-1 ≤ ψ ≤ 1	$\frac{1}{1,33 - 0,33\psi}$
	0,94
	0,90
	0,91
	0,86
	0,77
	0,82

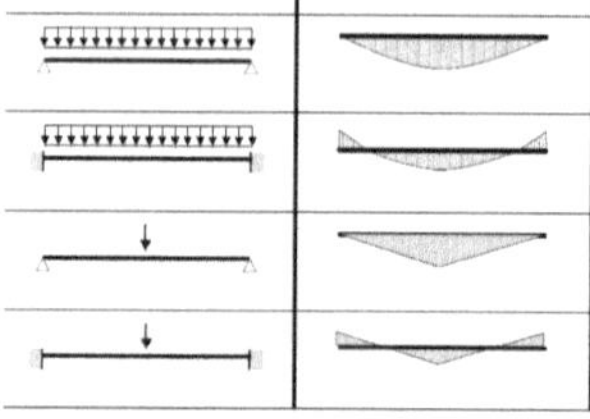

Loading and support conditions	Bending moment diagram	C_1	C_2
		1.127	0.454
		2.578	1,554
		1,348	0.630
		1,683	1,645

Kc= 0.94

$$f = 1 - 0.5 \times (1 - k_c) \times \left[1 - 2.0 \times \left(\bar{\lambda}_{LT} - 0.8\right)^2\right]$$

f 0.97

$\chi_{LT,mod}$	0.73	<1.0	So,$\chi_{LT,mod}$	0.73
C_{my}=C_{mLT}=	0.60	>	0.4	

$$M_{b.Rd} = \chi_{LT} W_y \frac{f_y}{\gamma_{M1}}$$ 714.71 kN.m O.k safe

For biaxial bending

MN,yrd= 974.14 kN.m

MN,zrd= 91.69 kN.m

a=	2.00	or	=1	conservartive
b=	0.00		=1	

- I and H sections: $\alpha = 2$; $\beta = 5n$ but $\beta \geq 1$
- circular hollow sections: $\alpha = 2$; $\beta = 2$
- rectangular hollow sections: $\alpha = \beta = \frac{1,66}{1 - 1,13n^2}$ but $\alpha = \beta \leq 6$

where $n = N_{Ed} / N_{pl,Rd}$.

$$\left[\frac{M_{y,Ed}}{M_{N,y,Rd}}\right]^{\alpha} + \left[\frac{M_{z,Ed}}{M_{N,z,Rd.}}\right]^{\beta} \leq 1$$ **0.8592** OK safe

Example 5.2

Consider the beam in Figure 5.16, which is supported by web cleats and subjected to two concentrated loads, $P = 80.0$ kN (design loads). Design the beam using a HEA profile (and alternatively an IPE profile) in S 235 steel ($E = 210$ GPa and $G = 81$ GPa), according to EC3-1-1. Consider free rotation at the supports with respect to the y-axis and the z-axis. Also, assume free warping at the supports, but note that the web cleats do not allow rotation around the axis of the beam (x-axis).

Assume:

Considering a HEA profile, a HEA 240 is adopted to account for the effect of lateral–torsional buckling. The required geometric characteristics of this cross section are $W_{pl,y} = 744.6\,\text{cm}^3$, $I_y = 7{,}763\,\text{cm}^4$, $I_z = 2{,}769\,\text{cm}^4$, $I_T = 41.55\,\text{cm}^4$, and $I_W = 328.5 \times 10^3\,\text{cm}^6$.

ELREEDYMAN www.elreedyman.com	Client :-		Page no.
	Project :-		REV. No :-
	Title :-		Calcs By :-
			Check. By :-
			Approve By :-
EC3 Beam column		Calculation sheet	Date:

Beam Calculation

INPUT DATA

MEMBER SIZE:	HEA240	
YIELD STRESS OF STEEL (Fy):	235	Mpa
ELASTIC MODULUS (E):	210,000	Mpa
UNBRACED LENGTH IN BENDING (l):	6.00	m
EFFECTIVE UNBRACED LENGTH ABOUT X AXIS (KLy):	6.00	m
EFFECTIVE UNBRACED LENGTH ABOUT Y AXIS (KLz):	6.00	m
γM0	1.00	
G	81000.00	MPa

DESIGN LOADING

ALLOWABLE STRESS DESIGN

BENDING MOMENT ABOUT X AXIS (MyED):	120.0	kN-m
BENDING MOMENT ABOUT Y AXIS (Mz):	0	kN-m
VED	80	kN

MEMBER PROPERTIES

GENERAL

A =	76.84	cm2				
d =	230.00	mm				
tw =	7.50	mm				
bf =	240.00	mm				
tf =	12.00	mm				
h = T =	230	mm				
It=	42	cm4	torsion constant			
hc =	230.00	mm				
Iw=	328500	cm6	Warping constant			

Y - AXIS

Iy =	7763	cm^4
Sy =	675.10	cm^3
ry =	10.05	cm
Wpl,y =	1307.00	cm^3

Z - AXIS

Iz =	2769	cm^4
Sz =	230.70	cm^3
rz =	7	cm
Wpl,z=	351.70	cm^3

e=	1.00			
cf/(tf ε)	8.63	<9	The flange is class 1	
cw/(twε)	24.08	<62	The web is class 1	
Mc,RD=	307.145	kN.m	>	120.00
Av=	2318.80	mm2		
Vc_rd=	314.61	kN	>	80.00 ok

Npl,Rd=AFy/γM0	1805.74	kN	
NED<Npl,Rd			
NED<0.25 Npl,Rd			
Mpl,y,Rd=Wpl,y fy/γM0	307.15	kN.m	
a=A-2bt/A=	0.25		
n=NED/Npl,R	0.00		
Mn,y,Rd=Mpl,y,Rd (1-n)/(1-0.5a)			
Mn,y,Rd=Mpl,y,Rd (1-n)/(1-0.5a)	351.10	kN.m	
My,ED=	120.00 <	Mn,y,Rd	oK

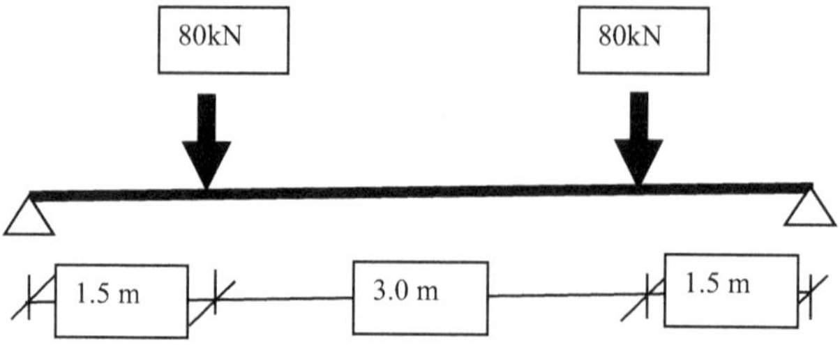

FIGURE 5.16 Example 2.

$$M_{cr} = C_1 \frac{\pi^2 EI_z}{(kL)^2}\left\{\sqrt{\left(\frac{k}{k_w}\right)^2 \frac{I_w}{I_z} + \frac{(kL)^2 GI_t}{\pi^2 EI_z} + (C_2 z_g)^2} - C_2 z_g\right\}$$

Cross-section	Limits	Buckling curve
Rolled I-sections	h/b ≤ 2	a
	h/b > 2	b
Welded I-sections	h/b ≤ 2	c
	h/b > 2	d
Other cross-sections	-	d

zg= 115.00 mm

Mcr= 231.57 kN.m

$$\bar{\lambda}_{LT} = \sqrt{\frac{W_y f_y}{M_{cr}}}$$

c1=	1.04
c2=	0.42

λLT= 1.15

h/b= 0.96 Curve b

α LT= 0.34

λLT,0= 0.40 maximum value

β= 0.75 minimum value

ϕ_{LT}= 1.13 $\Phi_{LT} = 0,5\left[1 + \alpha_{LT}\left(\bar{\lambda}_{LT} - \bar{\lambda}_{LT,0}\right) + \beta\bar{\lambda}_{LT}^2\right]$

χ_{LT} 0.61

$$\chi_{LT} = \frac{1}{\Phi_{LT} + \sqrt{\Phi_{LT}^2 - \beta\bar{\lambda}_{LT}^2}}$$

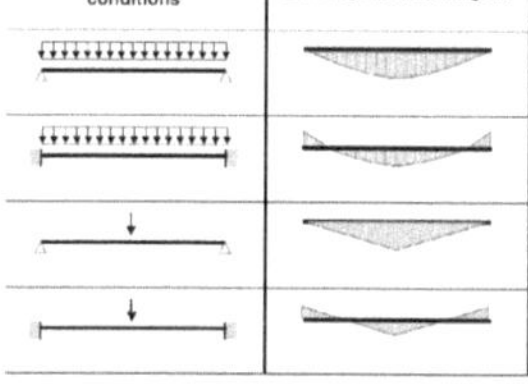

Loading and support conditions	Bending moment diagram	C_1	C_2
		1.127	0.454
		2.578	1.554
		1.348	0.630
		1.683	1.645

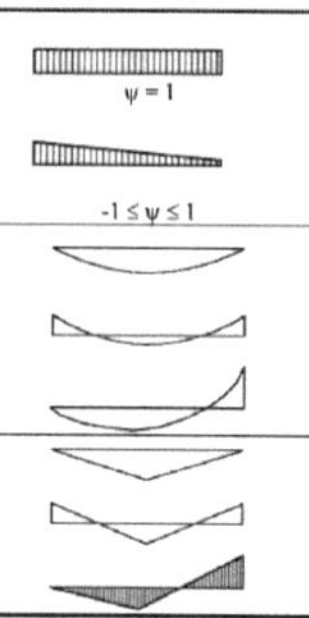

Moment distribution	k_c
ψ = 1	1,0
-1 ≤ ψ ≤ 1	$\frac{1}{1,33 - 0,33\psi}$
	0,94
	0,90
	0,91
	0,86
	0,77
	0,82

Kc= 0.91

$$f = 1 - 0.5 \times (1 - k_c) \times \left[1 - 2.0 \times (\bar{\lambda}_{LT} - 0.8)^2\right]$$

f 0.97

$\chi_{LT,mod}$ 0.63 <1.0 So,$\chi_{LT,mod}$ 0.63

$C_{my}=C_{mLT}$= 0.60 > 0.4

$$M_{b,Rd} = \chi_{LT} W_y \frac{f_y}{\gamma_{M1}}$$ 193.14 kN.m O.k safe

For biaxial bending

MN,yrd= 307.15 kN.m

MN,zrd= 54.21 kN.m

a = 2.00 or =1 conservartive

b= 0.00 =1

- I and H sections:
  $\alpha = 2\,;\ \beta = 5n$ but $\beta \ge 1$
- circular hollow sections:
  $\alpha = 2\,;\ \beta = 2$
- rectangular hollow sections:
  $\alpha = \beta = \frac{1,66}{1 - 1,13n^2}$ but $\alpha = \beta \le 6$

where $n = N_{Ed} / N_{pl,Rd}$.

$$\left[\frac{M_{y,Ed}}{M_{N,y,Rd}}\right]^{\alpha} + \left[\frac{M_{z,Ed}}{M_{N,z,Rd}}\right]^{\beta} \le 1$$ **0.3907** OK safe

Case 2:

In the case of a laterally braced beam at two points under load, the critical length for lateral–torsional buckling is 3 m in the lateral direction.

For the HEA 220 profile in S 235 steel (minimum section, governed by plastic resistance to bending), with $L = 3.00\,\text{m}$, $C_1 = 1.0$, $k_z = k_w = 1.0$, $I_z = 1{,}955\,\text{cm}^4$, $I_T = 28.46\,\text{cm}^4$, $I_W = 193.3 \times 10^3\,\text{cm}^6$, $W_{pl,y} = 568.5\,\text{cm}^3$, $E = 210\,\text{GPa}$, $G = 81\,\text{GPa}$ and $f_y = 235\,\text{MPa}$,

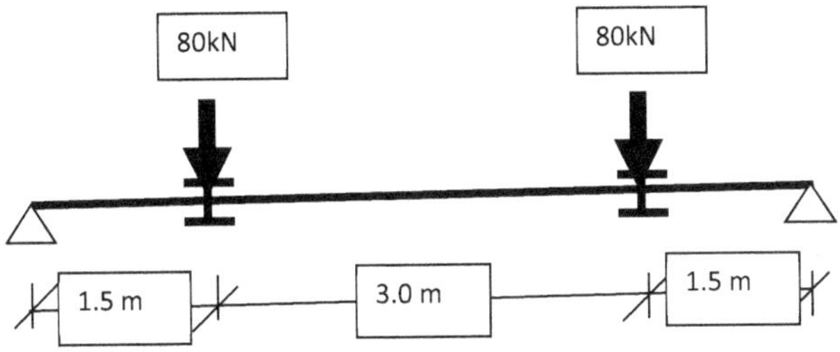

ELREEDYMAN
www.elreedyman.com
EC3 Beam column

Client :- | Project :- | Title :- | Calculation sheet
Page no. | REV. No :- | Calcs By :- | Check. By :- | Approve By :- | Date:

Beam Calculation

INPUT DATA

MEMBER SIZE:	HEA220	
YIELD STRESS OF STEEL (Fy):	235	Mpa
ELASTIC MODULUS (E):	210,000	Mpa
UNBRACED LENGTH IN BENDING (l):	3.00	m
EFFECTIVE UNBRACED LENGTH ABOUT X AXIS (KLy):	6.00	m
EFFECTIVE UNBRACED LENGTH ABOUT Y AXIS (KLz):	3.00	m
γM0	1.00	
G	81000.00	MPa

DESIGN LOADING

ALLOWABLE STRESS DESIGN

BENDING MOMENT ABOUT X AXIS (MyED):	120.0	kN-m
BENDING MOMENT ABOUT Y AXIS (Mz):	0	kN-m
VED	80	kN

MEMBER PROPERTIES

GENERAL

A =	76.80	cm2	
d =	210.00	mm	
tw =	7.00	mm	
bf =	220.00	mm	
tf =	11.00	mm	
h = T =	210	mm	
It =	28	cm4	torsion constant
hc =	210.00	mm	
Iw =	193300	cm6	Warping constant

Y - AXIS

Iy =	5410	cm^4
Sy =	515.20	cm^3
ry =	0.00	cm
Wpl,y =	568.50	cm^3

Z - AXIS

Iz =	1955	cm^4
Sz =	177.70	cm^3
rz =	0	cm
Wpl,z=	0.00	cm^3

e=	1.00				
cf/(tf ε)	8.53	<9	The flange is class 1		
cw/(twε)	23.23	<62	The web is class 1		
Mc,RD=	133.5975	kN.m	>	120.00	
Av=	3196.40	mm2			
Vc_rd=	433.68	kN	>	80.00	ok

Npl,Rd=AFy/γM0 1804.80 kN
NED <Npl,Rd
NED <0.25 Npl,Rd
Mpl,y,Rd=Wpl,y fy/γM0 133.60 kN.m
a=A-2bt/A= 0.37
n=NED/Npl,R 0.00
Mn,y,Rd=Mpl,y,Rd (1-n)/(1-0.5a)
Mn,y,Rd=Mpl,y,Rd (1-n)/(1-0.5a) 163.90 kN.m
My,ED= 120.00 < Mn,y,Rd oK

$$M_{cr} = C_1 \frac{\pi^2 EI_z}{(kL)^2} \left\{ \sqrt{\left(\frac{k}{k_w}\right)^2 \frac{I_w}{I_z} + \frac{(kL)^2 GI_t}{\pi^2 EI_z} + (C_2 z_g)^2} - C_2 z_g \right\}$$

Cross-section	Limits	Buckling curve
Rolled I-sections	h/b ≤ 2	a
	h/b > 2	b
Welded I-sections	h/b ≤ 2	c
	h/b > 2	d
Other cross-sections	-	d

zg=	115.00	mm

Mcr= 551.91 kN.m

$$\bar{\lambda}_{LT} = \sqrt{\frac{W_y f_y}{M_{cr}}}$$

c1=	1.00
c2=	0.00

λLT=	0.49	
h/b=	0.95	Curve b
α LT=	0.34	
λLT,0=	0.40	maximum value
β=	0.75	minimum value
φ_{LT}=	0.61	$\Phi_{LT} = 0{,}5\left[1 + \alpha_{LT}\left(\bar{\lambda}_{LT} - \bar{\lambda}_{LT,0}\right) + \beta\bar{\lambda}_{LT}^2\right]$
χ_{LT}	0.96	

$$\chi_{LT} = \frac{1}{\Phi_{LT} + \sqrt{\Phi_{LT}^2 - \beta\bar{\lambda}_{LT}^2}}$$

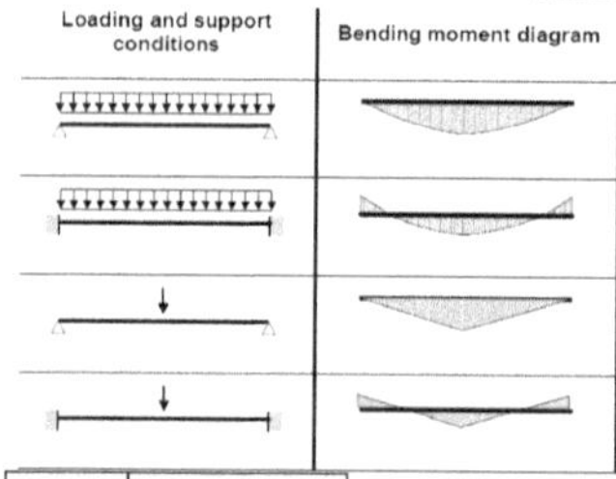

Loading and support conditions / Bending moment diagram	C_1	C_2
1	1.127	0.454
2	2.578	1.554
3	1.348	0.630
4	1.683	1.645

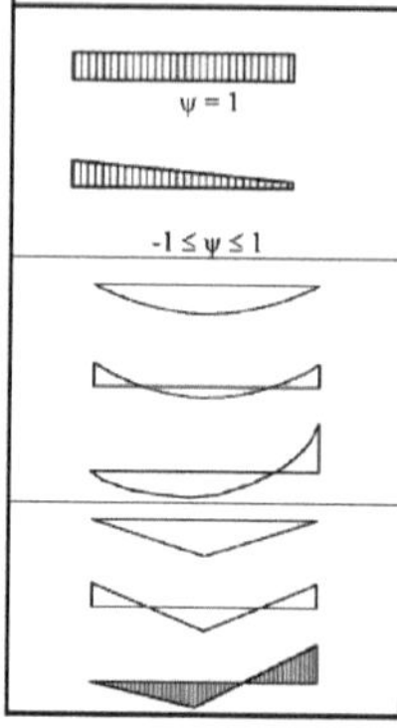

Moment distribution	k_c
ψ = 1	1,0
-1 ≤ ψ ≤ 1	$\frac{1}{1{,}33 - 0{,}33\psi}$
	0,94
	0,90
	0,91
	0,86
	0,77
	0,82

Kc=	1.00

$$f = 1 - 0.5 \times (1 - k_c) \times \left[1 - 2.0 \times (\bar{\lambda}_{LT} - 0.8)^2\right]$$

f 1.00

$\chi_{LT,mod}$	0.96	<1.0	So,$\chi_{LT,mod}$	0.96
$C_{my}=C_{mLT}$=	0.60	>	0.4	

$$M_{b,Rd} = \chi_{LT} W_y \frac{f_y}{\gamma_{M1}}$$ 128.72 kN.m O.k safe

For biaxial bending

MN,yrd= 133.60 kN.m
MN,zrd= 41.76 kN.m

a=	2.00	or	=1	conservartive
b=	0.00		=1	

- I and H sections:
  $\alpha = 2$; $\beta = 5n$ but $\beta \geq 1$
- circular hollow sections:
  $\alpha = 2$; $\beta = 2$
- rectangular hollow sections:
  $\alpha = \beta = \frac{1{,}66}{1 - 1{,}13n^2}$ but $\alpha = \beta \leq 6$

  where $n = N_{Ed} / N_{pl,Rd}$.

$$\left[\frac{M_{y,Ed}}{M_{N,y,Rd}}\right]^{\alpha} + \left[\frac{M_{z,Ed}}{M_{N,z,Rd.}}\right]^{\beta} \leq 1$$ **0.8982** OK safe

5.3 BEAM DESIGN BY AISC

As usual, the concept is the same, but the approach to design is different. For AISC 22, both LRFD and ASD are presented; this book will focus on LRFD as it is equivalent to the EC approach.

5.3.1 For Flexural Member

1. Like beams they take the reduction factor:

$$\varphi b = 0.90(\text{LRFD})$$

2. For singly symmetric members in single curvature and all doubly symmetric members, the lateral–torsional buckling modification factor, C_b, for non-uniform moment diagrams when both ends of the segment are braced is determined as follows:

$$C_b = \frac{12.5M_{\max}}{2.5M_{\max} + 3M_A + 4M_b + 3M_c}$$

where

M_{max} = absolute value of the maximum moment in the unbraced segment, kip-in. (N-mm)
M_A = absolute value of the moment at the quarter point of the unbraced segment, kip-in. (N-mm)
M_B = absolute value of the moment at the centreline of the unbraced segment, kip-in. (N-mm)
M_C = absolute value of the moment at the three-quarter point of the unbraced segment, kip-in. (N-mm)
C_b = 1.0 for the case of equal end moments of opposite sign (uniform moment) and in the case of cantilever if the free end is unbraced
C_b = 2.27 for the case of equal end moments of the same sign (reverse curvature bending),
C_b = 1.67 when one end moment equals zero.

3. Check for yield

$$M_n = M_p = F_y Z_x$$

where F_y = specified minimum yield stress of the type of steel being used, ksi (MPa), Z_x = plastic section modulus about the x-axis, in.^{3} (mm^3).

4. Check LTB
   a. When $L_b \leq L_p$, the limit state of lateral–torsional buckling does not apply.

      L_b = length between points that are either braced against lateral displacement of the compression flange or braced against twist of the cross section, in. (mm).

      L_p, the limiting laterally unbraced length for the limit state of yielding, in. (mm), is

$$L_p = 1.76 r_y \sqrt{\frac{E}{F_y}}$$

b. When $L_p < L_b \leq L_r$

L_r will be calculated from the following equation:

$$L_r = 1.95 r_{ts} \frac{E}{0.7F_y} \sqrt{\frac{Jc}{S_x h_o} + \sqrt{\left(\frac{Jc}{S_x h_o}\right)^2 + 6.76\left(\frac{0.7F_y}{E}\right)^2}}$$

where r_y = radius of gyration about the y-axis, in. (mm).

$$r_{ts}^2 = \frac{\sqrt{I_y Cw}}{S_x}$$

and the coefficient c is determined as follows: (a) For doubly symmetric I-shapes, $c = 1$, (b) For channels

$$c = \frac{h_o}{2}\sqrt{\frac{I_y}{C_w}}$$

where I_y = moment of inertia about the y-axis, in.^{4} (mm^4).

For doubly symmetric I-shapes with rectangular flanges, $C_w = I_y.(h_o)^2/4$, and, thus, r_{ts} will be calculated as follows:

$$(r_{ts})^2 = I_y h_o / 2S_x$$

After defining L_r of the unbraced length is less than or equal to L_r and greater than L_p

$$M_n = C_b\left[M_p - \left(M_p - 0.7F_y S_x\right)\left(\frac{L_b - L_p}{L_r - L_p}\right)\right] \leq M_p$$

c. When $L_b > L_r$

$$M_n = F_{cr} S_x \leq M_p$$

F_{cr} is the critical stress, ksi (MPa) and is calculated from the following equation:

$$= \frac{C_b \pi^2 E}{\left(\frac{L_b}{r_{ts}}\right)^2}\sqrt{1 + 0.078\frac{Jc}{S_x h_o}\left(\frac{L_b}{r_{ts}}\right)^2}$$

E = modulus of elasticity of steel = 29,000 ksi (200 000 MPa)
J = torsional constant, in.^{4} (mm^4)
S_x = elastic section modulus taken about the x-axis, in.^{3} (mm^3)
h_o = distance between the flange centroids, in. (mm).

As an approximate and conservative approach, F_{cr} can be calculated as follows:

$$F_{cr} = c_b (\pi)^2 E / (L_b / r_{ts})^2$$

Considering that it is equal to the previous edition, calculate the critical moment as per the following equation, which follows the same approach as per EC3:

$$M_{cr} = C_b \frac{\pi}{L_b} \sqrt{EI_y GJ + \left(\frac{\pi E}{L_b}\right)^2 I_y C_w}$$

5.3.2 I-Shaped Members and Channels Bent about their Minor Axis

This section applies to I-shaped members and channels bent about their minor axis. The nominal flexural strength, M_n, shall be the lower value obtained according to the limit states of yielding (plastic moment) and flange local buckling:

a. Check Yield

$$M_n = M_p = F_y Z_y \le 1.6 F_y S_y$$

where

S_y = elastic section modulus taken about the y-axis, in.^{3} (mm^3)
Z_y = plastic section modulus taken about the y-axis, in.^{3} (mm^3)

b. Flange Local Buckling
   1. For sections with compact flanges, the limit state of flange local buckling does not apply.
   2. For sections with non-compact flanges

$$M_n = M_p - \left(M_p - 0.70 F_y S_y\right)\left(\frac{\lambda - \lambda_{pf}}{\lambda_{rf} - \lambda_{pf}}\right)$$

$\lambda = b\ tf\ \lambda_{pf} = \lambda_p$, the limiting width-to-thickness ratio for a compact flange, as defined in Table B4.1b; $\lambda_{rf} = \lambda_r$, the limiting width-to-thickness ratio for a non-compact flange, as defined in Table B4.1b.

   3. For sections with slender flanges

$$M_n = F_{cr} S_y$$

$$F_{cr} = 0.70\ E/\left(b/t_f\right)^2$$

b = for flanges of I-shaped members, half the full flange width, b_f; for flanges of channels, the full nominal dimension of the flange, in. (mm), t_f = thickness of the flange, in. (mm).

5.3.3 Square and Rectangular HSS and Box Sections

This section applies to square and rectangular HSS and box sections bent about either axis, having compact, non-compact, or slender webs or flanges, as defined in Section B4.1 for flexure. The nominal flexural strength, M_n, shall be the lowest value obtained according to the limit states of yielding (plastic moment), flange local buckling, web local buckling, and lateral–torsional buckling under pure flexure.

a. Check Yielding

$$M_n = M_p = F_y Z$$

b. Flange Local Buckling
  a. For compact sections, the limit state of flange local buckling does not apply.
  b. For sections with non-compact flanges

$$M_n = M_p - \left(M_p - F_y S\right)\left(\frac{\lambda - \lambda_{pf}}{\lambda_{rf} - \lambda_{pf}}\right) \le M_p$$

  c. For sections with slender flanges, $M_n = F_y S_e$ (F7-3), where S_e = effective section modulus determined with the effective width, b_e, of the compression flange taken as follows:
    1. For HSS

$$b_e = 1.92 t_f \sqrt{\frac{E}{F_y}}\left(1 - \frac{0.38}{b/t_f}\sqrt{\frac{E}{F_y}}\right) \le b$$

    2. For box sections

$$b_e = 1.92 t_f \sqrt{\frac{E}{F_y}}\left(1 - \frac{0.34}{b/tf}\sqrt{\frac{E}{F_y}}\right) \le b$$

Web local buckling (a) for compact sections, the limit state of web local buckling does not apply, (b) for sections with non-compact webs

$$M_n = M_p - \left(M_p - F_y S\right)\left(\frac{\lambda - \lambda_{pw}}{\lambda_{rw} - \lambda_{pw}}\right) \le M_p$$

where $\lambda = h/t_w$

h = depth of web, as defined in Section B4.1b, in. (mm)
t_w = thickness of the web, in. (mm)
$\lambda_{pw} = \lambda_p$, the limiting width-to-thickness ratio for a compact web as defined in Table B4.1b $\lambda_{rw} = \lambda_r$, the limiting width-to-thickness ratio for a non-compact web as defined in Table B4.1b.

Check LTB

When $L_p < L_b \le L_r$

$$M_n = C_b\left[M_p - \left(M_p - 0.7 F_y S_x\right)\left(\frac{L_b - L_p}{L_r - L_p}\right)\right] \le M_p$$

When $L_b > L_r$

$$M_n = 2EC_b \frac{\sqrt{JA_g}}{L_b/r_y} \le M_p$$

where A_g = gross area of member, in.^{2} (mm^2)

L_p, the limiting laterally unbraced length for the limit state of yielding, in. (mm), is

$$L_p = 0.13Er_y \frac{\sqrt{JA_g}}{M_p}$$

L_r, the limiting laterally unbraced length for the limit state of inelastic lateral–torsional buckling, in. (mm), is

$$L_r = 2Er_y \frac{\sqrt{JA_g}}{0.7F_yS_x}$$

5.3.4 Round HSS

In case of *D/t* < 0.45*E/Fy*

Check yield

$$M_n = M_p = F_y Z$$

Local buckling

a. For compact sections, the limit state of flange local buckling does not apply.
b. For non-compact sections

$$M_n = \left(0.021E/(D/t) + F_y\right)S$$

c. For sections with slender walls $M_n = F_{cr} S$
   where D = outside diameter of round HSS, in. (mm).

$$F_{cr} = 0.33\ E/(D/t)$$

t = design wall thickness of HSS member, in. (mm).

REFERENCE

Boissonnade, N., Greiner, R., Jaspart, J.P., and Lindner, J. (2007). *Rules for Member Stability in EN1993-1-1 –Background Documentation and Design Guidelines, ECCS TC8 – Stability*, ECCS Report No. 119, ECCS, Brussels, Belgium.

FURTHER READING

CEN (2006). *Eurocode 3: Design of Steel Structures – Part 1-1: General Rules and Rules for Buildings*, EN 1993-1-1, UK.

Chen WF, Atsuta T (1977). *Theory of beam-columns*. McGraw-Hill, New York. https://fgg-web.fgg.uni-lj.si/~/pmoze/esdep/master/wg07/l0920.htm

Simoes da Silva, L., Simoes, R., and Gervasio, H. (2010) *Design of Steel Structure- Eurocode 3: Design of Steel Structures Part 1-1- General Rules and Rules for Building*, Ernst&Sohn, A Wiley Company.

6 Beam–Column

6.1 INTRODUCTION

Beam–columns exist in any frame structure in general. Given this, the choice of a suitable structural system is the engineer's responsibility during the preliminary engineering stage. The main structural systems include main girders, continuous beams, trusses, or frames.

The primary purpose of selecting a structural system is to identify the optimum solution from an economic perspective and functional use. To cover large spans in the past, we moved towards truss member systems, which are traditionally discussed in many chapters about steel trusses in textbooks.

For spans less than 10 m, it is reasonable to use a main girder rather than a frame structure, or to use a continuous beam for housing with spans less than 6.0 m, as shown in Figure 6.1.

Frame structures have been used worldwide since World War II. During this period, there was extensive research, leading to the fabrication of high-strength bolts, making the design and erection of moment connections easier and more economical. Moreover, frame structures can cover long spans, reaching up to 40 m or more. The frame structure is shown in Figure 6.2a and b.

As shown in Figure 6.3a, steel sections can be built-up to reduce the tonnage of steel, and many international companies specialize in this type of frame, as they can control the design in the workshop. Moreover, the connections will be adjusted to accommodate the increased bending moment effects on the connections. Figure 6.3b presents a simpler version of the frame structure that can be fabricated from rolled sections, with the sections being continuous along the member length, suitable for small factories and workshops.

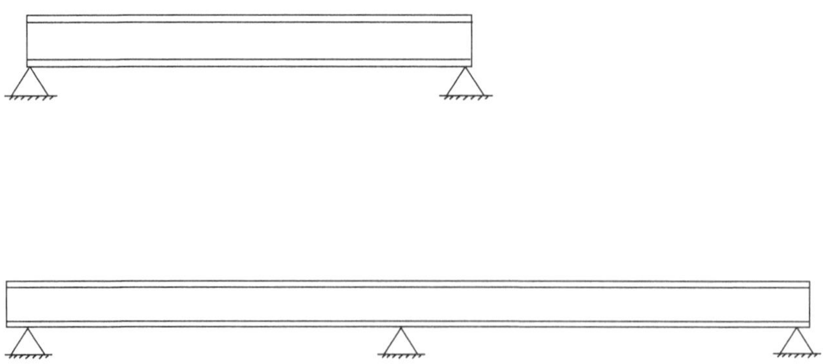

FIGURE 6.1 Main girder and continuous beam.

DOI: 10.1201/9781003307303-6

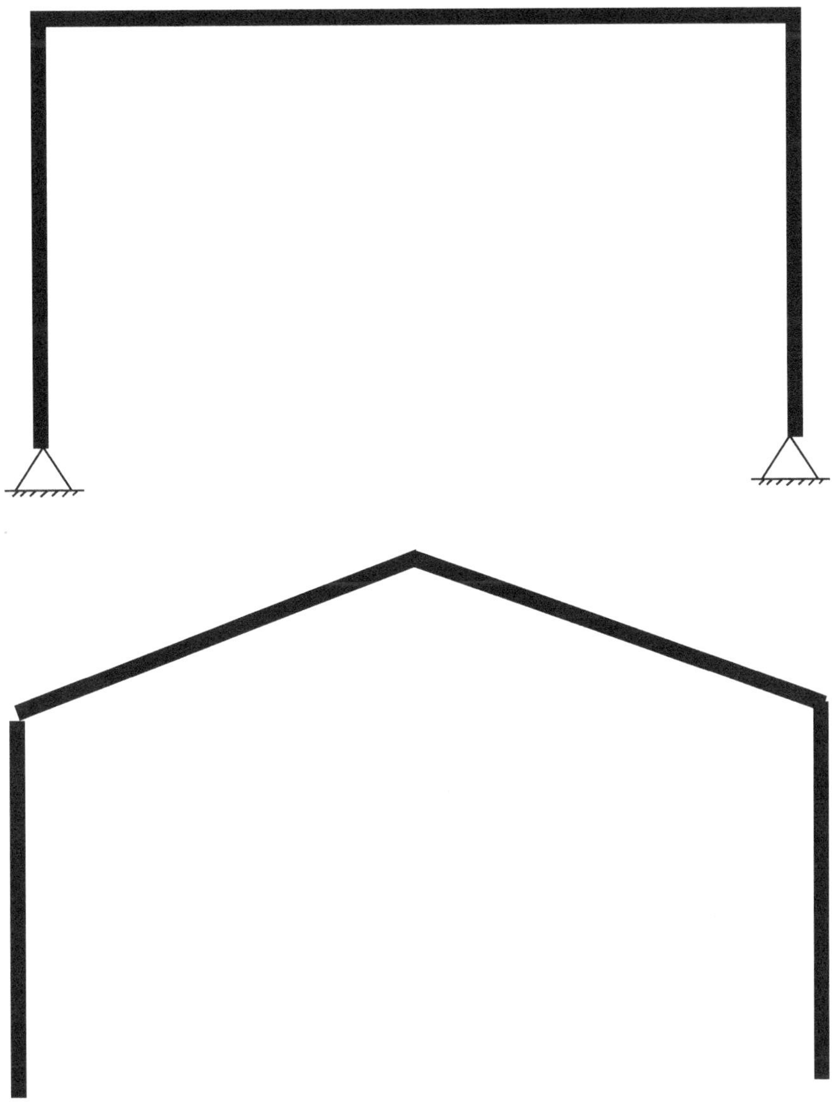

FIGURE 6.2 (a) Frame structure. (b) Frame structure with eaves.

The column truss system, as shown in Figure 6.4, was traditionally used instead of frame structures for wide spans. However, nowadays, the frame structure has superseded it, as it has lower manufacturing costs compared to truss fabrication and erection due to fewer connections. Figure 6.5 presents a different configuration for truss shapes; if all the members are under tension, this configuration will be the best choice. In steel structures, one of the main governing design considerations, which will be discussed in later chapters, is the buckling of the members. A truss system with all members in compression should also be considered.

For lateral loads due to earthquakes and wind, there are various systems to handle these loads, which will be discussed in detail in Chapter 7.

As the frame structure is the most traditional steel structural system, we need to focus on it in this chapter (Figure 6.6).

6.2 FRAME BEHAVIOUR

In general, the horizontal load in rigid frame structures is resisted by bending in the beams and columns. The columns, particularly in the lower storeys, must resist heavy moments, so their sections will be much larger than those in braced buildings.

The frame normally has H-section columns. It is rigid in one direction only, across the short span, and is braced longitudinally.

The connections are expensive, being welded or made with hunched beam ends and high-strength bolts. Frames that are rigid in both directions, with box section columns, have been constructed in areas subject to seismic loads.

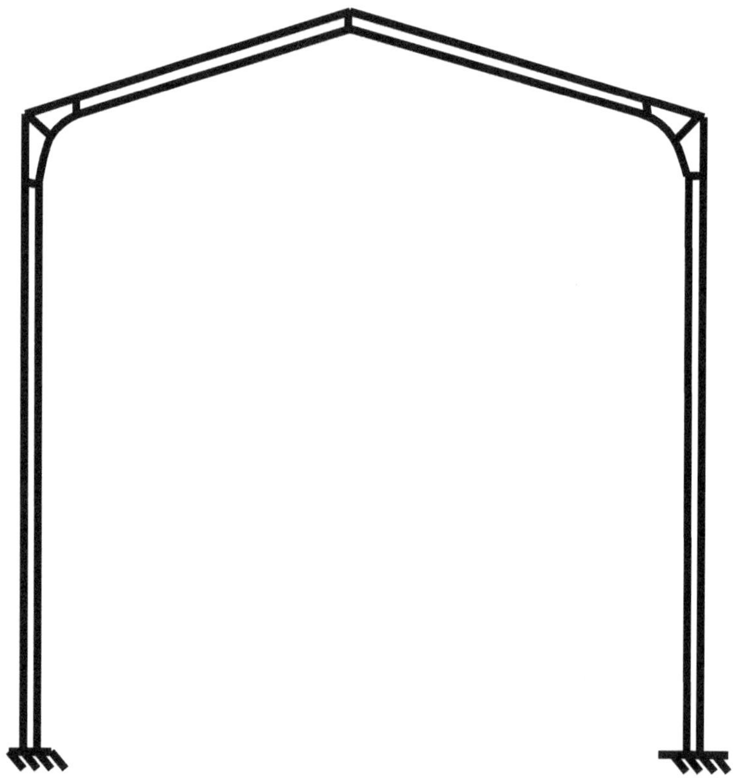

FIGURE 6.3 (a) Frame structure manufactured in workshop. (b) Frame structure with fixed section).

(Continued)

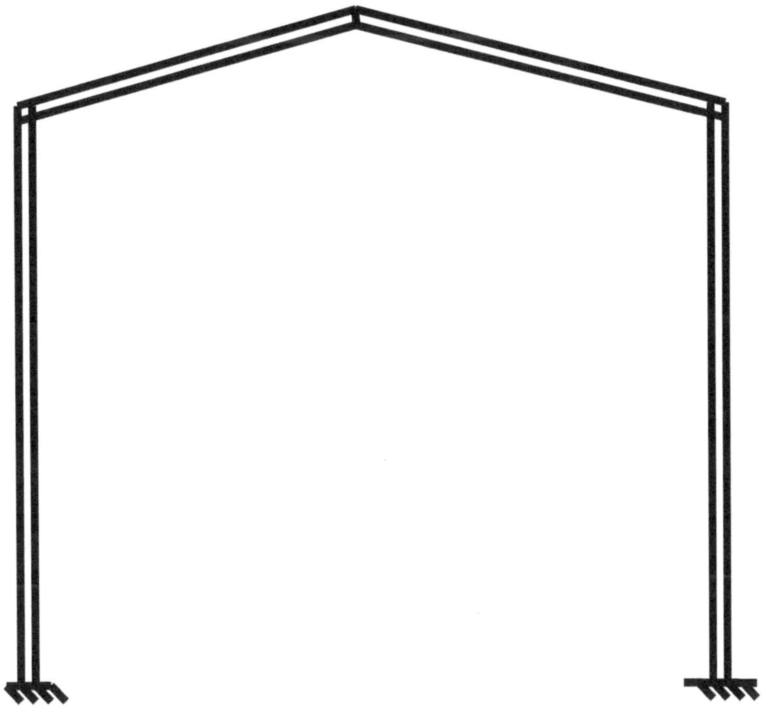

FIGURE 6.3 (*Continued*) (a) Frame structure manufactured in workshop. (b) Frame structure with fixed section).

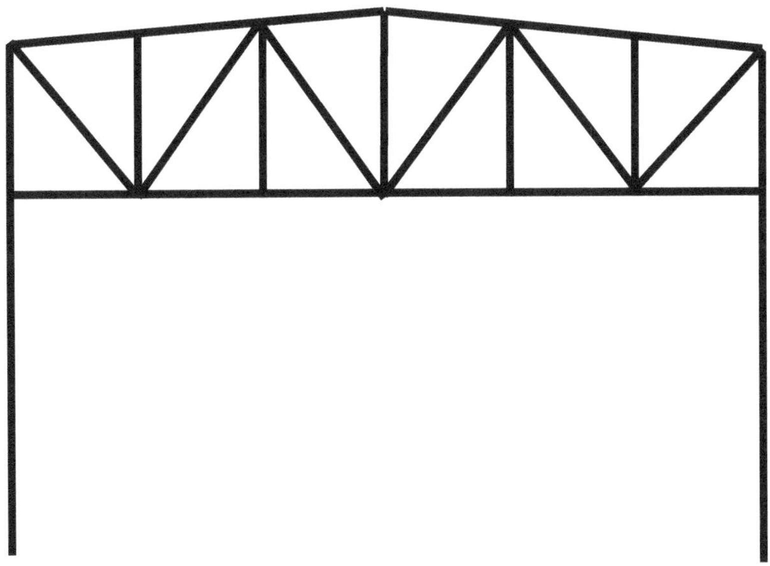

FIGURE 6.4 Truss column.

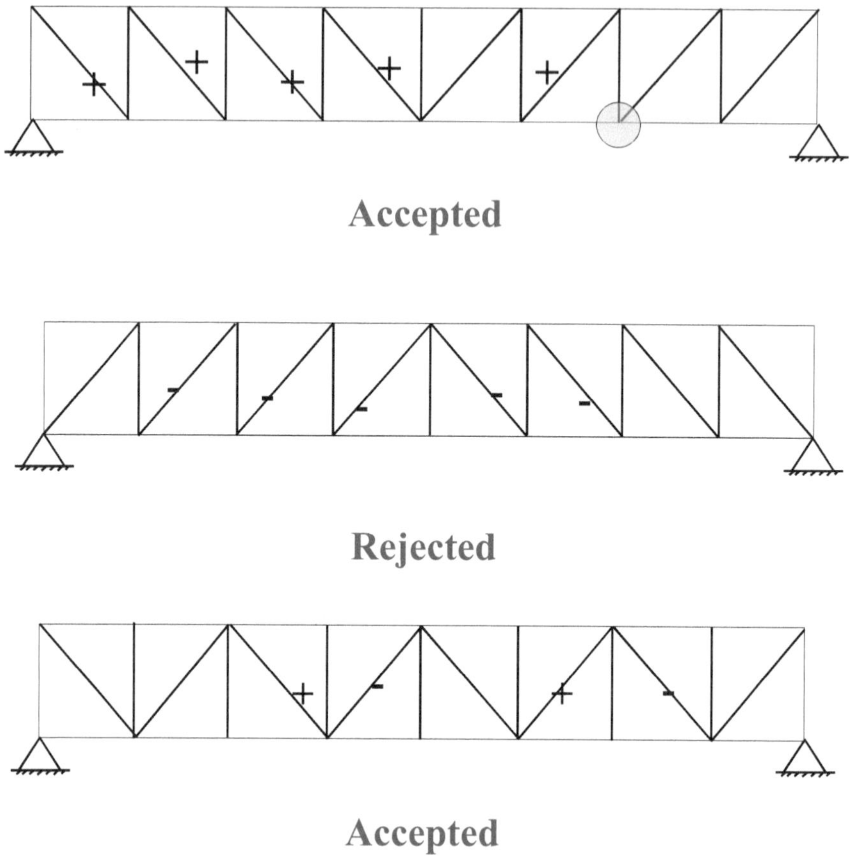

FIGURE 6.5 Truss configuration.

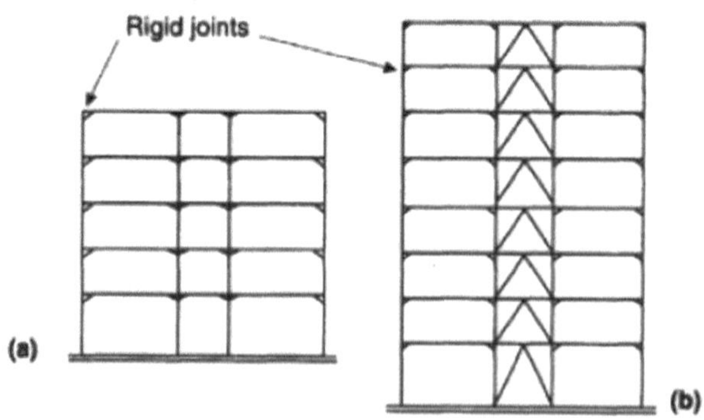

FIGURE 6.6 Rigid frame and rigid frame with bracing.

The rigid frame structure deflects more than a braced structure. The deflection comprises sway in each storey plus overall cantilever action. Due to excessive deflection, rigid frames are suitable only for low- or medium-rise buildings.

Global frame analysis is required to obtain the distribution of internal forces and the corresponding deformations in a structure subjected to specified loading.

Achieving the structural analysis requires the adoption of adequate models that incorporate assumptions about the behaviour of the structure, particularly regarding its members and joints.

6.2.1 Relationship between Load–Displacement in Frames

The frame behaviour is measured by monitoring the structural response during an increase in the applied load. This is represented by a graph illustrating the relationship between the applied load and significant displacement.

The load parameter is most often a multiplier, derived from a load factor, that is applied to all the load components. The displacement is considered in Figure 6.7 as the lateral displacement of the frame at the top floor level as per ECCS (1997). The resulting curve can be regarded as characterizing the overall structural behaviour. Based on this, the lateral stiffness of the frame structure is represented by the curve's slope.

It can be observed that the response of the structure is quasi-linear up to the linear limit point. When reaching the linear limit, the positive slope of the rising part of the curve gradually reduces due to the non-linear behaviour of geometric, joint or connection, and material properties.

Geometric non-linearity will affect the distribution of internal forces due to the actual deformation of the frame. The joints or connections may carry less load, while the materials exhibit non-linearity, transitioning from yield to plasticity of the section. Therefore, due to all these non-linear aspects, the frame experiences progressive failure as the load increases up to its maximum at failure.

Noting that the structure's stiffness, represented by the curve slope, is zero at the peak load and subsequently becomes negative, indicating that the structure is

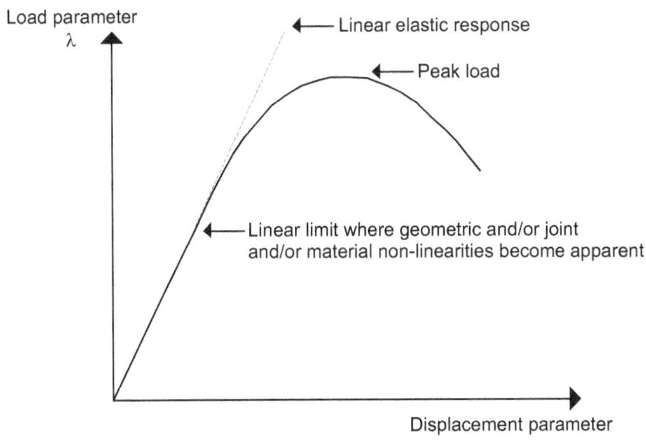

FIGURE 6.7 Load–displacement response of a framed structure.

unstable. The peak load, often termed the ultimate load, is the point of imminent structural collapse in the absence of load shedding.

Global analysis of frames is conducted on a model based on numerous assumptions, including those regarding the structural model, the geometric behaviour of the structure and its members, and the behaviour of the sections and joints.

Once the analysis is completed, a number of design checks for the frame members and joints must be performed. These checks depend on the type of analysis conducted and the type of cross-sectional verification and ultimate limit state criteria used.

Simplified models for buildings subjected to predominantly static loading are proposed as alternatives to more sophisticated models. However, this does not cover methods intended for seismic design or for "stressed skin" design, which are addressed in other specific Eurocodes, such as EC-8 for the earthquake resistance of structures and EC3 Part 1-3 for cold-formed members and sheeting.

6.2.2 Structural Conceptual Design

The first step in conceptual design is to develop a layout for the structure based on the requirements of the end user, including the resistance to loads that are likely to occur during the structure's lifetime.

It is necessary to identify the following categories of structural elements:

a. Main Structural Elements: These include main frames, joints, and foundations that form the path through which vertical and horizontal forces are transferred to the ground.
b. Secondary Structural Elements: These include secondary beams and purlins that transfer loads to the main structural elements.
c. Other Elements: These elements only transfer loads to the main or secondary elements. Examples include sheeting, roofing, and partitions.

In cases where these three categories of elements are subject to different safety requirements, they should be modelled separately, if necessary.

6.2.3 Frame Components

Any frame is composed of members and joints, . Members are elements whose length is greater than their depth, and joints are the points where more than one member is connected. Members are typically classified depending on the type of loading they carry. A member is referred to as a beam if bending predominates; as a column or tension member if axial load predominates; and as a beam–column if significant amounts of both bending and axial load are present. Frame structures mainly consist of beam and beam–column members. The truss or bracing structural system usually consists of tension and compression members.

Rather than having a separate category for members subject to both bending and axial tension, beam–columns can be generalized to cover this case.

Almost all frame members are beam–columns. A beam–column member is one that is subjected to both significant bending and axial compression.

Members subject to both significant bending and axial tension can be included in this category. Strictly speaking, most members are beam–columns; beams represent the limiting case where axial forces can be disregarded, and columns represent the limiting case where bending moments are not significant.

It is worth mentioning that if the structure's configuration makes it sensitive to eccentricity due to horizontal loading, relative to the centre of resistance to twisting of the structure, the effects of applying only part of the horizontal loading should also be taken into account.

6.2.4 Ground–Structure Interaction

Ground–structure interaction should be considered depending on its significance regarding the effects on the internal forces and moments in the structural elements due to foundation settlements resulting from loading on the ground and the soil characteristics.

The following procedure is important in examining ground–structure interaction:

- As a first step, the structure may be analysed assuming that the ground is rigid. From this analysis, the loading on the ground should be determined, and the resulting settlements should be calculated.
- The resulting settlements should be applied to the structure in the form of imposed deformations, and the effects on the internal forces and moments should be calculated.
- If the settlement effects are significant, the ground–structure interaction should be modelled using equivalent springs to represent the soil behaviour.

No criteria are provided in Eurocode 3 Part 1-1 to determine the significance of the ground–structure interaction. It is suggested that if they do not reduce the resistance of the structure by more than 5%, they can be considered insignificant and neglected in the design. This criterion is the same as that given for the classification of rigid joints.

6.2.5 Framing and Joints

In EC3, the term "framing" is used to differentiate between the various ways that joint behaviour can be considered for global analysis.

In general, due to joint deformations, the deflected shapes of members will be discontinuous at the joints. Depending upon the effects of this discontinuity, a distinction can be made between the following cases:

- The frame is considered continuous if the joints are assumed to be rigid, allowing the frame to be analysed as continuous.
- A frame is called simple framing if the joints are pinned, taking advantage of possible rotations without considering joint moment resistances.
- The discontinuity at the joints may also be accounted for by using ***semi-continuous*** framing. This involves a joint model regarded as semi-rigid, in which its moment rotation behaviour is taken into account more precisely.

The use of continuous and simple types of framing must be justified by an appropriate choice of joint type. While it is likely that, for the analysis of many typical frames, only one framing option will be used throughout the frame for beam-to-column joints, using different framing types for various parts of a given frame can be envisaged.

6.2.5.1 Modelling of Frames

To provide proper modelling for the structure, the following points should be considered:

1. The modelling of members and joints in the global analysis should reflect the expected behaviour of the structure under the applied actions. This is governed by the support criteria and restraint at the ends of the members
2. The basic geometry of a frame should be represented by the centrelines of the members.
3. It is normally sufficient to represent the members by linear structural elements located at their centrelines, disregarding the overlapping of the actual widths of the members.
4. Alternatively, account may be taken of the actual width of all or some of the members at the joints between members, including those involving special flexible joints.

6.2.6 Allowance for Imperfections

Appropriate allowance must be made for the effects of practical imperfections in the global analysis, in the analysis of bracing systems, and in member design.

Practical imperfections, including residual stresses, are geometrical imperfections such as lack of verticality, lack of straightness, lack of fit, and the unavoidable eccentricities present in practical joints.

Allowance for these imperfections may be achieved by incorporating suitable geometric imperfections with values that reflect all types of imperfection. EC 3 Part 1-1 addresses these by requiring that two kinds of imperfections be included in the global analysis of all frames: initial sway imperfections (frame imperfections) and member imperfections, where necessary.

It should be noted that for sway frames with slender columns, it may be required that member imperfections be incorporated in the analysis; however, the effects of member imperfections may be neglected when carrying out the analysis of non-sway frames.

6.3 BASIC CONCEPTS OF STRUCTURAL ANALYSIS

The first step in establishing a model of the structure to be analysed is to define the structural system configuration. The component members that make up typical structures are predominantly composed of straight members connected to other members.

After defining the structural system, the degrees of freedom of the structure will be determined. Additional information must be provided on the relaxation of constraints at specific locations within the structure. These are primarily related to the idealization of the support conditions and possibly the behaviour of joints in the structure, where relative movement or rotation is permitted.

Once the structural geometry has been established, the next step is to define the material properties and assign loads to each member. During the preliminary design phase, the member cross-sectional properties and joint properties are selected.

At the core of any structural analysis model is the structural element model, which represents the behaviour of members and joints. For the members, the model expresses the relationships between the displacements and rotations at the end joints of the member and the forces applied to it. These relationships satisfy the three basic principles of equilibrium, compatibility, and the constitutive laws at all sections along the member. The structural element representing the joint, when included, is developed along similar principles to that for a member, although the constitutive laws used are often based partly on empirical or semi-empirical relationships established through calibration with test results.

The structural element model employed in most computer analysis techniques to represent the behaviour of frame components takes the form of a stiffness matrix. Consequently, the solution to the analysis involves satisfying equilibrium and compatibility at the joints between the elements. More than one structural element can be used to represent a single frame component.

Many analysis programmes currently in use in the structural analysis and can assume joints to be either pinned or rigid. However, it is often possible to introduce joint elements (i.e. semi-continuous framing) and to provide the behavioural model for them. In particular, the influence of movements at the foundation can be investigated by assigning appropriate properties to the support joints. By treating the joints at the ends of a beam as an extension of the beam itself, it is possible to establish an equivalent beam element that incorporates the joint properties. This technique is particularly useful for hand calculations, especially when studying the in-plane stability of a frame or an individual member, including joint behaviour.

Most commonly used structural analysis software on the market is based on the stiffness method. This method relies on determining the displacements and rotations at each of the joints, which must satisfy compatibility. Compatibility implies that the member deformations must be geometrically compatible with the displacements and rotations of the joints and supports. It is important to note that the member and joint forces and deformations must satisfy the behavioural laws for the material from which they are constructed, taking into account Young's modulus value, the yield strength, and the ultimate capacity of the material. Then, through displacement and rotation, the forces are calculated to satisfy the equations of static equilibrium.

The traditional approach involves finding the member forces that satisfy equilibrium at each of the joints. This method, known as the force method, is well suited to manual analysis procedures such as the moment distribution method.

Due to the many simplifying assumptions upon which they are based, many contemporary manual and computer methods for structural analysis do not detect the onset of structural instability. Instability can arise from local buckling of part of a member's cross section or a joint, lateral–torsional buckling of a member, or buckling instability affecting part or all of the structure. Therefore, in addition to structural analysis, supplementary analysis and design measures typically need to be undertaken to guard against instability phenomena. These measures depend to some extent on the type and complexity of the analysis model used and on the types of members and joints involved.

Tables 6.1–6.3 can be used for hand calculations for a frame with one bay, both with and without eaves.

1. Frame with Hinge–Hinge Support

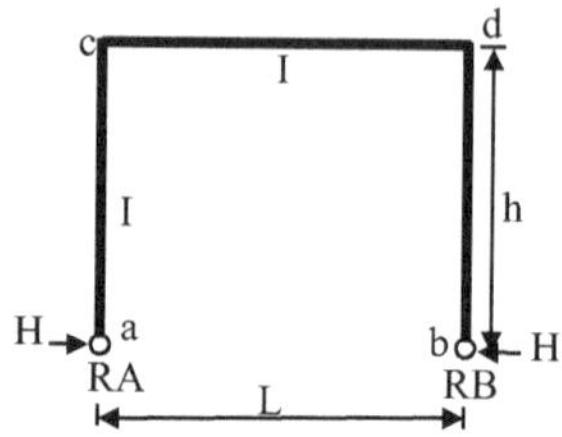

$K = (I_G/I_C).\ (h/L)$

$N = 2K + 3$

$M_c = -H_a.h$

$M_d = -H_b.h$

TABLE 6.1

Expression on Reduced Plastic Moment Resistance for Different Cross Sections

Cross Section	Shape	Expression for M_N
Rolled I or H	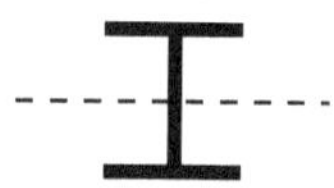	$M_{N,y} = 1.11M_{pl.y}(1-n)$
		$M_{N,z} = 1.56M_{pl.z}(1-n)(0.6+n)$
Square hollow section		$M_{N,y} = 1.26M_{pl}(1-n)$
Rectangular hollow section	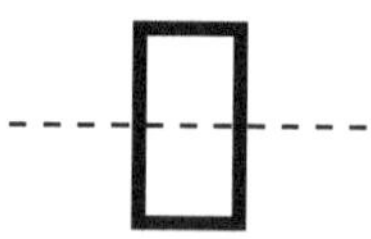	$M_{N,y} = 1.33M_{pl.y}(1-n)$
	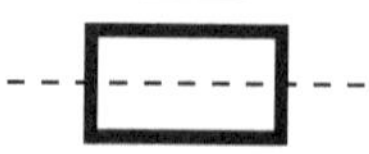	$M_{N,y} = M_{pl.z}\dfrac{1-n}{0.5+\dfrac{ht}{A}}$
Circular hollow section		$M_{N,y} = 1.04M_{pl}\left(1-n^{1.7}\right)$

Notation: $n = N_{Sd}/N_{pl.Rd}$.

TABLE 6.2
Equivalent Uniform Moment Factors β_M

Moment Diagram	Equivalent Uniform Moment Factor β_M
End moments 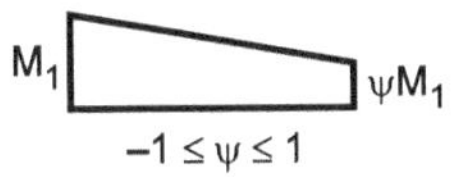	$\beta_{M,\psi} = 1.8 - 0.7\psi$
Moments due to in-plane lateral loads 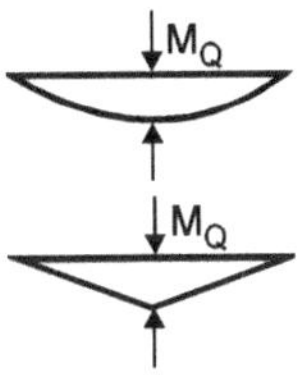	For uniformly distributed load, $\beta_{M,Q} = 1.3$ For concentrated load, $\beta_{M,Q} = 1.4$
Moments due to in-plane lateral loads plus end moments 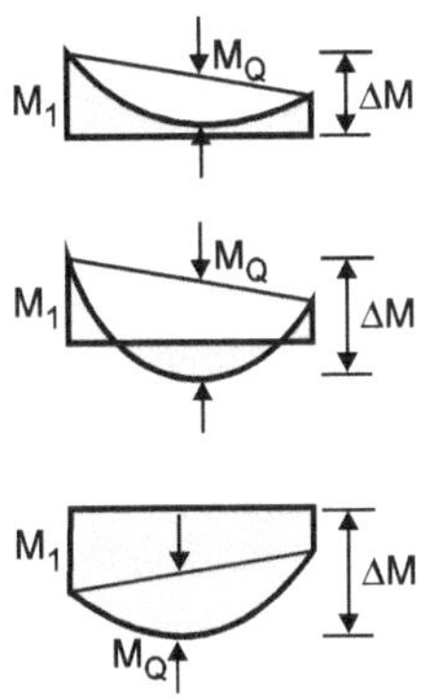	$\beta_M = \beta_{M,\psi} + \frac{M_Q}{\Delta M}(\beta_{M,Q} - \beta_{M,\psi})$ where $M_Q = \lvert \max M \rvert$ due to lateral load only and $\Delta M = \lvert \max M \rvert$ for moment diagram without change of sign $\Delta M = \lvert \max M \rvert + \lvert \min M \rvert$ where sign of moment diagram changes

TABLE 6.3
Values of α and β

Type of Cross Section	α	β
I- and H-sections	2	$5n$ but ≥ 1
Circular tubes	2	2
Rectangular hollow sections	$\frac{1.66}{1-1.33n^2}$ but ≤ 6	$\frac{1.66}{1-1.33n^2}$ but ≤ 6
Solid rectangles and plates	$1.73 + 1.8n^3$	$1.73 + 1.8n^3$

Notation: $n = N_{Sd}/N_{pl.Rd}$.

2. Uniform Load

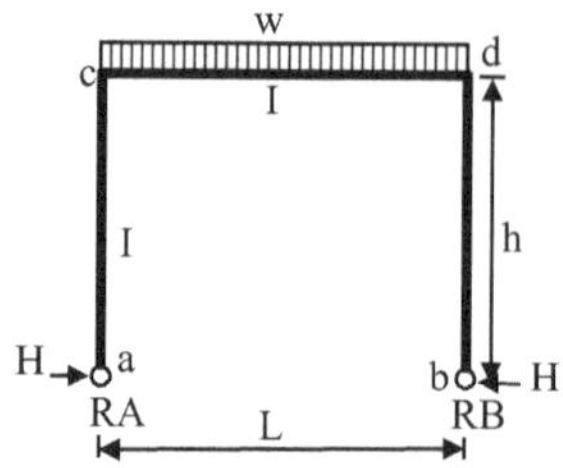

$H_a = H_b = wL^2/4hN$
$RA = RB = wL/2$

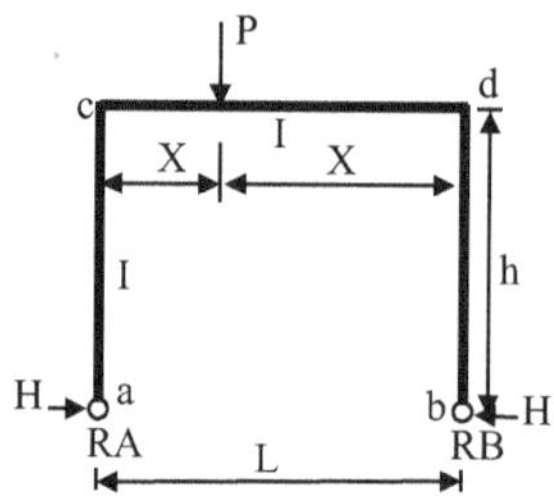

$H_a = H_b = 1.5\ P\ X_1.X_2/(hLN)$
$RA = PX_2/\mathrm{L}$
$RB = PX_1/\mathrm{L}$

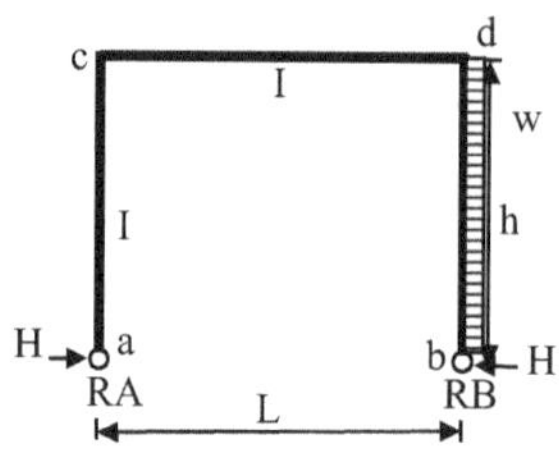

$H_a = (w.h/8).(5k + 6)/N$
$H_b = -(w.h/8).(11k + 18/N)$
$RA = RB = w.\ h^2/2L$

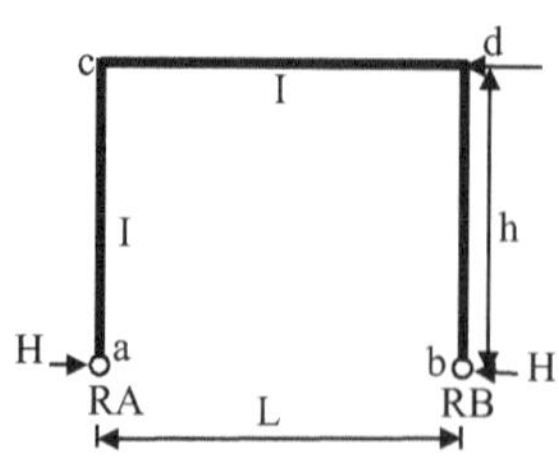

$H_a = -H_b = P/2$
$RA = RB = P.h/\mathrm{L}$

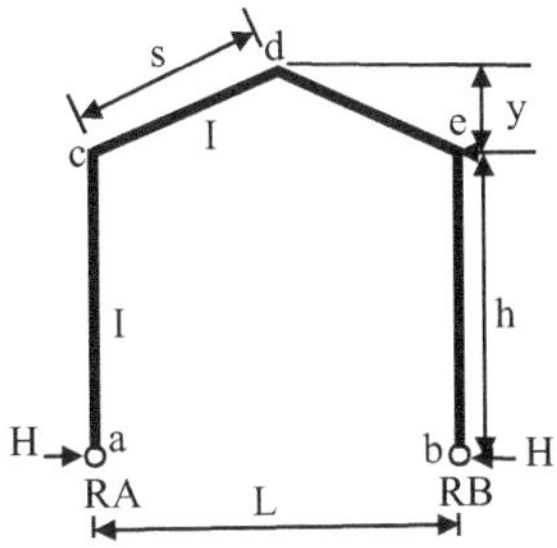

$K = (I_G/I_C).(h/s)$
$M = y/h$
$N = (K+3) + M(M+3)$

In case of unloaded column,

$M_c = -H_a.h$
$M_e = -H_b.h$
$M_d = M_o - H_a(h+y)$

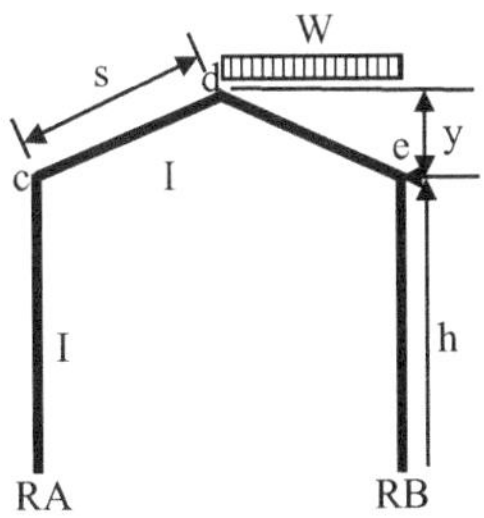

$H_a = H_b = wL^2/64h.((5M+8)/N)$
$RA = wL/8$
$RB = 3wL/8$

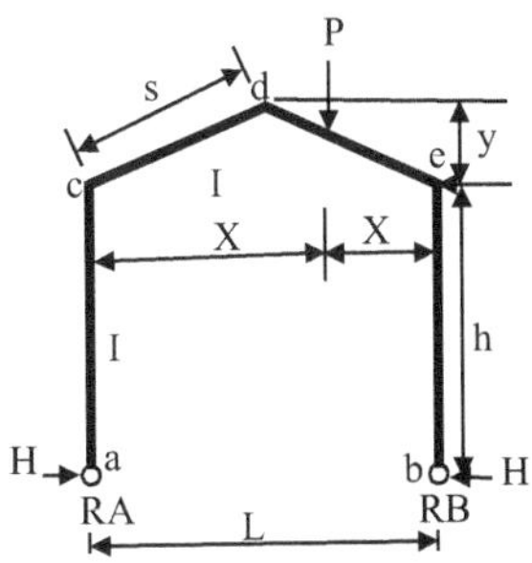

$H_a = H_b = (P.X_2/4h).[6X_1L + M(3L^2 - 4X_2^2)]/L^2.N$
$RA = P.X_2/L$
$RB = P.X_1/L$

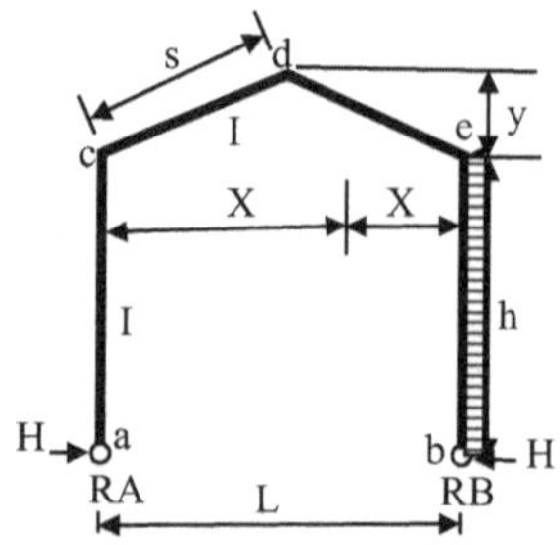

$H_a = (w.h/16).(5K + 6(M+2)/N)$
$H_b = H_a - w.h$
$RA = -RB = wh^2/(2L)$

3. Fixed Rectangular Frame

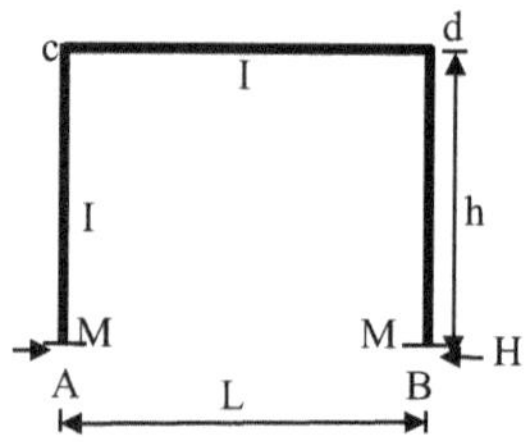

$K = (I_g/I_c).(h/L)$
$N_1 = k + 2$
$N_2 = 6k + 1$

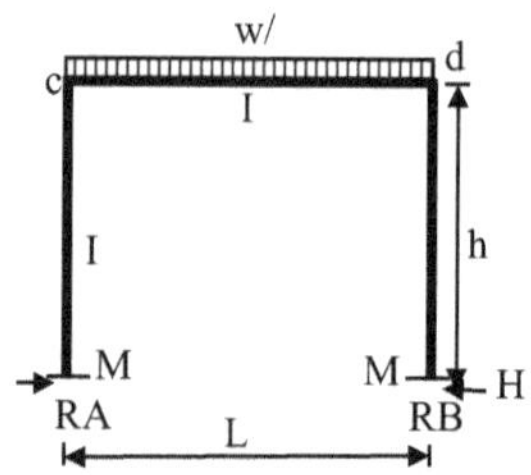

$H_a = H_b = wL^2/4hN_1$
$RA = RB = w.L/2$
$M_a = M_b = wL^2/12N_1 = H_a.h/3$
$M_c = M_d = wL^2/6N_1 = -2H_a.h/3$

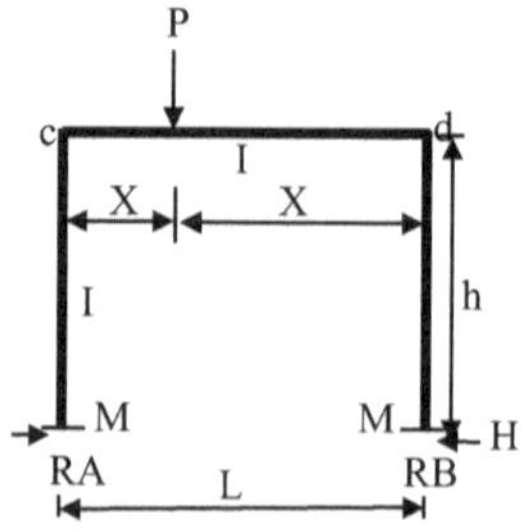

$H_a = H_b = 3P.X_1.X_2/2h.N_1.L$
$RA = P.X_2/L[1 + X_1(X_2 - X_1)/N_2L^2)]$
$RB = P - A$
$M_a = P.X_1.X_2/2L.\ (5k - 1 + 2.(X_1/L).N_1/(N_1.N_2)$
$M_c = M_a - H_a.h$
$M_b = (PX_1.X_2/2L).(7K + 3 - 2(X_1/L).N_1/(N_1.N_2)$
$M_d = M_b - H.h$

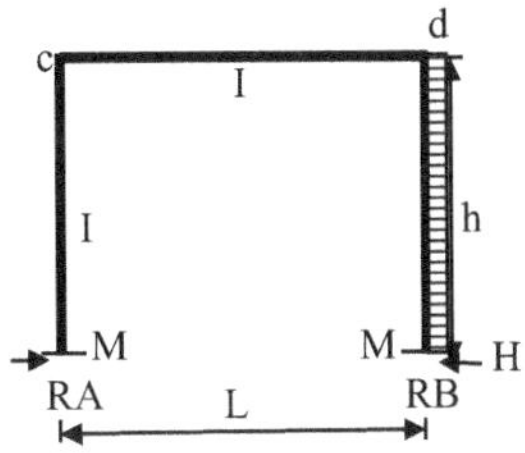

$H_a = (w.h/8).(2K + 3)/N_1$
$H_b = H_a - w.h$
$RA = -RB = wh^2K/L.N_2$
$M_a = (wh^2/24)[((5K + 9)/N_1) - (12K/N_2)$
$M_c = M_a - H_a.h$
$M_b = -wh^2/24[12 - ((5K + 9)/N1) - (12K/N_2)]$
$M_d = M_b - H.h + wh^2/2$

4. Fixed Support for Polygon Frame

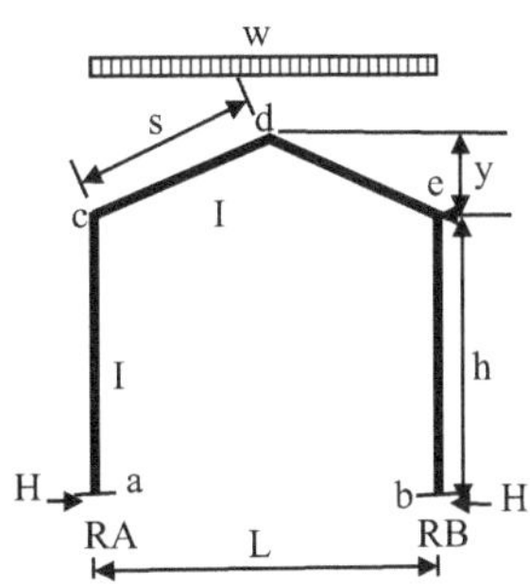

$K = I_G/I_C.h/S$
$M = y/h$
$N_1 = K(K + 4) + 2KM(3 + 2M) + M^2$
$N_2 = 3k + 1$
$H_a = H_b = (wL^2/8h).[4K + 5KM + M)/N_1]$
$RA = RB = wL/2$
$M_a = M_b = wL^2/48[(K(8 + 15M) + M(6 - M)]/N_1$
$M_c = M_e = -H_a.h + M_a$
$M_d = w.L^2/8 + M_a - H_a.(h + y)$

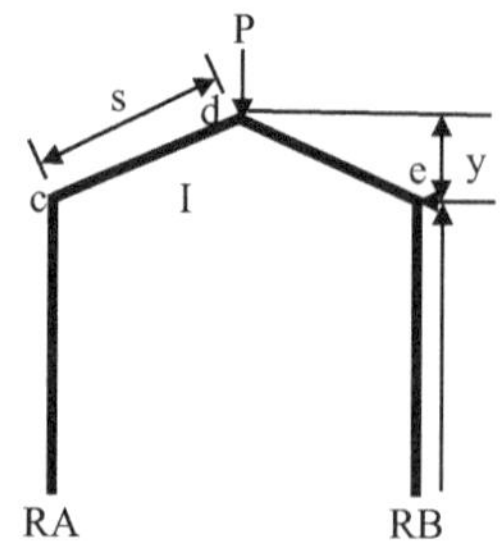

$H_a = H_b = (PL/4h)[(3K + 4KM + M)/N_1]$
$RA = RB = P/2$
$M_a = M_b = PL/4[(K+2KM + M)/N_1]$
$M_c = M_e = -H_a.h + M_a$
$M_d = P.L/4 + M_a - H_a\ (h + y)$

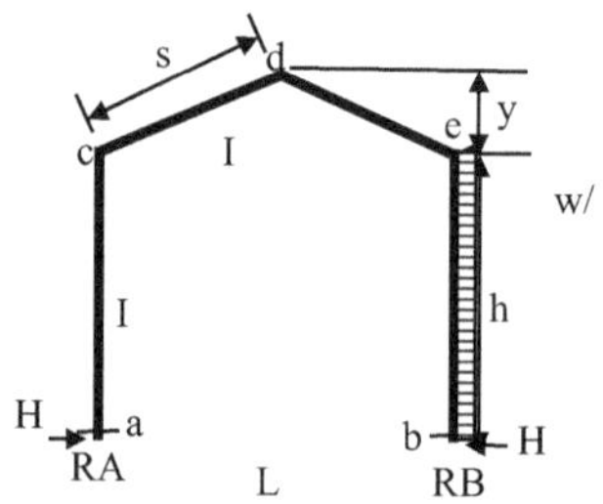

$H_a = (whk/4)[(K + 3 + 2M)/N_1]$
$RA = -RB = wh^2/2L - (M_b - M_a)/L$
M_a or $M_b = wh^2/24[-((K(K+6) + KM(15 + 16M) + 6M^2)/N_1) + 6(2K + 1)/N_2$
$M_c = -h_a.h + M_a$
$M_e = wh^2/2 + M_a - H_b.h$

6.4 METHODS OF GLOBAL FRAME ANALYSIS

The determination of the actual load–deformation response generally requires the use of a sophisticated analysis method. For practical purposes, assumptions regarding the frame, its component members, and joint models are made to allow for the determination of a safe bound for the ultimate load. Consequently, models range from simple elastic analyses or rigid–plastic analyses to more complex elastic–plastic analyses, which can provide a close representation of the real behaviour of the structure.

The elastic and plastic analysis methods are the primary structural methods. Elastic analysis can be used in all cases, while plastic analysis is applied with certain limitations. These methods are referred to as second-order theory and first-order theory, respectively. While second-order theory can be adopted in all cases, first-order theory may only be used when the displacement effects on the structural behaviour are negligible or when they can be incorporated in some other way.

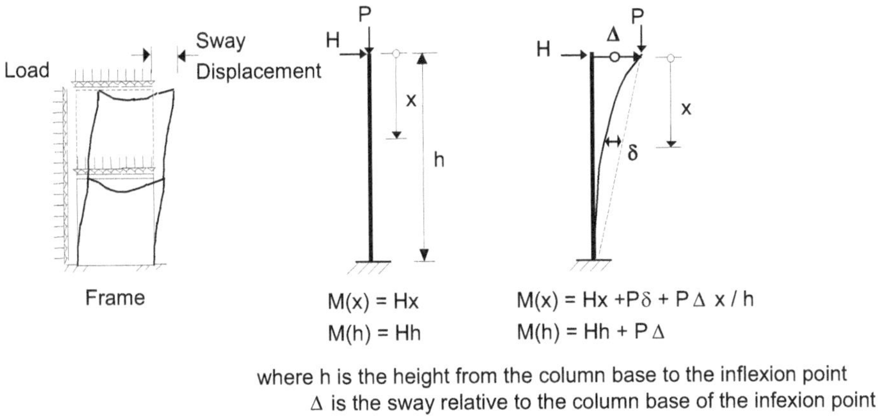

FIGURE 6.8 First and second order moments in a beam–column.

The following sections will present different methods of analysis, focusing only on two-dimensional frames that are subject to in-plane loading and deformations.

6.4.1 Second-Order Effects

The deflections caused by external loads modify the structural response and the distribution of internal forces. Therefore, it is necessary to evaluate their influence so that appropriate action can be taken to account for them if required.

In the case of frame structures, the most significant effect in linear analysis is the P-delta effect. This is well illustrated in Figure 6.8 as, where the frame is represented as a cantilever extending from the base to the end floor. This figure depicts two-storey floors. The delta effect results from building sway due to horizontal forces, such as wind loads, seismic activity, or wave loads in the case of offshore structures.

In the presence of axial load, the lateral displacement (Δ) is commonly referred to as drift when it results from wind or seismic action in multi-storey frames. At the top of the member and due to the curvature of the member itself, second-order effects in the form of secondary moments are induced along the member. Consequently, the actual deformations of the column under a given loading will be greater than those predicted by a first-order analysis if the secondary effects are neglected.

It can be seen that a global second-order moment, commonly referred to as the P-Δ effect, arises throughout the member due to the relative lateral displacement (Δ) between the top and the bottom of the member. In addition, a local second-order moment, commonly referred to as the P-δ effect, arises in the axially loaded member due to the deflections (δ) relative to the chord line connecting the member ends.

For frames, whilst the P-δ effect still arises when sway deformation is prevented, both the P-δ effect and the P-Δ effect arise when sway can occur. In the case of sway in typical frames, the P-Δ effect is usually much more significant than the P-δ effect. It is always necessary to evaluate whether sway effects in a frame are significant, whereas consideration of the P-δ effect needs to only be made for particularly slender members.

6.4.2 First-Order Elastic Analysis

A basic understanding of elastic frame analysis can be obtained by examining the slope–deflection equations for a simple beam or beam–column member. Starting from the equations for first-order theory, the introduction of second-order effects is then explained.

The basic slope–deflection equations express the moment at the end of a member as the superposition of the end moments due to external loads on the member with the ends assumed fixed, and of the moments caused by the actual end displacements and rotations. A set of simultaneous equations is written to express the equilibrium of the joints, in which the moments caused by joint displacements and rotations are expressed as linear functions of those same displacements and rotations. These equations are used to establish the terms related to bending and shear in the stiffness matrix for the member.

The slope–deflection method is used to analyse a sway frame, by considering the compatibility of the member rotations at the joints. In addition, equations expressing the lateral equilibrium of each floor are also established. The solution of the full system of equations provides the unknown joint displacements and rotations. When these are back substituted into the original slope–deflection equations for each member, the end moments can then be obtained. Subsequently, the end reactions can also be determined.

The reader is assumed to have some familiarity with the slope–deflection method, which is, in fact, an application of the more general displacement method of first-order elastic analysis to plane frames, in which the effects of axial and shear strain energy can be disregarded compared to the bending strain energy. This simplification is usually justified when analysing frames.

Figure 6.9 shows the sign convention adopted here, along with the symbols used for the moments, shear forces, rotations, and sway displacement.

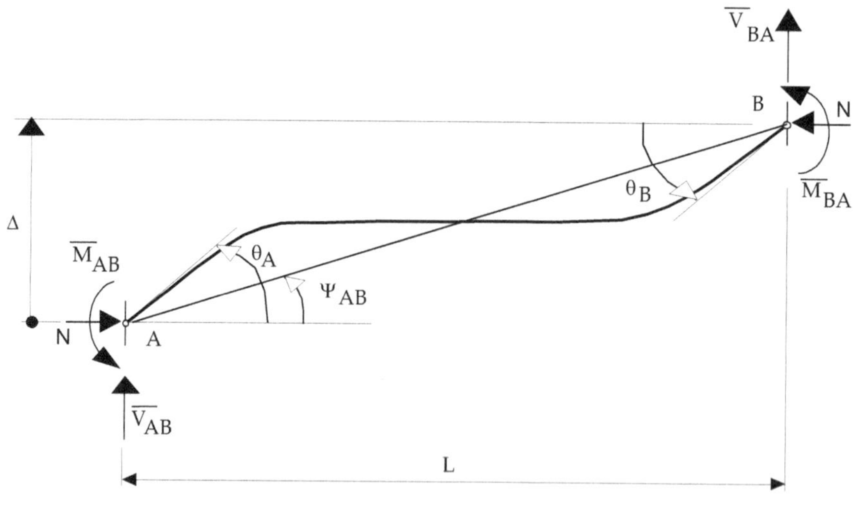

FIGURE 6.9 Slope–deflection methods.

The basic slope–deflection equations for the member may then be written as follows:

$$\bar{M}_{AB} = \left[\frac{EI}{L}\left(4\theta_A + 2\theta_B - 6\psi_{AB}\right) + M_{AB}\right]$$
$$\bar{M}_{BA} = \left[\frac{EI}{L}\left(2\theta_A + 4\theta_B - 6\psi_{AB}\right) + M_{BA}\right] \quad (6.1)$$

6.5 GLOBAL ELASTIC FRAME ANALYSIS

6.5.1 First-Order Theory

Linear elastic analysis implies an indefinite linear response of sections and joints, as presented in Figure 6.10. Equilibrium is expressed with reference to the undeformed configuration of the structure in a first-order analysis.

First-order elastic global analysis, with linear member and joint behaviour, results in a linear load–deflection curve, as shown in Figure 6.11 as presented in EUR 18563 report (1993).

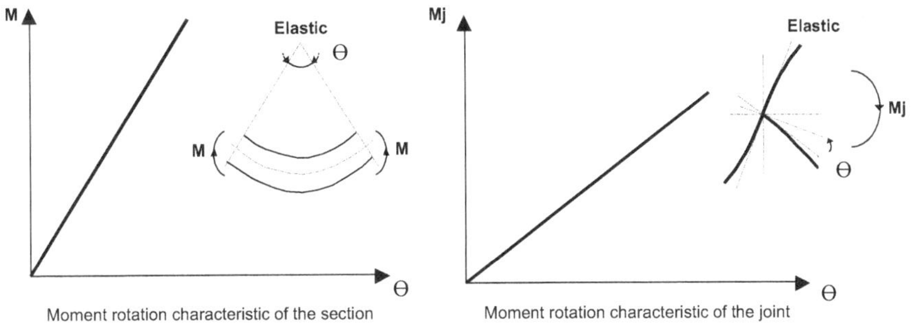

FIGURE 6.10 Characteristics of moment rotation for members and joints.

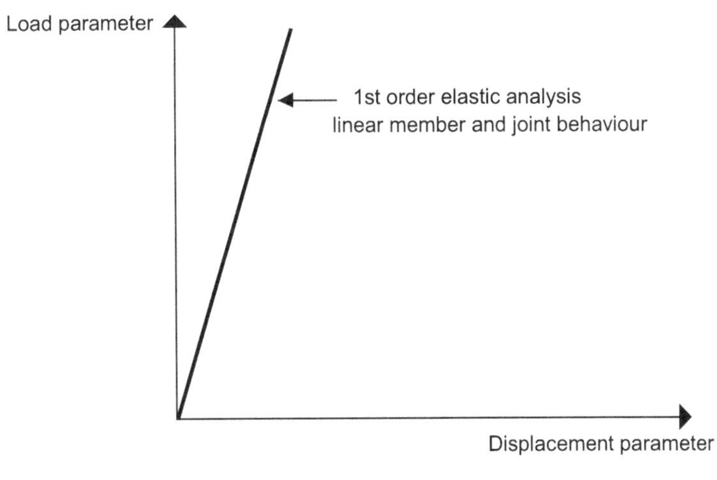

FIGURE 6.11 Relationships between load and displacement in first-order elastic analysis.

Frame imperfections must be accounted for in the analysis.

The use of tapered members is quite common for single-storey pitched-roof portal frames, and it is typical to model each such member using a sufficient number of elements of constant section. An alternative for single-bay pinned-base frames is to use a semi-graphical method based on an exact elastic model for tapered members, which is valid for both first- and second-order analyses.

Designers are generally familiar with first-order elastic analysis, which is the simplest of all possible types of analysis. Over the years, a variety of methods aimed at hand calculation have been developed, such as the slope–deflection method, the moment distribution method, as well as analytical formulae (sometimes presented in non-dimensional graphical form) for rapid analysis. These methods can be generalized to include the joint behaviour, as mentioned by Maquoi and Chabrolin (1998). The same applies to procedures based on matrix formulation, which have now almost entirely supplanted the hand methods as computer use has become common practice in design offices.

The principle of superposition of loading and load effects can be applied in the case of first-order elastic analysis.

For structures with low axial loads, first-order elastic analysis provides a safe basis for design, as long as the predicted response of the structure deviates only slightly from the actual response across a considerable range of loads.

It is mandatory to check the overall stability of the frame. The elastic critical buckling load for the frame may be evaluated using either a special analysis, such as the matrix method presented by Livesley (1969), or non-linear analysis presented by Gachon and Galea (1978).

For most frames, first-order elastic analysis generally serves as a good tool for predicting the response of the structure and its elements at the serviceability limit state. At this level of loading, the non-linear effects are relatively small.

Once the design forces such as axial forces, bending moments, and shear forces have been determined throughout the structure, the following checks should be performed:

- Ground–structure interaction effects and allowance for them when required.
- Second-order effects and allowance for them when required.
- The ultimate limit state resistance of the cross sections of the members.
- The ultimate limit state resistance of the joints.
- Instability phenomena: column buckling, lateral–torsional buckling, and partial or overall frame instability.
- Resistance to local buckling and resistance to concentrated loads, as well as fire resistance.
- The serviceability limit state requirements and vibrations.
- Resistance to fatigue

6.5.2 Second-Order Theory

In this type of elastic analysis, the indefinitely linear elastic response of sections and joints is still implied, as shown in Figure 6.12. The distribution of the internal forces

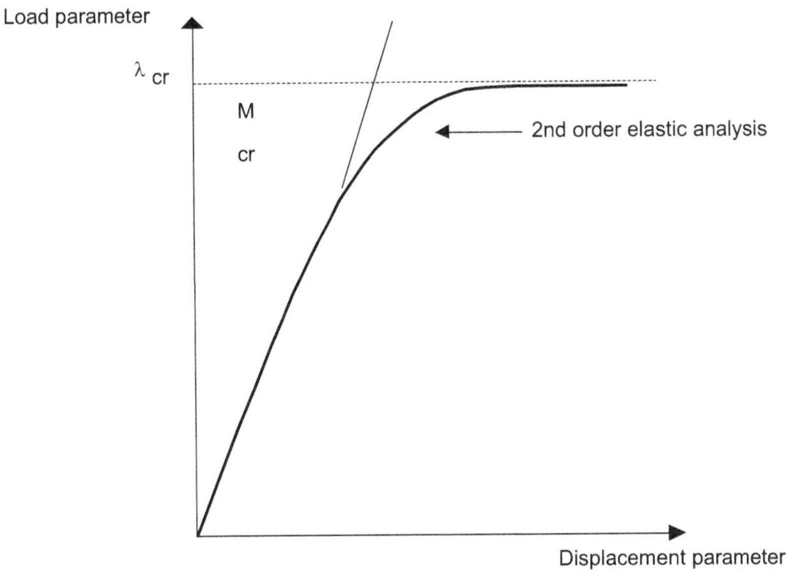

FIGURE 6.12 Relationship between load and displacement in second-order elastic analysis.

is now computed based on a second-order theory of the kinds outlined previously (although, for frames meeting certain conditions, a method involving simplified nominal corrections to a first-order analysis may be used). Equilibrium equations are formulated for the deformed structure (P-Δ effects), taking into account member imperfections and the decrease in member stiffness due to axial compressive forces (*P-δ* effects) when necessary.

In contrast to a first-order elastic analysis, one can no longer superimpose the results from the analysis of individual load cases. It is advisable to analyse each load combination case separately.

Figure 6.12 illustrates the load deflection response resulting from a second-order elastic analysis in which all the loads are increased monotonically by the same load multiplier. The load deflection curve, which now includes geometric non-linearity, approaches asymptotically a horizontal value corresponding to a peak value of *Mcr* (critical load multiplier). This value of *Mcr* corresponds to the elastic critical buckling load of the frame, relevant to the specified load case combination.

If the *P-δ* effects are disregarded, the computed peak load may be higher than the actual one. It is noted that the more slender the compression members are, the more significant the *P-δ* effects become.

The elastic critical buckling load is an important reference, as it represents the highest theoretical load that the frame can withstand in the absence of any material yielding. When yielding occurs, as is often the case, it results in the actual maximum load being lower – usually significantly lower – than the elastic critical buckling load.

Unlike first-order elastic analysis, second-order analysis provides internal forces and moments at the ends of the members, which include the second-order effects, thus eliminating the need for further examination.

The elastic critical buckling load of the frame may also be obtained from a second-order elastic analysis, but only if the load multiplier is increased sufficiently. However, if the frame is designed to remain elastic, there is no need to check in-plane frame stability. Furthermore, since the P-δ effects are not usually significant, there is no need to check the in-plane stability of the members. When curved or very slender members are used, member imperfections and P-δ effects should be included in the analysis. Nevertheless, unless the members are modelled using enough elements to provide the moments and forces along their length, an in-plane stability check of the members is advisable

6.6 METHODS OF GLOBAL PLASTIC FRAME ANALYSIS

The plastic analysis method can be used only if there are minimum requirements for steel ductility, member cross-sectional ductility, joint ductility, and lateral support at hinge locations, as follows:

1. The ratio between the specified minimum tensile strength, f_u, and the specified minimum yield strength, f_y, is equal to or higher than 1.2.
2. The elongation at failure over a gauge length of $5.65\sqrt{A_0}$ is not less than 15%, where A_0 is the original cross-sectional area.
3. The ultimate strain, ε_u, corresponding to the ultimate strength, f_u, is equal to or higher than 20 times the yield strain, ε_y, corresponding to the yield strength, f_y
4. There is a lateral restraint at all plastic hinge locations under any load case. The restraint should be provided within a distance along the member from the theoretical plastic hinge location that does not exceed (1/2) the member depth.
5. The member section classifications must generally be Class 1. Classes 2 and 3 may also be allowed where hinges do not occur. Class 2 sections may be used at a hinge location only when a large rotation capacity is not required there.
6. Where cross sections of the members vary along their length, restrictions are placed on the distances from a hinge at which reductions of web thickness and changes in web and compression flange classification can be affected.

6.6.1 Elastic–Perfectly Plastic Analysis (Second-Order Theory)

In the elastic–perfectly plastic analysis, it is assumed that any section or joint remains elastic up to the attainment of the plastic moment resistance, at which point it becomes ideally plastic. Plastic deformations are assumed to be concentrated at the plastic hinge locations, which are assumed to have an infinite rotational capacity. The actual rotation capacities must usually be checked later to ensure they are sufficient to meet the requirements.

The influence of the normal force or shear force on the plastic moment resistance of the sections may either be accounted for directly or checked later at the design verification stage. Note that the plastic hinge moment resistance to be adopted is the design value denoted by $M_{pl,Rd}$.

FIGURE 6.13 Behaviour of members and joints.

The load–displacement curve of the frame can be determined. The computation of plastic rotations at the plastic hinges may also be carried out to permit the check that the required rotation capacity is available.

The pushover analysis principle is presented in Figure 6.14, which is based on the second-order elastic–perfectly plastic analysis. In this method, the load is applied in increments, during which plastic hinges are formed sequentially, considering that more than one may occur simultaneously.

One starts with the displacements from an elastic second-order analysis, as shown in Figure 6.14 for stage 1, while monitoring the bending moments in the frame at each incremental increase in the loads. The load at which the first hinge occurs is identified as the load at which the plastic moment resistance is attained in a section or joint. Subsequent analyses are conducted for further incremental loads, during which the frame behaves differently due to the introduction of a pinned joint at the location of the first plastic hinge (stage 2). It is important to note that a plastic hinge is assumed to be unable to resist any further moment increase and possesses sufficient ductility to undergo the necessary rotation. The joint introduced at the plastic hinge behaves as a pin only for the later incremental increases in loading, while continuing to transfer a moment equal to the plastic moment resistance. A new plastic hinge forms after a further increase in the load level, and this process is repeated.

Figure 6.14 illustrates the result of the second-order elastic–perfectly plastic analysis as the solid curve. Branch 1 corresponds to a fully elastic frame, and this curve becomes asymptotic to the elastic buckling load of the frame only if infinite elastic behaviour is assumed. A first hinge is formed, and the frame subsequently behaves under further load increments as if only one hinge exists (branch 2) until the next hinge develops. If unlimited elastic behaviour is assumed after the first hinge has formed, then branch 2 continues and becomes asymptotic to the buckling load of the "deteriorated" frame, which includes a pin introduced at the location of the first hinge. This process continues with new hinges forming (and corresponding pins being introduced) at each step, until the structure becomes unstable (mechanism or frame instability). The maximum load of the second-order elastic–plastic analysis corresponds to this load level, indicated as the reference load multiplier M_2 in Figure 6.14.

No additional design checks of the resistance of sections and joints are required if the influence of the normal force or shear force is accounted for. As the rotations at the plastic hinges have been calculated, this allows for verification of the required rotation capacity, if necessary.

The in-plane frame stability is addressed by the structural analysis when second-order theory is applied. Furthermore, in this case, the in-plane stability of the

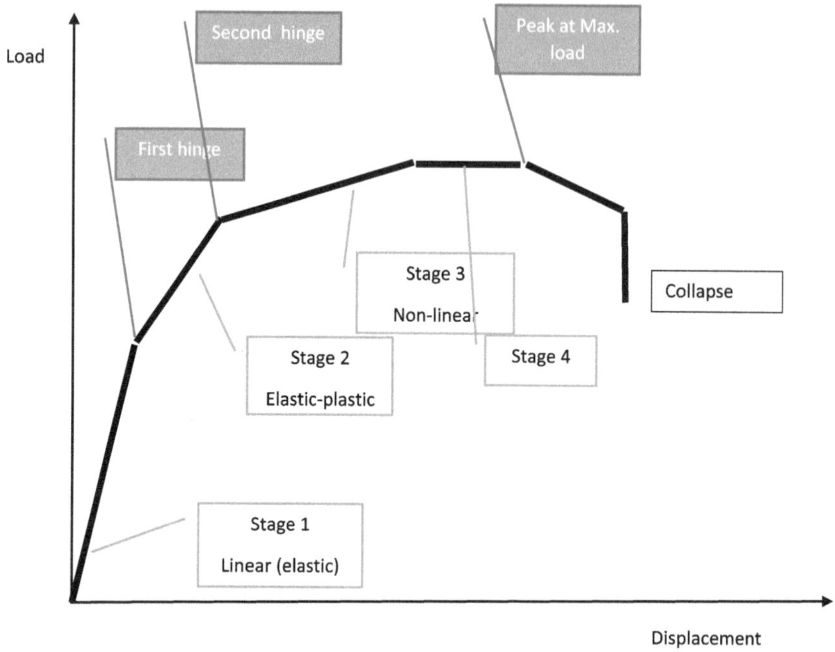

FIGURE 6.14 Relationship between load and displacement for second-order elastic–perfectly plastic analysis.

members need not be verified, provided that local member imperfections have been taken into account in the initial analysis when required.

It is clear that the number of remaining design tasks has been reduced compared to those required when an elastic first- or second-order analysis has been used.

6.6.2 Elastic–Plastic Analysis (Second-Order Theory)

Using a second-order elastic–plastic analysis provides a better estimation of the structural response compared to that provided by a first- or even a second-order elastic–perfectly plastic analysis, for instance.

Yielding of members and joints is a progressive process; therefore, the transition from elastic to plastic behaviour is a gradually phenomenon. Once yielding commences, as the moment in the member cross section continues to increases, the plastic zone extends partially along the member and through the depth of the cross section. This behaviour is considered by the plastic zone theory, noted by Chen et al. (1996) and Clarke (1994).

The ductility requirements for the members and joints, as well as the procedure for analysis and checks, are the same as those discussed for second-order elastic–perfectly plastic analysis.

Because of its complexity, the elastic–plastic method is not used for practical design purposes and is restricted to computer research applications only.

6.6.3 Rigid–Plastic Analysis (First-Order Theory)

This method is entirely different from elastic–plastic analysis, as the elastic deformations of members, joints, and foundations are small in comparison to the plastic deformations and are therefore ignored in this method.

In the elastic–perfectly plastic analysis, the plastic deformations are concentrated in sections and joints where plastic hinges are likely to occur. These sections and joints are assumed to have infinite rotational capacity.

Figure 6.15 shows the idealized rigid–plastic response of the sections and joints adopted for this type of analysis. As per Maquoi and Chabrolin (1993), the design moment resistance values for sections and joints, as well as the structural configuration and the loading, are the only parameters that affect rigid–plastic analysis.

The ductility requirements for the members and joints are the same as those outlined for the elastic–perfectly plastic analysis. Useful information on the minimum limits to be satisfied when using first-order rigid–plastic design for single-bay hinged-base frames is provided in ECCS (1991). It is important to note that this method is not suitable for second-order analysis.

Given that the maximum loading a structure can sustain corresponds to the loading at which collapse occurs due to a realistic plastic mechanism, the analysis involves identifying the critical mechanism.

The collapse load can be obtained by directly applying the fundamental theorems of simple plastic design. These fundamental theorems are the lower-bound and upper-bound theorems, also known as the static and kinematic theorems, respectively. A third theorem, the uniqueness theorem, states that when the conditions

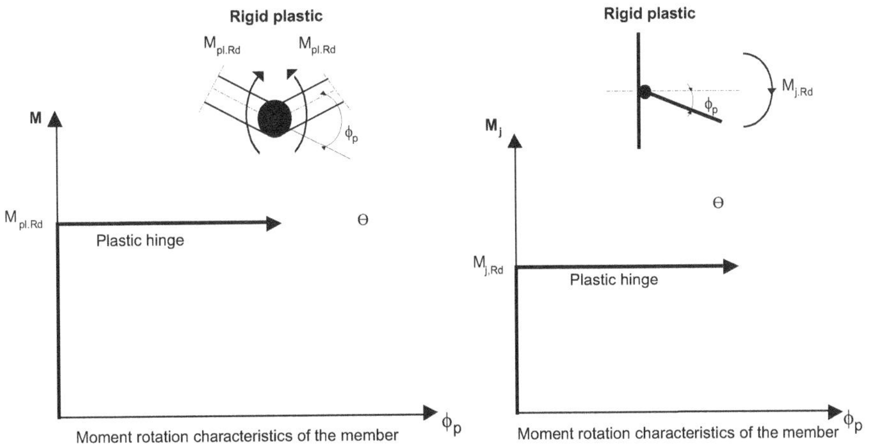

FIGURE 6.15 Characteristics of moment rotation for members and joints in rigid–plastic analysis.

of both the static and kinematic theorems are satisfied simultaneously by a given mechanism for the same loading case, the collapse load has been established, as stated by Neal (1956).

According to this theorem, for a given structure and loading, any arbitrarily assumed plastic collapse mechanism occurs at a load multiplier value that is greater than or equal to the value of the collapse load multiplier. By examining the various possible mechanisms, one can identify the collapse mechanism for which the load multiplier value is least and which is both statically and plastically admissible.

6.7 IN-PLANE BEHAVIOUR OF BEAM–COLUMNS

Beam–columns are defined as members subject to combined bending and compression. The most famous example of a beam–column is found in frame structures, where all the members experience both normal force and bending. In principle, all members in frame structures can be considered beam–columns, as per the definition of beams – steel members with nearly zero axial force – and columns – where the bending moment is approximately zero; these two cases represent extremes.

The possible responses of a member depend on factors such as the specific manner in which the applied loading is transferred, the type of support offered, and the cross-sectional shape of the member.

This chapter discusses the main concepts relating to the interaction between bending and compression effects by concentrating on uniaxial and in-plane behaviour from an EC3 perspective. Additionally, it addresses beam–columns to cover the cases of biaxial bending and out-of-plane buckling.

Beam–columns will be illustrated in three cases

i. The restrained beam–column;
ii. Uniaxial bending and compression of the unrestrained beam–column;
iii. Biaxial bending and compression of the unrestrained beam–column.

In case (i) is governed by the type of cross section, which will be discussed later. Case (ii) is influenced by the interaction of the cross-sectional behaviour with in-plane column buckling, with or without lateral–torsional buckling. For case (iii), it is governed by the same factors as stage (ii), but the moment about the other axis must be included in the equation of stresses.

The simplest scenario involves bending applied about one principal axis only, with the member responding by bending solely in the plane of the applied moment.

When the deformation of the beam–column is confined to the plane of bending, and as there is lateral restraint against out-of-plane bending, its behaviour shows an interaction between beam bending and compression member buckling.

The limits of the member that to reach plasticity in case of beam scenario and the limit of buckling in case of pure axial force. Considering there will be an additional moment, (N,y) which this will reduce the moment capacity in case of full plasticity.

6.8 CROSS-SECTIONAL BEHAVIOUR

6.8.1 Bending and Axial Force for Class 1 and 2 Cross Sections

In these sections, full plasticity can be achieved without local buckling of the flange or web, so the stress distribution will be as shown in Figure 6.16. The combination of axial load and moment that results in this condition will be

a. For *neutral axis in web* $y_n \le (h - t_f)/2$

$$N_M = 2 f_y t_w y_n$$

$$M_N = f_y b t_f (h - t_f) + f_y \left[\left(\frac{h - 2t_f}{2} \right)^2 - y_n^2 \right] t_w \tag{6.2}$$

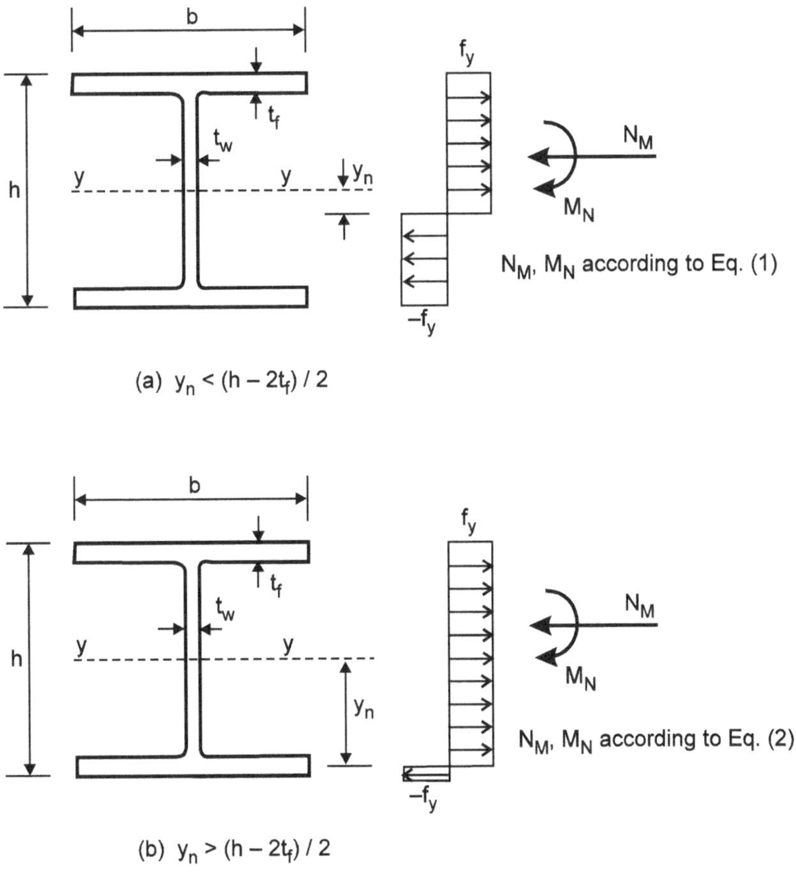

FIGURE 6.16 Full plasticity under axial load and moment.

b. For the *neutral axis in flange* $y_n > \left(h - t_f\right)/2$

$$N_M = f_y\left[t_w\left(h-2t_f\right)+2b\left(t_f-\frac{h}{2}+y_n\right)\right]$$

$$M_N = f_y b\left(\frac{h}{2}-y_n\right)\left(h-y_n\right)t_f \tag{6.3}$$

It is found from Figure 6.17 a slightly differences that equations 6.2 and 6.3 applied on H-section with the approximation used in Eurocode 3 of

$$M_{Ny.Rd} = M_{pl.y}\left(1-n\right)/\left(1-0,5a\right) \text{ but } M_{Ny.Rd} \le M_{ply.Rd} \tag{6.4}$$

in which $n = N_{Sd}/N_{pl.Rd}$ is the ratio of axial load to "squash" load (f_y A), and $a = \left(A - 2bt_f\right)/A \le 0.5$.

In case of cross sections without bolt holes, the following approximate equation can be used for moments about the z axis:

for $n \le a : M_{Nz.Rd} = M_{pl.z.Rd}$

for $n > a : M_{Nz.Rd} = M_{pl.z.Rd} \cdot \left[1-\left(\frac{n-a}{1-a}\right)^2\right]$

where $n = N_{Sd}/N_{pl.Rd}$ and $a = \left(A - 2bt_f\right)/A$ but $a \le 0.5$.

Further simplifications and approximations for a range of common cross-sectional shapes are presented in Table 6.1. In all cases, the value of M_N should be less than or equal to M_{pl}.

6.8.2 Bending and Axial Force for Class 3 Cross Sections

The behaviour of column stresses under axial compression and bending moment for I beams is presented in Figure 6.18. The bending moment about the y-axis produces the uniform and varying stress distributions shown in Figure 6.18a and b.

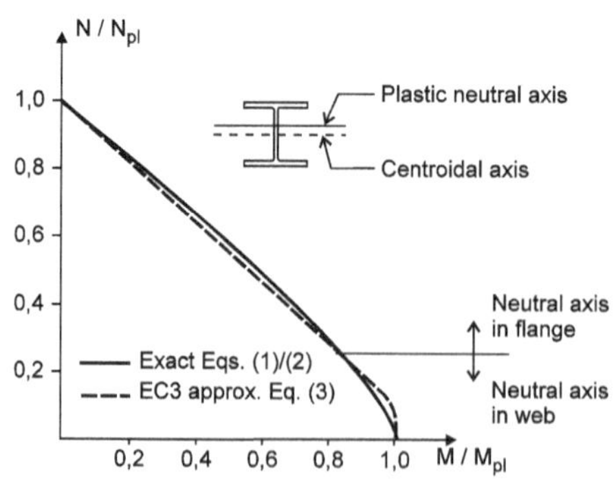

FIGURE 6.17 Full plasticity interaction (Adapted from EC3).

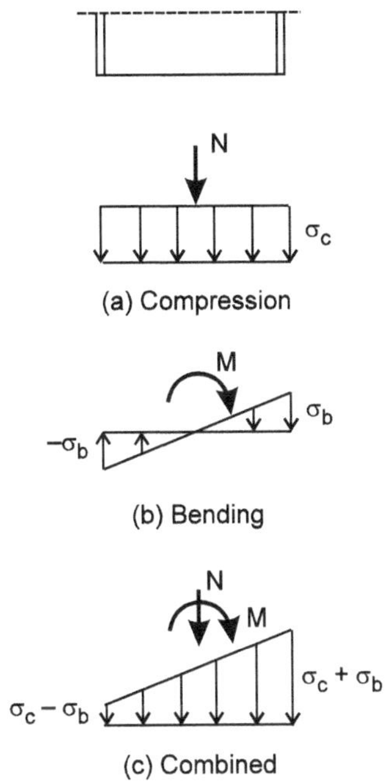

FIGURE 6.18 Elastic behaviour of cross section in compression and bending.

For elastic behaviour, the principle of superposition may be used to simply add the two stress distributions, as illustrated in Figure 6.18c. First yield will therefore develop at the edge where the maximum compressive bending stress occurs, corresponding to the condition:

$$f_y = \sigma_c + \sigma_b$$

where

f_y = the yield stress
$\sigma_c = N/A$ is the stress due to the compressive load N
$\sigma_b = \dfrac{Mh/2}{I}$ is the maximum compressive stress due to the moment M, h is the overall depth of the section, and I is the second moment of inertia about the y-axis

In the case of Class 3 cross sections, it will be satisfactory if the maximum longitudinal stress $\sigma_{x.Ed}$ satisfies the criterion:

$$\sigma_{x.Ed} \le f_{yd}$$

where $f_{yd} = f_y/\gamma_{M0}$.

6.8.3 Bending and Axial Force for Class 4 Cross Sections

In the case of Class 4 cross sections, they will be safe if the maximum longitudinal stress $\sigma_{x.Ed}$ calculated using the effective widths of the compression elements, as per paragraph 5.3.2.(2) in EC3, is valid for the following criterion:

$$\sigma_{x.Ed} \leq f_{yd}$$

where $f_{yd} = f_y / \gamma_{M0}$.

6.9 OVERALL STABILITY

The cross-sectional behaviour discussed in the previous section does not consider the exact manner in which the moment M at the particular cross section under consideration was generated. Figure 6.19 shows a beam–column with lateral deflection due to the effects of axial compression and applied moments at the ends.

The moment at any point along the length can conveniently be regarded as being composed of two parts:

- primary moment M
- secondary moment Nv.

In 1988, Trahair and Bradford performed an analysis for this case elastically by using strut theory, which gives the maximum deflection at the centre as follows:

$$v_{\max} = \frac{M}{N} \sec \frac{\pi}{2} \sqrt{\frac{N}{P_{E_y}}} - 1$$

where $P_{Ey} = \dfrac{\pi^2 EI_y}{L^2}$ is the Euler critical load for major axis buckling, and the maximum moment is as follows:

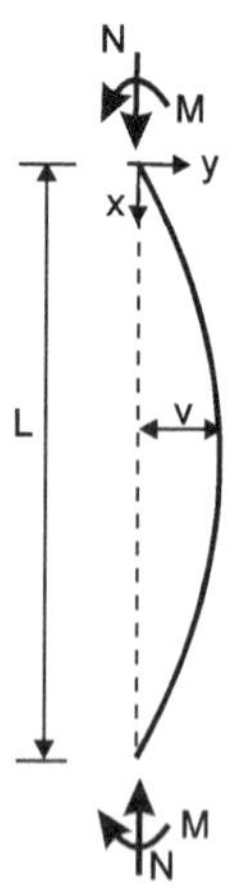

FIGURE 6.19 Primary and secondary moments.

$$M_{\max} = M \sec \frac{\pi}{2}\sqrt{\frac{N}{P_{Ey}}}$$

In both equations, the secant term may be replaced by noting that the first-order deflection created as a result of the end moments, M, acting alone and the first-order moment, M, obtained by ordinary beam theory, is approximately amplified by the term:

$$\frac{1}{1 - N/P_{Ey}}$$

Thus,

$$v_{\max} = \frac{ML^2}{8EI_y}\frac{1}{1 - N/P_{Ey}}$$

$$M_{\max} = M\frac{1}{1 - N/P_{Ey}}$$

Since the maximum elastic stress will be

$$\sigma_{\max} = \sigma_c + \sigma_b \frac{M_{\max}}{M} \tag{6.5}$$

Equation 6.5 may be rewritten as

$$\frac{\sigma_c}{f_y} + \frac{\sigma_b}{f_y\left(1 - N/P_{Ey}\right)} = 1.0 \tag{6.6}$$

Equation 6.6 may be solved for values of σ_c and σ_b that just cause yield, taking different values of P_{Ey} which is a function of the slenderness L/r_y. If the bending moment is zero then σ_c tends to the value of yield strength f_y in case of zero axial load and then σ_b tends to the value of yield strength f_y.

The main issue in equation 6.6 is that it does not account for the case of column buckling before reaching yield, as the stresses due to buckling will be explained by this equation.

$$\sigma_{Ey} = \frac{P_{Ey}}{A} = \frac{\pi^2 EI_y}{AL^2} = \frac{\pi^2 E}{\lambda_y^2} \tag{6.7}$$

The use of both equations 6.6 and 6.7 ensures that both conditions are covered, as shown in Figure 6.20.

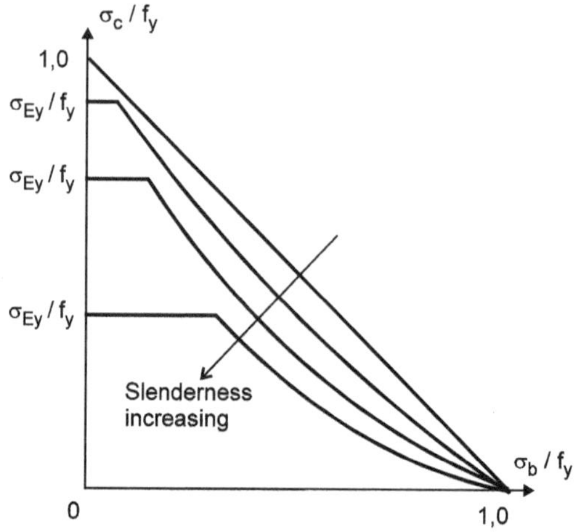

FIGURE 6.20 Combination of equations 6.6 and 6.7.

6.10 CALCULATION BASED ON EUROCODE 3

In general, equations 6.6 and 6.7 will be reformulated to check the member's safety against failure due to yield or elastic buckling. Considering that EC3 follows the limit states design code approach, the ultimate load will be considered regarding member capacity under static load.

Several factors are included, though they are somewhat hidden in the equations, such as initial lack of straightness, residual stresses, and others. These aspects are detailed in the book on structural reliability by Elreedy M.A. (2014).

6.10.1 Members with Class 1 and 2 Cross Sections

As per the direction of Eurocode 3, it is assumed that bending about the y-axis is to use

$$\frac{N_{Sd}}{\chi_y A f_y} + \frac{k_y M_{y.Sd}}{W_{pl.y} f_y} \leq 1 \tag{6.8}$$

in which χ_y is the reduction factor for column buckling, and

$$k_y = 1 - \frac{\mu_y N_{Sd}}{\chi_y A f_y} \text{ but } k_y \leq 1.5 \tag{6.9}$$

where k_y is a modification factor

$$\mu_y = \bar{\lambda}_y \left(2\beta_{My} - 4\right) + \frac{W_{pl,y}}{W_{el,y}} - 1 \text{ but } \mu_y \leq 0.90$$

where β_{My} is an equivalent uniform moment factor accounting for the non-uniformity of the moment diagram, as per Table 6.2, the moment diagram is about the y axis and the restraints in the z-direction.

6.10.2 Members with Class 3 Cross Sections

Members with class 3 cross sections subject to bending and axial load shall satisfy

$$\frac{N_{Sd}}{\chi_y A f_y} + \frac{k_y M_{y.Sd}}{W_{el.y} f_y} \leq 1 \tag{6.10}$$

where k_y and χ_y are defines in equation 6.9, with

$$\mu_y = \bar{\lambda}_y \left(2\beta_{My} - 4\right) \text{ but } \mu_y \leq 0.90 \tag{6.11}$$

6.10.3 Members with Class 4 Cross Sections

Steel members with class 3 cross sections subject to bending and axial load should satisfy the following equation:

$$\frac{N_{Sd}}{\chi_y A_{eff} f_y} + \frac{k_y \left(M_{y.Sd} + N_{Sd} e_{N.z}\right)}{W_{eff.y} f_y} \leq 1 \tag{6.12}$$

where

- k_y and χ_y are calculated from equation 6.9, and μ_y is derived from in equation 6.11.
- $A_{eff.y}$ is the effective cross-sectional area for pure compression
- $W_{eff.y}$ is the effective cross-sectional modulus for pure bending
- $e_{N.z}$ is the shift in the neutral axis, comparing the full cross section with the effective cross section, considering there is pure compression, which is used to account for local buckling

6.10.4 The Role of K_Y

The value of k_y, as shown by equation 6.9, depends in a rather complex way on

- the level of axial load, as measured by the ratio $\dfrac{N_{Sd}}{\chi_y A f_y}$
- the member slenderness λ_y
- the margin between the cross section's plastic and elastic section moduli W_{pl} and W_{el} (for class 1 and class 2 cross sections only)
- the pattern of primary moments.

In the worst-case scenario for all these parameters, k_y is 1.5. The role of k_y is to account for the secondary bending effect described earlier, as well as the effects of non-uniform moment and spread of yield.

Figure 6.21 illustrates the situation for end moments M and $\psi\, M$, where ψ can take values between +1 (uniform single curvature) and –1 (double curvature). The particular case shown corresponds to a value $\psi \cong -0.5$.

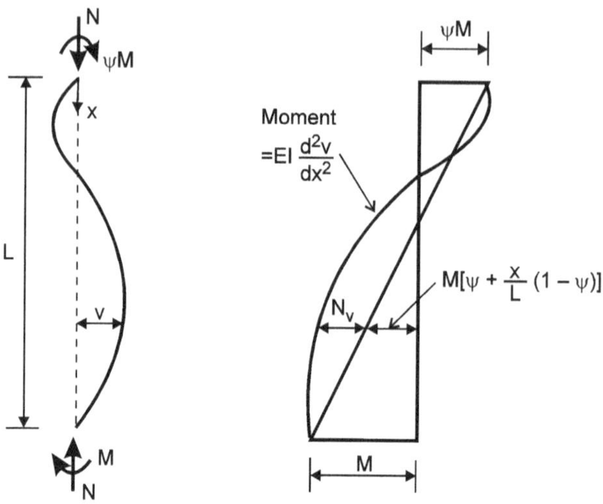

FIGURE 6.21 Non-uniform moment case.

For the case illustrated, the maximum moment still occurs within the member length; however, the situation is clearly less severe than that depicted in Figure 6.19, assuming all conditions are identical apart from the value of ψ. It is customary in design to recognize this by reducing the contribution of the moment term to the interaction relationship. Thus, in Eurocode 3, k_y in equation 6.9 depends on the ratio ψ.

Since the case of uniform single curvature moment is the most severe, it is a safe simplification to always use the procedure for $\psi = 1.0$.

Returning to Figure 6.21, it is possible for the point of maximum moment to be at the end where the larger primary moment is applied. This typically occurs if the axial load is small and/or the slenderness is low, resulting in relatively slight secondary bending effects. In such cases, design will be controlled by the need to ensure adequate cross-sectional resistance at this end. Therefore, the formula from Table 6.2, for the particular shape of the cross section being used, should be employed. In scenarios where only the uniform moment ($\psi = 1.0$) arrangement is being considered, the overall buckling check in equation 6.9 will always be more severe than (or, in the limit, equal to) the cross-sectional check, and therefore this latter check need not be performed separately (Figure 6.22).

6.11 LATERAL–TORSIONAL BEHAVIOUR OF BEAM–COLUMNS

In the case of an unrestrained beam–column bent about its major axis, it may buckle by deflecting laterally and twisting at a load that is significantly less than the maximum load predicted by an in-plane analysis. This lateral–torsional buckling may occur while the member is still elastic (see curve 1 of Figure 6.23), or after some yielding (curve 2) due to in-plane bending and compression.

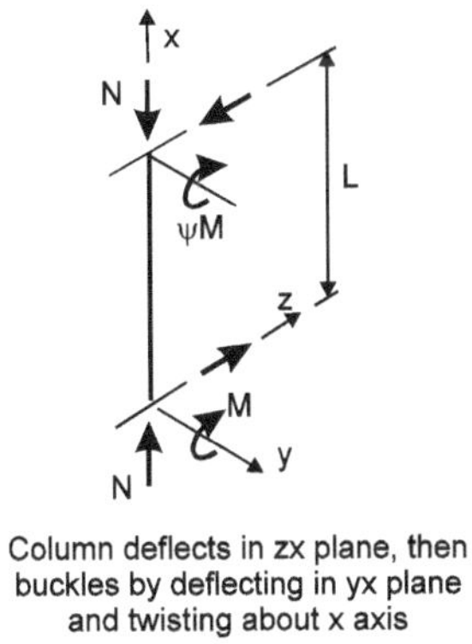

FIGURE 6.22 Lateral–torsional behaviour.

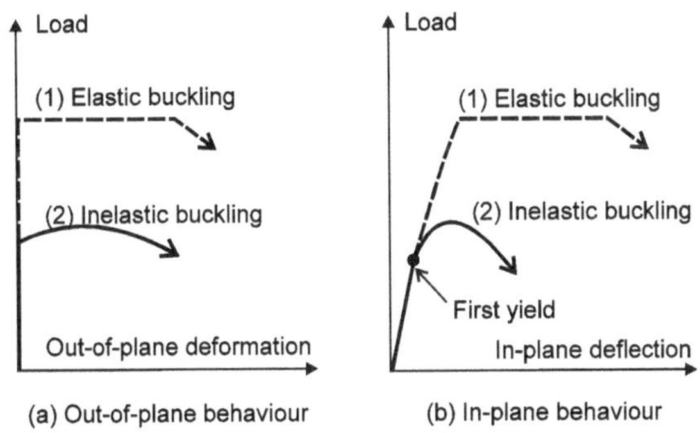

FIGURE 6.23 Lateral–torsional buckling.

6.11.1 Lateral–Torsional Buckling

For an unrestrained I-beam–column bent about the major axis, lateral–torsional buckling will occur. Assuming elastic behaviour and considering the arrangement of applied loading and support conditions, the critical combinations of N and M can be obtained from the solution presented by Chen and Atsuta (1976):

$$\frac{M^2}{i_0^2 P_{Ez} P_{E0}} = \left(1 - \frac{N}{P_{Ez}}\right)\left(1 - \frac{N}{P_{E0}}\right) \tag{6.13}$$

in which

- the polar radius of gyration; $i_0 = \sqrt{\dfrac{I_y + I_z}{A}}$
- the minor axis critical load; $P_{Ez} = \dfrac{\pi^2 E I_z}{L^2}$
- the torsional buckling load; $P_{E0} = \dfrac{G I_t}{i_0^2}\left(1 + \dfrac{\pi^2 E I_w}{G I_t L^2}\right)$

Equation 6.13 reduces to the buckling of a beam when the axial force (N) reaches zero and to the buckling of a column in either flexure (P_{Ez}) or torsion (P_{E0}) when the bending moment (M) reaches zero. In the first case, the critical value of M is given by

$$M_{cr} = \frac{\pi}{L}\sqrt{EI_z GI_t}\sqrt{1+\frac{\pi^2 EI_w}{L^2 GI_t}} \tag{6.14}$$

in which

- EI_z is the minor axis flexural rigidity
- GI_t is the torsional rigidity
- EI_w is the warping rigidity.

In deriving equation 6.13, no allowance was made for the amplification of the in-plane moments M by the axial load acting through the in-plane deflections. This may be approximated as $\frac{M}{1-N/P_{Ey}}$. Equation 6.13 can, therefore, be modified to

$$\frac{M^2}{i_0^2 P_{Ez} P_{E0}} = \left(1-\frac{N}{P_{Ey}}\right)\left(1-\frac{N}{P_{Ez}}\right)\left(1-\frac{N}{P_{E0}}\right) \tag{6.15}$$

Noting the relative magnitudes of P_{Ey}, P_{Ez} and P_{E0}, and re-arranging gives the following approximation:

$$\frac{N}{P_{Ez}} + \frac{1}{1-N/P_{Ey}} \frac{M}{i_0\sqrt{P_{Ez}P_{E0}}} = 1 \tag{6.16}$$

$$\frac{N}{P_{Ez}} + \frac{1}{1-N/P_{Ey}} \frac{M}{M_{cr}} = 1 \tag{6.17}$$

6.11.2 The Design Process in Eurocode 3

6.11.2.1 Members with Class 1 and 2 Cross Sections

Eurocode 3 provides the following interaction equation:

$$\frac{N_{Sd}}{\chi_z A f_y} + \frac{k_{LT} M_{y.Sd}}{\chi_{LT} W_{pl.y} f_y} \leq 1 \tag{6.18}$$

in which χ_z is the reduction factor for column buckling around the minor axis, χ_{LT} is the reduction factor for lateral–torsional beam buckling, and

$$k_{LT} = 1 - \frac{\mu_{LT} N_{Sd}}{\chi_z A f_y} \text{ but } k_{LT} \leq 1.0$$

with

$$\mu_{LT} = 0.15\left(\bar{\lambda}_z 2\beta_{M,LT} - 1\right) \text{ but } \mu_{LT} \leq 0.90$$

where $\beta_{M,LT}$ is a factor accounting for the non-uniformity of the moment diagram, see Table 6.2 (moment diagram about y-axis and restraints in the y direction).

6.11.2.2 Members with Class 3 Cross Sections

Members with Class 3 cross sections should satisfy the following criterion:

$$\frac{N_{Sd}}{\chi_z A f_y} + \frac{k_{LT} M_{y.Sd}}{\chi_{LT} W_{el.y} f_y} \leq 1 \tag{6.19}$$

6.11.2.3 Members with Class 4 Cross Sections

Members with class 3 cross sections should satisfy the following criterion:

$$\frac{N_{Sd}}{\chi_z A f_y} + \frac{k_{LT} M_{y.Sd} + N_{Sd} e_{N,z}}{\chi_{LT} W_{eff.y} f_y} \leq 1 \tag{6.20}$$

6.11.2.4 Role of k_{LT}

The value of k_{LT}, as shown by the equations explaining equation 6.18, depends on

- the level of axial load as measured by the ratio $\dfrac{N_{Sd}}{\chi_z A f_y}$
- the member slenderness λ_z
- the pattern of primary moments.

For the most severe combination k_{LT} adopts the value of unity, corresponding to a linear combination of the compressive and bending terms. This reflects the reduced scope for amplification effects in this case, since the value of N_{Sd} cannot exceed $\chi_z A f_y$, which will, in turn, be significantly less than the elastic critical load for in-plane buckling P_{Ey}.

It is, of course, also necessary to guard against the possibility of in-plane failure due to excessive deflection in the plane of the web at a lower load than that given by equation 6.18. This might occur, for example, in situations where differing bracing and/or support conditions are provided in the xy and xz planes, as illustrated in Figure 6.24. Such cases should be treated by checking, in addition to equation 6.18, an in-plane equation of the form:

$$\frac{N_{Sd}}{\chi_{\min} A f_y} + \frac{k_y M_{y.Sd}}{W_{pl.y} f_y} \leq 1 \tag{6.21}$$

in which χ_{min} depends on the in-plane conditions. Usually, however, equation 6.18 will govern.

6.12 BIAXIAL BENDING OF BEAM–COLUMNS

Analysis for the full three-dimensional case, even for the simple elastic version, is extremely complex, and closed-form solutions are not available. Rather than starting

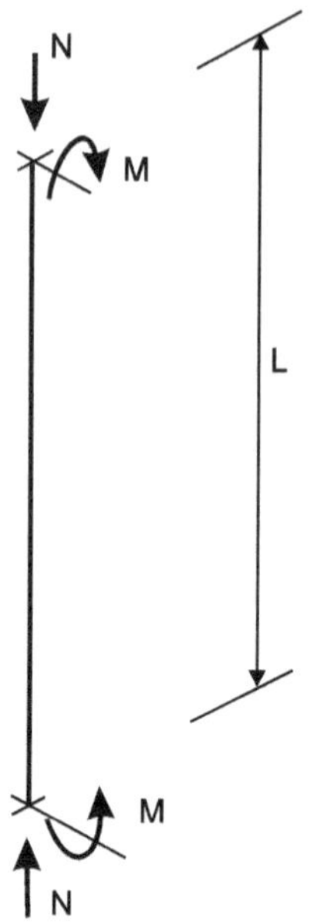

FIGURE 6.24 Column with different support conditions in xy and xz planes.

analytically, it is more convenient to approach the question of a suitable design method from behavioural considerations and the use of the methods already derived for the simpler cases of Figure 6.25.

Figure 6.26 presents a diagrammatic version of the design requirement. The N-M_z and N-M_y correspond to the two uniaxial cases that have already been examined. Interaction between the two moments, M_z and M_y, corresponds to the horizontal plane. When all three load components – N, M_y, and M_z – are present, the resulting interaction plots exist somewhere in the three-dimensional space represented by the diagram. Any point falling within the boundary corresponds to a safe combination of loads.

Assuming proportional loading, any combination may be regarded as a straight line starting at the origin, the orientation of which depends upon the relative sizes of the three load components. Increasing the loads extends this line from the origin until it just reaches and then exceeds the boundary. Non-proportional loading would correspond to a series of lines.

In each case, the axes have been taken as the ratio of the applied component to the member's resistance under the load component alone, e.g. $N_{Sd}/\chi_{min}\,Af_y$ in the case of compressive loading. Thus, Figure 6.26 actually represents the situation for one particular example with specific values of cross-sectional properties, slenderness, and load arrangement. Changing some or all of these will alter the shape of the interaction surface shown, but not the general principle involved.

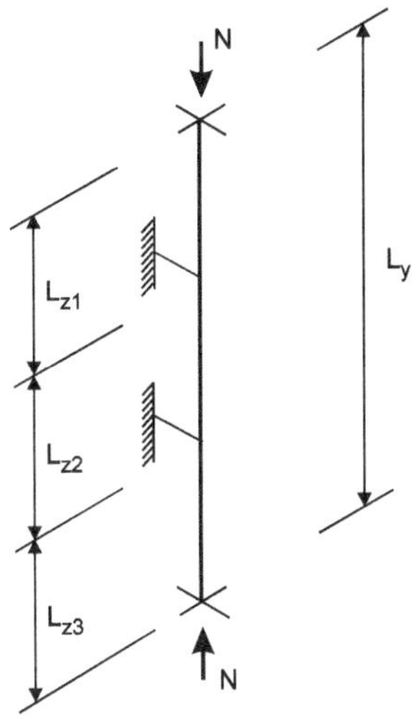

FIGURE 6.25 Biaxial bending.

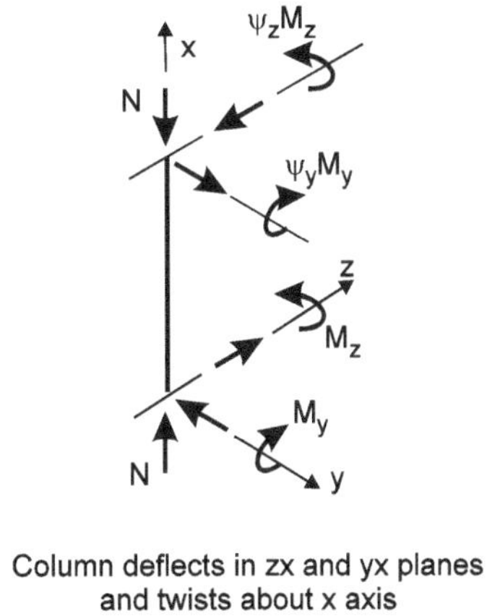

FIGURE 6.26 Interaction diagram for biaxial bending.

6.12.1 Design for Biaxial Bending and Compression

Members with class 1 and 2 cross sections subject to combined biaxial bending and axial compression should satisfy the following criterion:

$$\frac{N_{Sd}}{\chi_{\min} A f_y} + \frac{k_y M_{y,Sd}}{W_{pl,y} f_y} + \frac{k_z M_{z,Sd}}{W_{pl,z} f_y} \leq 1 \tag{6.22}$$

where k_z is a factor similar to k_y, see equation 6.9.

Members with class 1 and 2 cross sections subject to combined biaxial bending and axial compression where lateral–torsional buckling is relevant should also satisfy the following criterion:

$$\frac{N_{Sd}}{\chi_z A f_y} + \frac{k_{LT} M_{y,Sd}}{\chi_{LT} W_{pl,y} f_y} + \frac{k_z M_{z,Sd}}{W_{pl,z} f_y} \leq 1 \tag{6.23}$$

Members with class 3 cross sections subject to combined biaxial bending and axial compression should satisfy the following criterion:

$$\frac{N_{Sd}}{\chi_{\min} A f_y} + \frac{k_y M_{y,Sd}}{W_{el,y} f_y} + \frac{k_z M_{z,Sd}}{W_{el,z} f_y} \leq 1 \tag{6.24}$$

Members with class 3 cross sections subject to combined biaxial bending and axial compression where lateral–torsional buckling is relevant, should also satisfy the following criterion:

$$\frac{N_{Sd}}{\chi_z A f_y} + \frac{k_{LT} M_{y,Sd}}{\chi_{LT} W_{el,y} f_y} + \frac{k_z M_{z,Sd}}{W_{el,z} f_y} \leq 1 \tag{6.25}$$

Members with class 4 cross sections subject to combined biaxial bending and axial compression should satisfy the following criterion:

$$\frac{N_{Sd}}{\chi_{\min} A_{eff} f_y} + \frac{k_y \left(M_{y,Sd} + N_{Sd} e_{Nz}\right)}{W_{eff,y} f_y} + \frac{k_z \left(M_{z,Sd} + N_{Sd} e_{Ny}\right)}{W_{eff,z} f_y} \leq 1 \tag{6.26}$$

Members with class 4 cross sections subject to combined biaxial bending and axial compression where lateral–torsional buckling is relevant, should also satisfy the following criterion:

$$\frac{N_{Sd}}{\chi_z A_{eff} f_y} + \frac{k_{LT} M_{y,Sd} + N_{Sd} e_{Ny}}{\chi_{LT} W_{eff,y} f_y} + \frac{k_z \left(M_{z,Sd} + N_{Sd} e_{Nz}\right)}{W_{eff,z} f_y} \leq 1 \tag{6.27}$$

An important point to note from the definition of A_{eff} and W_{eff} above is that the calculation of cross-sectional properties, and thus the classification of cross sections, should be undertaken separately for each of the three load components N, M_y, and M_z. This does mean that the same member may be classified as (for example) class 1 for major axis bending, class 2 for minor axis bending, and class 3 for compression. In such cases, the safe design approach is to conduct all beam–column checks using the procedures for the least favourable class.

6.12.2 Cross-Sectional Checks

If allowances have been made when determining the k factors (through the use of β_M) for the less severe effect of moment patterns other than uniform single curvature bending, it is necessary to further check that the cross section is capable of locally restraining the combination of compression and primary moment(s) present at any point.

The following equation is presented in Eurocode 3 for biaxial bending as follows:

$$\left(\frac{M_{y.Sd}}{M_{Ny.Rd}}\right)^{\alpha} + \left(\frac{M_{z.Sd}}{M_{Nz.Rd}}\right)^{\beta} \leq 1 \tag{6.28}$$

in which the values of α and β depend upon the type of cross section as indicated in Table 6.3. The following equation is simple but more conservative.

$$\frac{N_{Sd}}{N_{pl.Rd}} + \frac{M_{y.Sd}}{M_{Ny.Rd}} + \frac{M_{z.Sd}}{M_{Nz.Rd}} \leq 1 \tag{6.29}$$

6.12.3 Cross-Sectional Requirements for Plastic Global Analysis

At plastic hinge locations, the cross section of the member containing the plastic hinge should have a rotation capacity not less than that required at the plastic hinge location.

In a uniform member, sufficient rotation capacity may be assumed at a plastic hinge if both of the following requirements are satisfied:

a. the member has Class 1 cross sections at the plastic hinge location;
b. Where a transverse force that exceeds 10% of the shear resistance of the cross section, as described in Section 6.2.6, is applied to the web at the plastic hinge location, web stiffeners should be provided within a distance along the member of $h/2$ from the plastic hinge location, where, h is the height of the cross section at this location.

In case of the member's cross section varies along its length, additional criteria should be considered as follow:

a. Adjacent to plastic hinge locations, the thickness of the web should not be reduced for a distance of at least $2d$ along the member from the plastic hinge location, where d is the clear depth of the web at the plastic hinge location.
b. Adjacent to plastic hinge locations, the compression flange should be Class 1 for a distance of not less than the greater of
   $2d$, where d is the web clear depth
   the distance to the adjacent point at which the moment in the member has fallen to 0.8 times the plastic moment resistance at the point concerned.
c. Elsewhere in the member, the compression flange should be class 1 or class 2, and the web should be class 1, class 2, or class 3.

In the case of a bolted connection, the location of bolts near plastic hinge locations must satisfy the following conditions regarding their holes in tension.

The bolts holes in the tension flange may be ignored, provided that for the tension flange:

$$\frac{A_{f,net}0,9f_u}{\gamma_{M2}} \geq \frac{A_f f_y}{\gamma_{M0}} \tag{6.30}$$

where A_f is the area of the tension flange.

Example

Safety check of a beam–column of the first storey of the building. The member, composed of a HEB 320 cross section in steel S 355, has a length of 4.335.

HEB 320 cross section are $A=161.3\,cm^2$; $W_{pl,y}=2{,}149\,cm^3$, $I_y=30{,}820\,cm^4$, $i_y=13.82\,cm$; $I_z=9{,}239\,cm^4$, $i_z=7.57\,cm$; $I_T=225.1\,cm^4$ and $I_W=2{,}069\times10^3\,cm^6$. The mechanical characteristics of the material are $f_y=355\,MPa$, $E=210\,GPa$ and $G=81\,GPa$.

Design values are $N_{Ed}=1{,}704$ kN; $M_{y,Ed}=24.8$ kNm

i. Cross-sectional classification: As the compression force is high, the cross section is classified under compression only (conservative approach). Since the HEB 320 section is stocky, it is classified as class 1, even under this load condition.
ii. Cross-sectional resistance

The design internal forces: $M_{y,Ed}=24.8$ kNm and $N_{Ed}=1{,}704$ kN (compression).

$$N_{pl,Rd} = Af_y/\gamma_{M0} = 161.3\times10^{-4}\times355\times10^3/1.0 = 5{,}726.2\text{ kN}$$

As, N_{Ed} = 1,704 kN < $N_{pl,Rd}$ = 5,726.2 kN, the axial force resistance is verified.

Since, N_{Ed} = 1,704 kN > 0.25 $N_{pl,Rd}$ = 1,431.5 kN, in accordance with clause 6.2.9.1(4) of EC3-1-1, it is necessary to reduce the plastic bending resistance (to $M_{n,y,Rd}$)

$$M_{pl,y,Rd} = W_{pl,y}f_y/\gamma_{M0} = 2{,}149\times10^{-6}\times355\times10^3/1.0 = 762.9\text{ kNm}$$

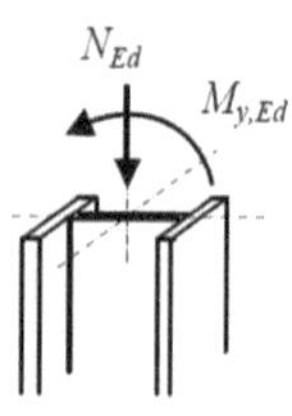

$$a = A - 2bt/A = 161.3 - 2\times30\times2.05/161.3 = 0.24$$

$$n = N_{ED}/N_{pl,Rd} = 1{,}704/5{,}726.2 = 0.30$$

$$M_{n,y,Rd} = M_{pl,y,Rd}(1-n)/(1-0.5a)$$

As $M_{y,Ed}=24.8$ kNm < $M_{n,y,Rd}=606.9$ kNm, the bending resistance is verified

iii. Steel member stability

In this example, Method 2 is applied. Since the member is susceptible to torsional deformations due to its thin-walled open cross section, it is assumed that lateral–torsional buckling constitutes the relevant instability mode. Since $M_{z,Ed}=0$, the following conditions must be verified:

$$\frac{N_{Ed}}{\chi_y N_{Rk}/\gamma_{M1}} + K_{yy}\frac{M_{y,Ed}}{\chi_{LT} M_{y,Rk}/\gamma_{M1}} \leq 1.0$$

$$\frac{N_{Ed}}{\chi_z N_{Rk}/\gamma_{M1}} + K_{zy}\frac{M_{y,Ed}}{\chi_{LT} M_{y,Rk}/\gamma_{M1}} \leq 1.0$$

Step 1: Characteristic resistance of the cross section

$N_{Rk}=A\,f_y=161.3\times10^{-4}\times355\times10^{3}=5{,}726.2$ kN

$M_{y,Rk}=W_{pl,y}f_y=2{,}149\times10^{-6}\times355\times10^{3}=762.9$ kNm

Step 2: Calculation of the χ_{LT} using the alternative method applicable to rolled or equivalent welded sections as per clause 6.3.2.3 of EC3-1-1

The length between braced sections is $L=4.335$ m. The critical moment M_{cr}, assuming a linear diagram, is obtained in this example:

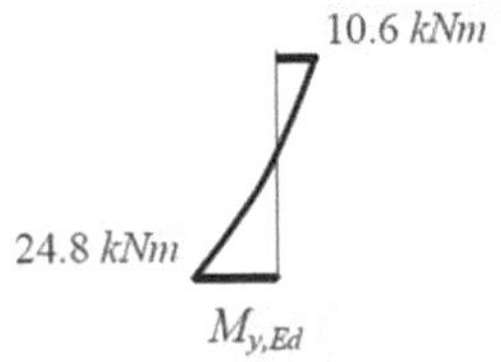

$$M_{cr}=5{,}045.1 \text{ kNm}$$

$$\lambda'_{LT}=\left[2419(10)^{-6}\,(355)(1000/5945.1)\right]^{0.5}=0.39$$

Rolled I- or H-sections with $h/b=320/300=1.07<2$.

Curve b and $\alpha_{LT}=0.34$.

Taking $\lambda'_{LT,0}=0.4$ and $\beta=0.75$.

$$\phi_{LT}=0.5\left[1+0.34(0.39-0.40)+0.75(0.39)^2\right]=0.56$$

$$\chi_{LT}=\frac{1}{0.56+\left((0.56)^2-0.75(0.39)^2\right)^2}=0.99$$

Section	Limits	Buckling Curve (EC3-1-1)
I- or H-sections rolled	$h/b\leq2$	B
	$h/b>2$	C
I- or H-sections welded	$h/b\leq2$	C
	$h/b>2$	D

Step 3: Calculation of χ_{LT} using the alternative method applicable to rolled or equivalent welded sections as per clause 6.3.2.3 of EC3-1-1.

The correction factor k_c, according to Table 6.6 of EC3-1-1, with $\boldsymbol{I} = 10.6/(-24.8) = -0.43$, is given by

$$k_c = \frac{1}{1.33 - 0.33\Psi} = \frac{1}{1.33 - 0.33 \times (-0.43)} = 0.68$$

$$f = 1 - 0.5 \times (1 - k_c) \times \left[1 - 2.0 \times \left(\bar{\lambda}_{LT} - 0.8\right)^2\right]$$

$$= 1 - 0.5 \times (1 - 0.68) \times \left[1 - 2.0 \times (0.39 - 0.8)^2\right] = 0.89$$

$$K_c = 1/(1.33 - 0.33\lambda) = 1/1.33 - 0.33(-0.43) = 0.68$$

$$f = 1 - 0.5(1 - k_c)\left[1 - 2(\lambda'_{LT} - 0.8)^2\right]$$

$$= 1 - 0.5(1 - 0.68)\left[1 - 2(0.39 - 0.8)^2\right] = 0.89$$

The modified lateral–torsional buckling reduction factor is given by

$$\chi_{LT,mod} = 0.99/0.89 = 1.11 > 1.0$$

So, $\chi_{LT,mod} = 1.00$ must be adopted.

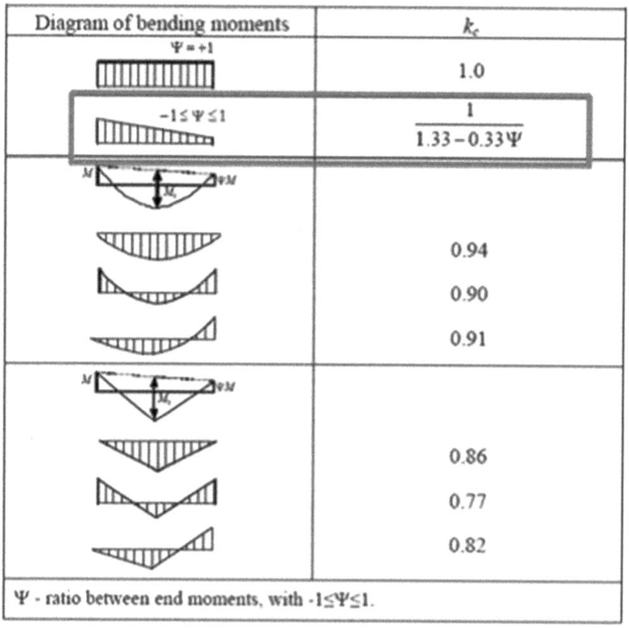

Diagram of bending moments	k_c
$\Psi = +1$	1.0
$-1 \leq \Psi \leq 1$	$\frac{1}{1.33 - 0.33\Psi}$
M, ΨM, M_0	0.94 0.90 0.91
M, ΨM, M_0	0.86 0.77 0.82
Ψ - ratio between end moments, with $-1 \leq \Psi \leq 1$.	

Step 4: Interaction factors k_{yy} and k_{zy}.

The equivalent factors of uniform moment C_{my} and C_{mLT} are obtained based on the bending moment diagram between braced sections according to the z direction in case of C_{my} and laterally in case of C_{mLT}. Assuming the member braced in the z direction and laterally just at the base and top cross sections, the factors C_{my} and C_{mLT} must be calculated based on the bending moment diagram along the total length of the member.

Since the bending moment diagram is assumed linear, it can be defined by $M_{y,Ed,base} = -24.8$ kNm;
$M_{y,Ed,top} = 10.4$ kNm, from Table B.3 of EC3-1-1, is obtained:

Table B.3: Equivalent uniform moment factors C_m in Tables B.1 and B.2

Moment diagram	range	C_{my} and C_{mz} and C_{mLT}	
		uniform loading	concentrated load
M 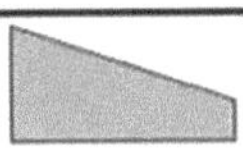 ψM	$-1 \le \psi \le 1$	$0.6 + 0.4\psi \ge 0.4$	

$\Psi = M_{y,Ed,top}/M_{y,Ed,base} = 10.6/(-24.8) = -0.43$
$C_{my} = C_{mLT} = 0.60 + 0.40(-0.43) = 0.43 > 0.4$

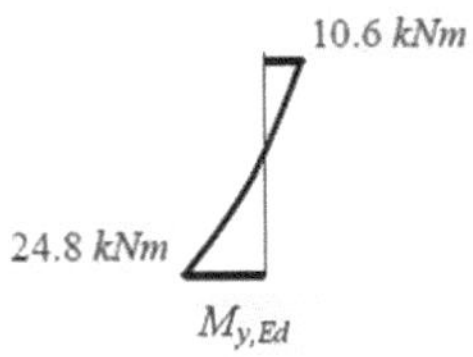

Because the member is susceptible to torsional deformations, the interaction factors k_{yy} and k_{zy} are obtained from Table B.2 of EC3-1-1, through the following calculations:

$$k_{yy} = C_{my}\left[1 + \left(\lambda'_y - 0.2\right)\frac{N_{Ed}}{x_y N_{RK}/\gamma_{M2}}\right]$$

$$= 0.43\left[1 + (0.41 - 0.2)\frac{1704}{0.92 \times 5,726.2/1}\right] = 0.46$$

$$\text{As } k_{yy} \le C_{my}\left(1 + 0.8\frac{N_{Ed}}{\chi_y N_{RK}/\gamma_{M1}}\right) = 0.54,\ k_{yy} = 0.46$$

$$k_{ZY} = \left[1 - \frac{0.1\bar{\lambda}_Z}{(C_{mLT} - 0.25)}\frac{N_{Ed}}{\chi_Z N_{Rk}/\gamma_{M1}}\right]$$

$$= \left[1 - \frac{0.1 \times 0.75}{(0.43 - 0.25)}\frac{1704}{0.69 \times 5726.2/1.0}\right] = 0.82$$

$$As \quad k_{ZY} \ge \left[1 - \frac{0.1}{(C_{mLT} - 0.25)}\frac{N_{Ed}}{\chi_Z N_{RK}/\gamma_{M1}}\right] = 0.76$$

Then $K_{zy} = 0.82$

Step 5: Finally, the verification of equations 6.61and 6.62 of EC3-1-1 yields

$1{,}704/(0.92 \times 5{,}726.2/1.0 + 0.46 \times (24.8/(1.0 \times 762.1/1.0)) = 0.34 < 1.0$

$1{,}704/(0.69 \times 5{,}726.2/1.0 + 0.82 \times (24.8/(1.0 \times 762.1/1.0)) = 0.46 < 1.0$

You can contact us to send you the excel sheet www.elreedyman.com

ELREEDYMAN www.elreedyman.com	Client :-		Page no.	
	Project :-		REV. No :-	
	Title :-		Calcs By :-	
			Check. By :-	
			Approve By :-	
EC3 Beam column		Calculation sheet	Date:	

Beam-Column Calculation

INPUT DATA

MEMBER SIZE:	HEB320	
YIELD STRESS OF STEEL (Fy):	355	Mpa
ELASTIC MODULUS (E):	210,000	MPa
UNBRACED LENGTH IN BENDING (l):	4.34	m
EFFECTIVE UNBRACED LENGTH ABOUT X AXIS (KLy):	4.34	m
EFFECTIVE UNBRACED LENGTH ABOUT Y AXIS (KLz):	4.34	m
γM0	1.00	
G	80770	MPa

DESIGN LOADING

ALLOWABLE STRESS DESIGN

AXIAL LOAD (COMPRESSION) (N_{ED}):	1704	kN
BENDING MOMENT ABOUT X AXIS (MyED):	24.8	kN-m
BENDING MOMENT ABOUT Y AXIS (Mz):	660	kgf-m

MEMBER PROPERTIES

GENERAL

A =	161.30	cm2
d =	320.00	mm
tw =	11.50	mm
bf =	300.00	mm
tf =	20.50	mm
h = T =	320	mm
h0 =	0	mm
hc =	320.00	mm
kc=	0.76	

Y- AXIS

Iy =	30820	cm^4
Wpl,y =	2149.00	cm^3
ry =	13.82	cm
Zy =	0.00	cm^3

Z - AXIS

Iz =	9239	cm^4
Wpl,z =	37.59	cm^3
rz =	7.57	cm

$N_{pl},Rd=AFy/\gamma M0$ 5726.15 kN

N_{ED}<Npl,Rd

N_{ED}<0.25 Npl,Rd

Mpl,y,Rd=Wpl,y fy/M0 762.90 kN.m

a=A-2bt/A= 0.24

n=N_{ED}/Npl,R 0.30

Mn,y,Rd=Mpl,y,Rd (1-n)/(1-0.5a)

Mn,y,Rd=Mpl,y,Rd (1-n)/(1-0.5a) 608.06 kN.m

My,ED= 24.80 < Mn,y,Rd oK

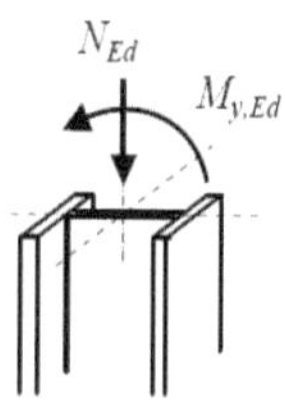

$$M_{cr} = C_1 \frac{\pi^2 EI_z}{L^2} \sqrt{\frac{I_w}{I_z} + \frac{L^2 GI_t}{\pi^2 EI_z}}$$

Section	Limits	Buckling curve (EC3-1-1)
I or H sections rolled	$h/b \le 2$	b
	$h/b > 2$	c
I or H sections welded	$h/b \le 2$	c
	$h/b > 2$	d

Mcr= 5045.10 kN.m

λLT= 0.39

h/b= 1.07 Curve b

α_{LT}= 0.34

λLT,0= 0.40

ϕ_{LT}=	0.55		
χ_{LT}	1.00		

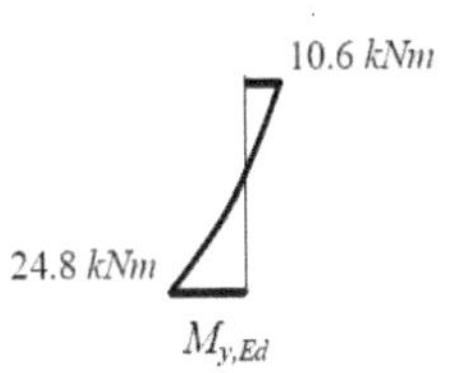

I =	=10.6/-24.8	-0.43	
Kc=	=1/(1.33-0.33ī)= 0.68		

$$f = 1 - 0.5 \times (1 - k_c) \times \left[1 - 2.0 \times (\bar{\lambda}_{LT} - 0.8)^2\right]$$

f	0.89		
$\chi_{LT,mod}$	1.12 >1.0	So,$\chi_{LT,mod}$	1.00
C_{my}=C_{mLT}=	0.43 >	0.4	

$$k_{yy} = C_{my}\left[1 + (\bar{\lambda}_y - 0.2)\frac{N_{Ed}}{\chi_y N_{Rk}/\gamma_{M1}}\right]$$

Kyy=	0.46
χ_y	0.92

$$k_{zy} = \left[1 - \frac{0.1\bar{\lambda}_z}{(C_{mLT} - 0.25)}\frac{N_{Ed}}{\chi_z N_{Rk}/\gamma_{M1}}\right] =$$

Kzy=	0.82
χ_z	0.69

$$\text{As } k_{zy} \geq \left[1 - \frac{0.1}{(C_{mLT} - 0.25)}\frac{N_{Ed}}{\chi_z N_{Rk}/\gamma_{M1}}\right] = 0.76$$

$$\frac{N_{Ed}}{\chi_y N_{Rk}/\gamma_{M1}} + k_{yy}\frac{M_{y,Ed}}{\chi_{LT} M_{y,Rk}/\gamma_{M1}} \leq 1.0 \qquad 0.34 \qquad \text{O.k safe}$$

$$\frac{N_{Ed}}{\chi_z N_{Rk}/\gamma_{M1}} + k_{zy}\frac{M_{y,Ed}}{\chi_{LT} M_{y,Rk}/\gamma_{M1}} \leq 1.0 \qquad 0.46 \qquad \text{O.k safe}$$

6.13 DESIGN BEAM–COLUMN BY AISC

The main concept is the same, but there are differences in the applied equations for calculating the member capacity under axial and bending forces.

6.13.1 Doubly and Singly Symmetric Members Subjected to Flexure and Compression

The member strength for the interaction of flexural and compressive forces in doubly symmetric members and singly symmetric members constrained to bend about a geometric axis (x and/or y) shall be calculated using the following equations:

- In case of $P_r/P_c \geq 0.2$,

$$\frac{P_r}{P_c}+\frac{8}{9}\left(\frac{M_{rx}}{M_{cx}}+\frac{M_{ry}}{M_{cy}}\right)\leq 1.0 \tag{6.31}$$

- In case of $P_r/P_c > 0.2$,

$$\frac{P_r}{2P_c}+\left(\frac{M_{rx}}{M_{cx}}+\frac{M_{ry}}{M_{cy}}\right)\leq 1.0 \tag{6.32}$$

where P_r is the required compressive strength, P_c is the available compressive strength, M_r is the required flexural strength, M_c is the available flexural strength, $\phi_b M_n$, x is the subscript relating to major-axis bending, and y is subscript relating to minor-axis bending.

From this curve the required flexural can be higher than the flexural capacity to reach to 6% more in case of minor axis stresses and axial force to axial strength less than 0.05 Figure 6.27 presents the relation between axial force to strength ratio with respect to minor and major axis ratio. Figure 6.28 presents the intraction curve as presented by Bernuzzi and cordova (2016).

6.13.2 Doubly and Singly Symmetric Members Subjected to Flexure and Tension

In the case of interaction between flexural and axial tension forces, doubly symmetric members and singly symmetric members constrained to bend about a geometric axis (x and/or y) shall be calculated using equations 6.31 and 6.32, as in the above section. But in this case, P_r and P_c are the required and available tensile strengths, respectively.

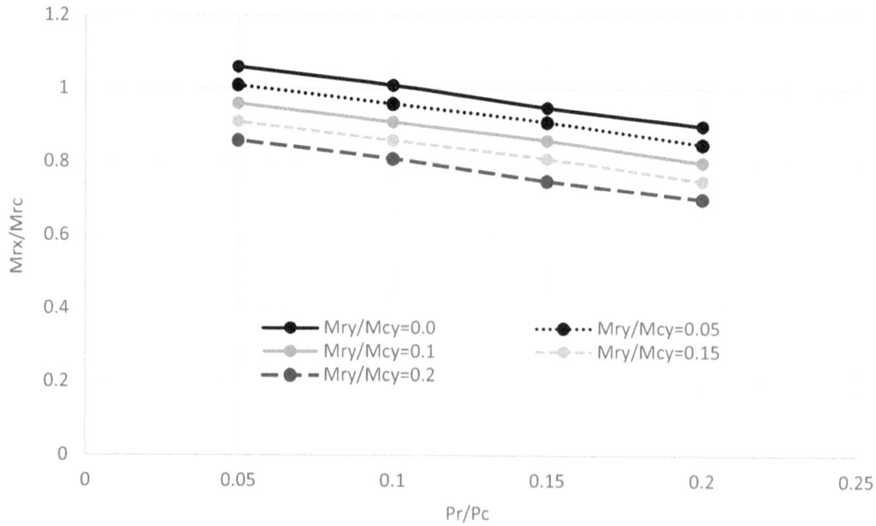

FIGURE 6.27 The relation between axial force to strength ration w.r.t minor and major axis.

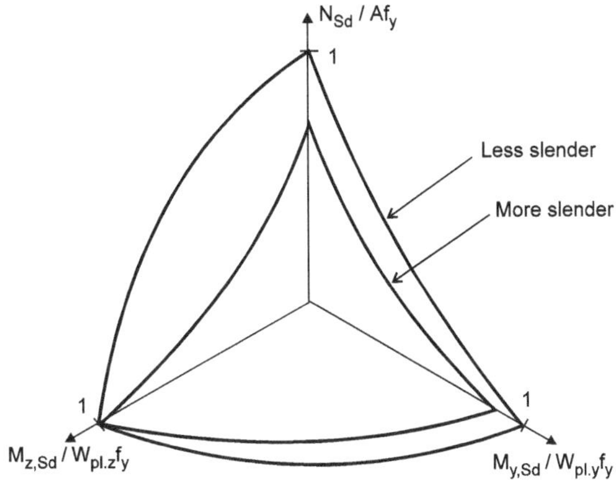

FIGURE 6.28 Relation between axial force to strength ratio with respects to minor and major axis ratios.

In case of doubly symmetric members, C_b is permitted to be multiplied by the following factor $(1+\alpha P_r/P_{ey})^{0.5}$,

where

$$P_{ey}=(\pi)^2 EI_y/(L_b)^2$$

$\alpha = 1.0$

E = modulus of elasticity of steel = 29,000 ksi (200,000 MPa)

I_y = moment of inertia about the y-axis,

L_b = length between points that are either braced against lateral displacement of the compression flange or braced against twist of the cross section.

6.13.3 Doubly Symmetric Rolled Compact Members Subjected to Single-Axis Flexure and Compression

In the case of doubly symmetric rolled compact members, with the effective length for torsional buckling less than or equal to the effective length for y-axis flexural buckling, and simultaneously subjected to flexure and compression with moments primarily about their major axis, it is permissible to address the two independent limit states: in-plane instability and out-of-plane buckling or lateral–torsional buckling.

So in this case, it is required to check the yields and buckling at the same time.

- For the limit state of in-plane instability, equations 6.36 and 6.37 shall be used with P_c taken as the available compressive strength in the plane of bending and M_{cx} taken as the available flexural strength based on the limit state of yielding.
- For the limit state of out-of-plane buckling and lateral–torsional buckling

$$\frac{P_r}{P_{cy}}\left(1.5-0.5\frac{P_r}{P_{cy}}\right)+\left(\frac{M_{rx}}{C_b M_{cx}}\right)^2 \le 1.0$$

where C_b is the lateral–torsional buckling modification factor determined as per Chapter 4, 1 M_{cx} is the lateral–torsional strength for major-axis flexure determined using C_b taken 1.0, and P_{cy} is the compressive strength out of the plane of bending, about the minor axis.

REFERENCES

Annex H [Informative] (1992). *Modelling of Building Structures for Analysis*, Amendment A2 to Eurocode 3: ENV 1993-1-1:1992/A2.

Caludio Bernuzzi and Benedetto Cordova. *Structural steel design to Eurocode3 and AISC specifications*, Wiley Blackwell, India, 2016.

CEN (1993). Eurocode 3 Part 1-3 *General Rules - Supplementary Rules for Cold Formed Thin Gauge Members and Sheeting*, ENV 1993-1-3.

CEN (1998). Eurocode 8 *Design Provisions for Earthquake Resistance of Structures*, ENV 1998.

Chabrolin, B., Maquoi,R. (1993). *Technical steel reserach: Frame design including joint behaviour*, EUR 18563 Report (1993), http://europa.eu.int

Chen, W.K., Goto, Y., and Liew, J.Y.R. (1996). *Stability Design of Semi-Rigid Frames,* John Wiley & Sons, Inc., New York.

Clarke, M.J. (1994). Plastic Zone Analysis of Frames, in *Advanced Analysis of Steel Frames: Theory, Software and Applications,* Chen, W.F. and Toma, S., eds., Boca Rotan, FL, CRC Press, pp. 259–274.

ECCS (1991). *Practical Analysis of Single-Storey Frames*, ECCS -Technical Committee 8- Structural Stability, Technical Working Group 8.1/8.2 Skeletal Structures, ECCS Publication No. 61.

Gachon, H. and Galea, Y. (1978). *Model of Non-Linear Analysis for Plane Frames* (In French), Construction Métallique No. 2.

Galea, Y. (1978). *Non-Linear Elasto-Plastic Analysis Programme for Plane Frames* (In French), Construction Métallique No. 4.

Livesley, R.K. (1969). *Matrix Methods of Structural Analysis*, Pergamon Press, Oxford.

Maquoi, R. and Chabrolin, B. (1998). *Frame Design Including Joint Behaviour*, Report No. EUR 18563 EN, ECSC Contract No 7210-SA/212/320, European Commission.

Neal, B.G. (1956). *Plastic Methods of Structural Analysis,* Chapman and Hall, London.

The Steel Construction Institute (1992). *Steel Designers Manual,* 5th edition, Blackwell Scientific Publications, Oxford.

Elreedy, MA (2014) " reinforced concrete structure reliability", CRC press , USA

7 Composite Section

7.1 INTRODUCTION

The direction of using composite sections derived from steel and reinforced concrete aims to capitalize on the advantages offered by both materials. Concrete provides protection to steel against corrosion to some extent, and incorporating steel sections helps to minimize the dimensions of composite members, which is advantageous in high-rise buildings. Studies have demonstrated that the total slab thickness, including ribs, can be utilized to determine effective slab width, as per Grant et al. (1977) and Fisher (1970).

In this chapter, I will discuss an overview of the design of composite sections, focusing particularly on the use of EC4, which presents the design in detail. It is also essential to provide aligns with BS 5950-4 for ASD and has been modified to conform to EC 1994-1-1, along with AISC360-22, which adheres to ASD and LRFD principles. The reinforced concrete design will follow the corresponding codes and standards applicable to the design of steel structures.

7.2 COMPOSITE SLAB DESIGN

A critical step is to carry out calculations for both the construction phase and the composite phase, which represents the final configuration. Therefore, the load calculations will be as follows:

A. Construction stage
  A.1 Permanent load
    - Deck and mesh
    - Steel beam
  A.2 Variable actions
    - Wet concrete

- Construction loading about 0.75 kN/m^2, as per SCI AD-346, recommending the addition of 10% of the slab self-weight in areas of 3 m^2.

B. Composite stage
  B.1 Permanent actions
    - Deck and mesh
    - Steel beam
    - Dry concrete
    - Finishes
    - Services
    - Ponding

DOI: 10.1201/9781003307303-7

B.2 Variable actions
- Occupancy floor load based on usage
- Moveable partitions

The slab can consist of reinforced concrete resting on the steel plate, as presented in the case study for the construction section; in most cases, it will include haunching, as shown in Figure 10.1b, rest directly on the steel beam, as shown Figure 7.1a

The haunch is particularly important in cases of high shear stresses, and details regarding this haunch are provided in Figure 7.2. This detail is consistent match with ECP and EN 1994-1-1 (CEN, n.d.).

The concrete slab should be connected to the steel beam, enabling it to carry the stresses as a single unit. This is achieved by using a shear connector; one type of shear connector is a stud, and its construction is outlined in the chapter on steel construction. Details of this connector and its integration with the reinforced concrete are presented in Figure 7.3.

They are fastened to the top flange of the steel beam and embedded in the concrete slab to transmit the longitudinal shear and prevent any slippage between the concrete slab and the steel beam. Furthermore, they prevent slab uplift.

There are mainly three types of shear connectors: the stud, a series of bolts welded to the steel beam or steel channel section, or a steel angle section; these types are presented in Figure 7.4.

In composite beams where the longitudinal spacing of shear connectors has been varied according to the intensity of shear, duplicate beams with the required number of

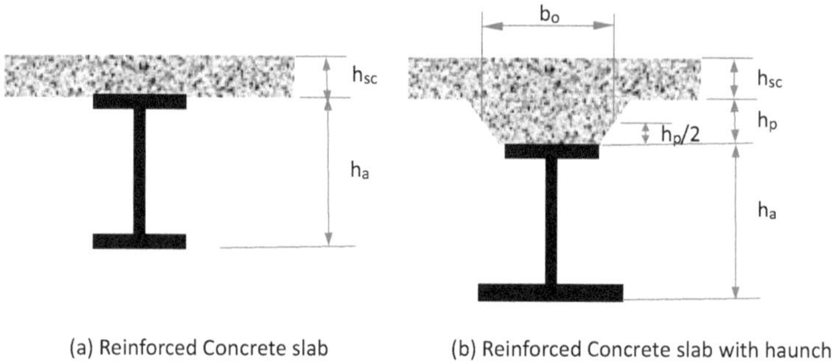

FIGURE 7.1 Slab composite section.

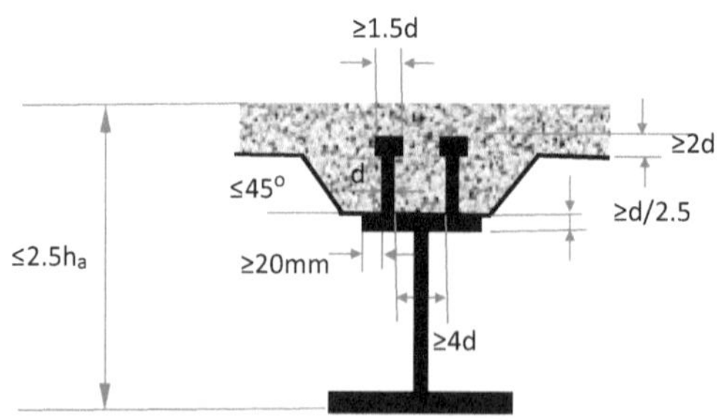

FIGURE 7.2 Detail for slab composite section.

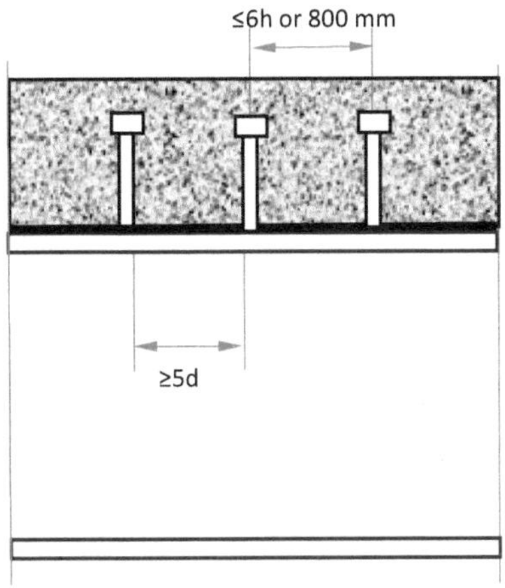

FIGURE 7.3 Connection between steel beam and reinforced concrete slab.

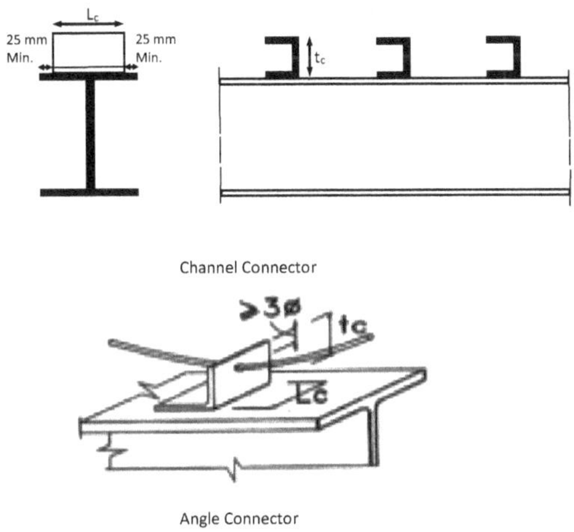

FIGURE 7.4 Different types of shear connector.

connectors uniformly spaced have exhibited the same ultimate strength and the same amount of deflection under normal working loads. Only a slight deformation in the concrete and the more heavily stressed shear connectors is needed to redistribute the horizontal shear to the other, less heavily stressed connectors. An important consideration is that the total number of connectors must be sufficient to develop the shear V on either side of the point of maximum moment. The provisions of the AISC ASD or any other code and specification are based upon this concept of composite action.

Studies have defined stud shear connection strength Q in terms of normal weight and lightweight aggregate concretes, as a function of both concrete modulus of elasticity and concrete strength (McGarraugh and Baldwin, 1971 and Ollgaard et al., 1971). The following equation is still used in AISC 360-22 (AISC, n.d.).

$$Q_u = 0.5\,A_S\left(f_c'\,E_c\right)^{0.5}$$

A_s = cross-sectional area of stud
f'_c = concrete compressive strength
E_c = concrete modulus of elasticity

As per AISC 360-22, the value of $Q_u \leq R_g R_p A_s F_u$.
f_u = specified minimum tensile strength of a steel headed stud anchor, ksi (MPa)
R_g = 1.0 in the following cases:

a. One steel headed stud anchor welded in a steel deck rib with the deck oriented perpendicular to the steel shape
b. Any number of steel headed stud anchors welded in a row directly to the steel shape
c. Any number of steel headed stud anchors welded in a row through a steel deck with the deck oriented parallel to the steel shape and the ratio of the average rib width to rib depth ≥1.5

R_g = 0.85 for the following cases:

a. Two steel headed stud anchors welded in a steel deck rib with the deck oriented perpendicular to the steel shape
b. One steel headed stud anchor welded through a steel deck with the deck oriented parallel to the steel shape and the ratio of the average rib width to rib depth < 1.5

R_g = 0.7 for three or more steel headed stud anchors welded in a steel deck rib with the deck oriented perpendicular to the steel shape

R_p = 0.75 in case of

a. Steel headed stud anchors welded directly to the steel shape
b. Steel-headed stud anchors welded into a composite slab with the deck oriented perpendicular to the beam and $e_{mid\text{-}ht \geq}$ 2 in. (50 mm)
c. Steel-headed stud anchors welded through the steel deck or steel sheet used as girder filler material, and embedded in a composite slab with the deck oriented parallel to the beam.

= 0.6 for steel-headed stud anchors welded in a composite slab with the deck oriented perpendicular to the beam and $e_{mid\text{-}ht} < 2$ in. (50 mm)

Here, $e_{mid\text{-}ht}$ = distance from the edge of steel headed stud anchor shank to the steel deck web, measured at mid-height of the deck rib, and in the load bearing direction of the steel headed stud anchor (in other words, in the direction of maximum moment for a simply supported beam), in. (mm)

7.3 SHEAR CONNECTION AS PER EN1994

As per EN1994, the design shear resistance of a headed stud, automatically welded in accordance with EN 14555 should be determined from

A. Check for stud failure mode from the equation:

$$P_{Rd} = 0.8 f_u \pi d^2 / 4\gamma_v \tag{7.4}$$

B. Check for concrete failure mode from the equation

$$P_{Rd} = 0.29 \alpha d^2 \left(f_{ck} E_{Cm}\right)^{0.5} / 4\gamma_v \tag{7.5}$$

where

γ_v is the partial factor equal to 1.25 and refer to your annex that you follow
d is the diameter of the shank of the stud, $16\text{ mm} \leq d \leq 25\text{ mm}$
f_u is the specified ultimate tensile strength of the stud materials equal or less than 500 Mpa
f_{ck} is the concrete characteristics cylinder compressive strength at the age considering and with concrete density not less than $1{,}750\text{ kg/m}^3$
h_{sc} is the stud nominal height.

- In case of $3 \leq h_{sc}/d \leq 4$,

$$\alpha = 0.2\left(h_{sc}/d + 1\right) \tag{7.6}$$

- In case of $h_{sc}/d > 4$,

$$\alpha = 1$$

B_o is defined in Figure 7.5. Considering that $h_{sc} \leq h_p + 75\text{ mm}$.

A. In case of *profiled steel sheeting spanning parallel to the supporting beam*
In general, the design of a single headed stud connector in a rib that is parallel to the supporting beam in sheeting that is continuous across the beam is determined from
$K_t P_{rd}$, where P_{rd} will be calculated using the above equations and k_t from this equation.

$$k_t = 0.6 \frac{b_o}{h_p}\left(\frac{h_{sc}}{h_p} - 1\right) k_t \leq 1.0 \tag{7.7}$$

b_o is the width of a trapezoidal rib at mid-height of the profile (h_p), or the minimum width of the rib for re-entrant profiles.
h_p is the height of the steel sheeting measured to the shoulder of the profile.
h_{sc} is the as-welded height of the stud, but not greater than $h_p + 75\text{ mm}$.

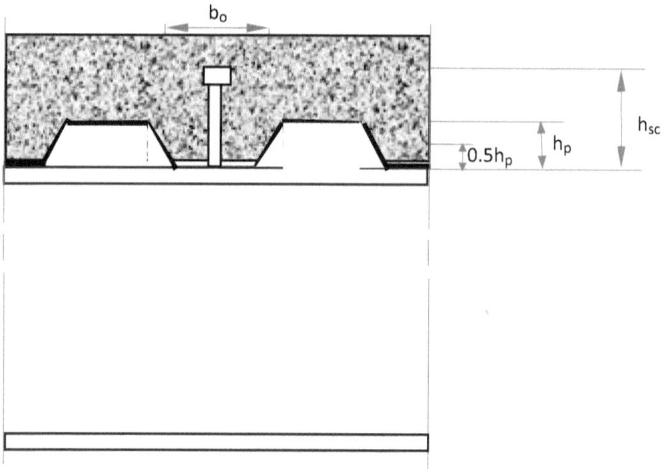

FIGURE 7.5 Shear connection detail.

B. In case of profiled steel sheeting spanning transverse to the supporting beam.
C. In this case, the reduction factor will be applied using the following equation, taking into account the following criteria:

- The studs are placed in ribs with $hp \leq 85$ mm and $bo \geq hp$.
- For shear studs welded through the steel sheeting, $d \leq 20$ mm.
- For holes provided in the sheeting, $d \leq 22$ mm.

$$k_t = \frac{0.7}{\sqrt{n_r}} \frac{b_o}{h_p} \left(\frac{h_{sc}}{h_p} - 1 \right) \tag{7.8}$$

where n_r is the number of stud connectors in one rib at a beam intersection, and this must not be greater than 2.

h_p, b_o, and h_{sc} are defined above, and the values for k_{max} presented in Table 7.1.

7.4 COMPOSITE BEAM RESISTANCE CALCULATION

As a rule of thumb, the steel flange provides higher strength than the concrete flange. There are three cases to calculate section resistance based on the location of the natural axis (NA), as per SCI-P359.

Case 1: NA in the concrete

This case is presented in Figure 7.6 and occurs when the sheeting is perpendicular to the secondary beam axis

TABLE 7.1
Values of k_{max}

Number of Stud Connector per Rib	Sheeting Thickness (mm)	Studs Welded through Sheeting $d \leq 20$mm	Sheeting with Holes $d = 19$ or 22 mm
1	≤1.0	0.85	0.75
1	>1.0	1.0	0.75
2	≤1.0	0.7	0.6
2	>1.0	0.8	0.6

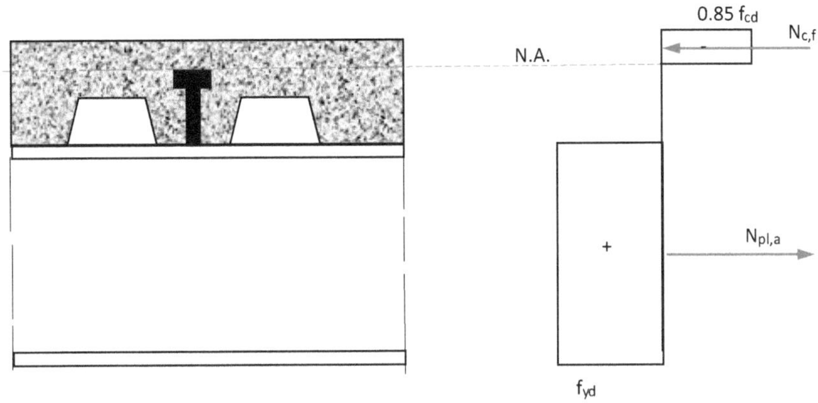

FIGURE 7.6 Plastic natural axis in concrete.

$$M_{Pl,Rd} = N_{Pl,a}\left(\frac{h_a}{2} + h_S - \left[\frac{N_{pl,a}}{N_{c,f}}\frac{h_c}{2}\right]\right) \tag{7.9}$$

$N_{pl,a}$ is the axial resistance of the steel section
$N_{c,f}$ is the resistance of the effective area of the concrete flange acting compositely with the steel section= $(0.85 f_{cd}\, b_{eff}\, x_c)$
h_a is the depth of the steel section, as shown in Figure 7.7
h_s is the depth of the concrete above the steel sheeting profile
h_p is the depth of the steel sheeting profile measured to the shoulder of the profile
h_s is the overall depth of the steel sheeting profile, including the height of the top dovetail stiffener, if present. In the case of a deck without a stiffener, $h_d = h_p$

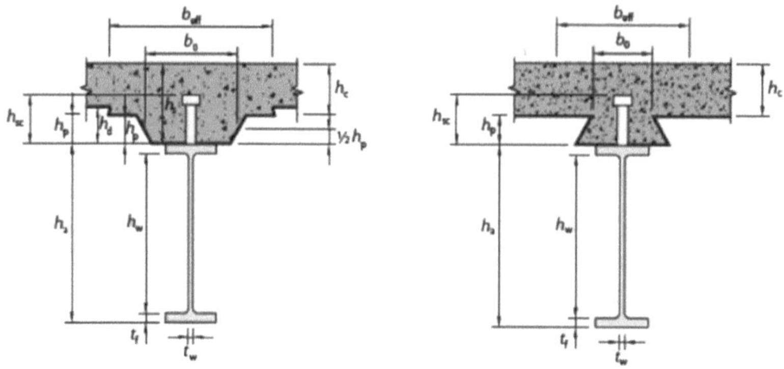

FIGURE 7.7 Details of composite section.

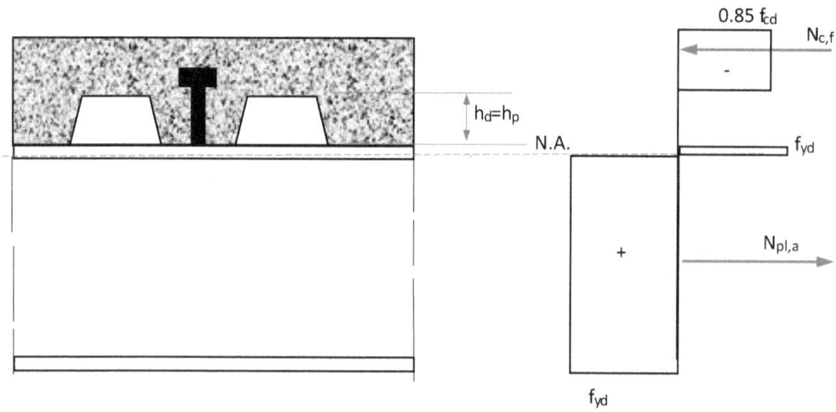

FIGURE 7.8 Plastic natural axis in the top flange of the steel section.

Case2: NA in the top flange

This case, along with the next, occurs in the case of a primary beam; in these two cases, $N_{pl,a} > N_{c,f}$. This case is presented in Figure 7.8.

$$M_{Pl,Rd} = N_{pl,a}\left(\frac{h_a}{2}\right) + N_{c,f}\left(\frac{h_c}{2} + h_d\right) \tag{7.10}$$

Case 3: NA in the Web

This case is presented in Figure 7.9.

$$M_{Pl,Rd} = M_{pl,aRd} + N_{c,f}\left(\frac{h_c + 2h_d + h_a}{2}\right) - \left(\frac{N_{c,f}^2}{N_w}\frac{h_a}{4}\right) \tag{7.11}$$

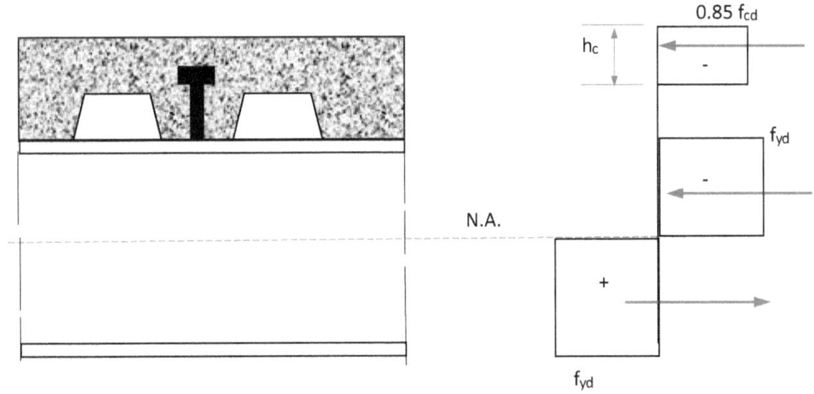

FIGURE 7.9 Plastic natural axis in the web of the steel section.

where

$M_{pl,aRd} = W_{ply} f_y / \gamma_{Mo}$
$N_w = 0.95 f_y \, t_w \, h_w$
$h_w = h_a - 2t_f$

Considering that b_{eff} will be calculated as follows:

$$b_{eff} = b_o + \sum b_{el} \tag{7.12}$$

where b_{ei} is the effective width at midspan of the concrete flange on each side of the steel section, taken as $L_e/8$, and L_e is the effective length of the span, which is taken equal to the system length for a simply supported beam.

7.4.1 Plastic Bending Resistance with Partial Shear Connection

The composite beam will be designed with partial shear connection. If the full compression resistance of the concrete flange ($N_{c,f}$) is not required for the bending resistance of the composite beam, then the shear connectors are not required to transfer a force equal to $N_{c,f}$.

$$M_{Rd} = M_{pl,a,Rd} + \left(M_{pl,Rd} - M_{pl,a,Rd}\right)\frac{N_c}{N_{c,f}} \tag{7.13}$$

The aim of the minimum degree of shear connection in buildings is to ensure that the shear studs have adequate deformation capacity, based on a characteristic slip capacity of 6 mm. Tests conducted on shear connectors with greater values of slip capacity have been determined from tests in the UK and are given in NCCI PN002a-GB[26].

For steel sections with equal flanges, the following empirical equation can be used:

In case of $L_e \leq 25$,

$$\eta \geq 1-\left(\frac{355}{f_y}\right)(0.75-0.03\ L_e) \geq 0.4 \tag{7.14}$$

In case of $L_e > 25$ $\eta \geq 1.0$,

where, L_e is the distance (m) between points of zero bending moment (beam span for simply supported beams).

7.4.1.1 Slab Steel Reinforcement

There will be longitudinal reinforcement in most cases, which will consist of a top and bottom mesh around the shear stub.

Transverse reinforcement is required to ensure that the longitudinal shear is transferred into the concrete without failure of the concrete flange.

The transverse reinforcement per unit length should satisfy

$$\frac{A_{sf} f_{yd}}{s_f} \geq V_{Ed} \frac{h_f}{\cot \theta_t} \tag{7.15}$$

where

s_f is the spacing of the reinforcement bars
A_{sf} is the area of reinforcement per unit length
f_{yd} is the design yield strength of the reinforcement
V_{Ed} is the design value of the transverse shear force
h_f is the depth of the flange, which is equal to hc for a composite slab
θ_f is the angle of dispersion of the force from the shear connector, taken as, $26.3° \leq \theta_f \leq 45°$ for compression flanges

7.4.2 Vertical Shear Resistance

As per SCI, the resistance of the composite beam to shear is normally taken by the resistance of the steel section, while ignoring the concrete slab, which is a conservative approach.

$$V_{pl,Rd} = A_v\left(f_y/1.73\right)/\gamma_{Mo} \tag{7.16}$$

$$A_v = A - 2b\ t_f + (t_w + 2r)t_f \tag{7.17}$$

7.4.3 Lateral–Torsional Buckling for Composite Beam

As per EC4, if the following requirements are met, a direct calculation is not needed. In the case of a continuous beam (or a beam within a frame that is composite throughout its length) with Class 1, 2, or 3 cross sections, it may be designed without additional lateral bracing when the following conditions are satisfied:

a. Adjacent spans do not differ in length by more than 20% of the shorter span. Where there is a cantilever, its length does not exceed 15% of that of the adjacent span.
b. The loading on each span is uniformly distributed, and the design permanent load exceeds 40% of the total design load.
c. The top flange of the steel member is attached to a reinforced concrete or composite slab by shear connectors.
d. The same slab is also attached to another supporting member approximately parallel to the composite beam considered, to form an inverted-U frame, as presented in Figure 7.10.
e. If the slab is composite, it spans between the two supporting members of the inverted-U frame considered.
f. At each support of the steel member, its bottom flange is laterally restrained, and its web is stiffened. Elsewhere, the web may be unstiffened.
g. If the steel member is an IPE section or an HE section that is not partially encased, its depth h does not exceed the limit given in Table 6.1.
h. If the steel member is partially encased in concrete according to 5.5.3(2), its depth h does not exceed the limit given in Table 6.1 by more than 200 mm for steel grades up to S355 and by 150 mm for grades S420 and S460 (Table 7.2).

7.4.4 Calculation of Deflection

The deflection calculation mainly depends on the second moment of area; therefore, it is necessary to calculate the second moment of area, also known as the second moment of inertia, from the following equation:

$$I \cong \frac{A_a\left(h_c + 2h_d + h_a\right)^2}{4(1+nr)} + \frac{b_{eff}h_c^3}{12n} + I_a \tag{7.18}$$

where

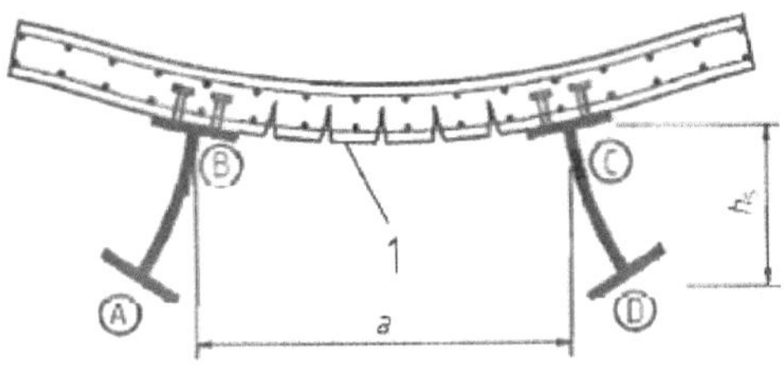

FIGURE 7.10 Concrete slab deformation.

TABLE 7.2
Max. Depth, *h* (mm) of Uncased Steel Member

	Nominal Steel Grade			
Steel Member	S235	S275	S355	S420 & S460
IPE	600	550	400	270
HE	800	700	650	500

A_a is the area of the steel section
I_a is the second moment of area of the steel section
h_c is the depth of the concrete above the steel sheeting profile $(h_s - h_d)$
h_a is the depth of the steel section
h_d is the overall height of the steel sheeting profile
r is the ratio of the cross-sectional area of the steel section relative to the concrete section, $A_a/(b_{eff}\cdot h_c)$
b_{eff} is the effective width.
n is the modular ratio for the design situation considered. As per Sci, it is recommended to calculate this from

$$n = 2/3n_o + 1/3n_l$$

where n_o and n_l are the modular ratios for short and long-term loading, respectively.

$$n_o = E_a/E_{cm}$$

E_a is the modulus of elasticity of structural steel (given as 210,000 MPa)
E_{cm} is the secant modulus of elasticity of concrete for short-term loading, calculated from EN1992 as follows:

$$E_{cm} = 22\left(\frac{f_{cm}}{10}\right)^{0.3} \text{MPa} \tag{7.19}$$

This can also be obtained from the table for normal and lightweight concrete as presented in Table 7.3.

For long-term modular

$$n_l = n_o(1 + \psi_L \phi_t) \tag{7.20}$$

ϕ_t is creep coefficient $\phi(t,t_0)$, as per EC2 which may be taken as 1.05 E_{cm}.
t is the age of the concrete in days at the time considered in the design
t_0 is the age of the concrete in days at loading. This may be modified to account for different cement classes.

TABLE 7.3
Normal and Light Concrete Grade with Its Characteristics

	Normal Concrete		Lightweight Concrete	
	C25/30	C30/37	LC25/28	LC30/33
Concrete Grade			(1,800 kg/m^3)	(1,800 kg/m^3)
Characteristic cylinder compressive strength, f_{ck} (MPa)	25	30	25	30
Mean value of cylinder compressive strength, f_{cm} (MPa)	33	38	33	38
Secant elastic modulus, E_{cm} (GPa)	31	33	21	22

ψ_L is the creep multiplier, which, depending on the type of loading, is taken as
$\psi_L = 1.1$ for permanent loads
$\psi_L = 0.55$ for primary and secondary effects of shrinkage

For deflection due to occupancy loads, the modular ratio may be taken as a proportion of the long- and short-term ratios, as noted above

Where the neutral axis is within the concrete slab, the concrete below that level is neglected during resistance verifications. However, that concrete will contribute some stiffness, so should be included when determining the second moment of area.

Eurocode 4 *does not specify deflection limits for composite beams. Therefore, the deflection limits for composite beams should be specified for each project, depending on the sensitivity of the finishes, visual appearance, etc., to meet the client's needs.*

For composite slabs, a span/depth ratio criterion may be substituted for a deflection calculation. A limit of L/350 or 20 mm is recommended for the deflection of a composite slab due to imposed loads. The deflection due to the total load, less the deflection due to the self-weight of the slab, should be limited to $L/250$.

7.4.5 Vibration

In the UK, the traditional approach used to determine the sensitivity of a floor to vibrations has been to establish the natural frequencies of the primary and secondary supporting beams. If the natural frequencies are found to be greater than 4 Hz, the floor is considered acceptable for normal use (e.g. offices).

The design procedures for determining the dynamic performance of a composite floor include the following steps:

- Determine the natural frequency.
- Determine the modal mass for the floor.
- Evaluate the response of the floor.
- Verify the response of the floor against the requirements

7.5 COMPOSITE BEAM

7.5.1 Encased Beam

The composite section of the beam is only represented by encased beams, as the steel beam will be inserted into a reinforced concrete beam, and certain dimension regulations should be followed, as shown in Figure 7.11.

This case is presented in AISC and ECP, but not clearly defined in EC4. The AISC provides the following limitations:

- The strength reduction factor, $\phi = 0.9$
- The nominal flexural strength, M_n, shall be determined using one of the following methods

1. The superposition of elastic stresses on the composite section, considering the effects of shoring for the limit state of yielding (yield moment).
2. The plastic stress distribution in the steel section alone, for the limit state of yielding (plastic moment) in the steel section.
3. The plastic stress distribution in the composite section or the strain-compatibility method, for the limit state of yielding (plastic moment) in the composite section. For concrete-encased members, steel anchors shall be provided.
   - The total cross-sectional area of the steel core shall comprise at least 1% of the total composite cross section.
   - Concrete encasement of the steel core shall be reinforced with continuous longitudinal bars and transverse reinforcement (stirrups, ties, hoops, or spirals). Detailing of longitudinal reinforcement, including bar spacing and concrete cover requirements, shall conform to ACI 318. Transverse reinforcement consisting of stirrups, ties, or hoops shall be a minimum of either a No. 3 (10 mm) bar spaced at a maximum of 12 in. (300 mm) on centre, or a No. 4 (13 mm) bar or larger spaced at a maximum of 16 in. (400 mm) on centre. Deformed wire or welded wire reinforcement of equivalent area is permitted.

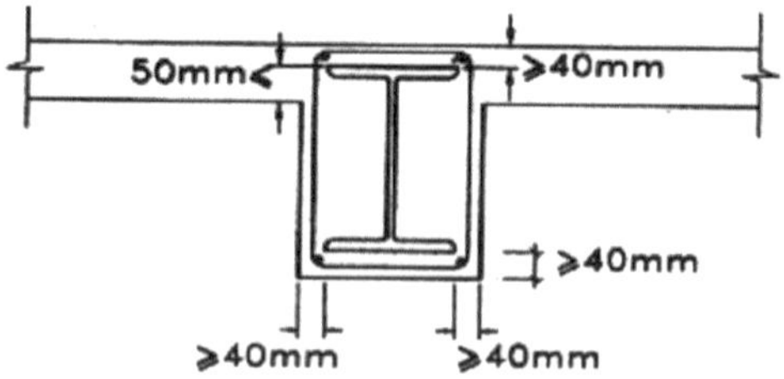

FIGURE 7.11 Encased beam detail.

– The minimum reinforcement ratio for continuous longitudinal reinforcement, ρsr, shall be 0.004, where ρ_{sr}

$$\rho_{sr} = A_{sr}/A_g \tag{7.21}$$

where A_g = gross area of the composite member, in.² (mm²)

A_{sr} = area of continuous longitudinal reinforcing bars, in.² (mm²)

- Clear spacing between the steel core and the longitudinal reinforcing steel shall be a minimum of 1.5 reinforcing bar diameters, but not less than 1.5 in. (38 mm).
- For shear strength, the design shear strength, φVn, which is related to the nominal shear strength of the structural steel section, as defined in Chapter G, plus the nominal strength of the transverse reinforcement, as defined by ACI 318, with a combined resistance or safety factor of

$$\phi_v = 0.75$$

7.5.2 Filled Composite Beam

For filled composite members, the total cross-sectional area of the structural steel section shall comprise at least 1% of the total composite cross section.

Flexural strength can be calculated as follows:

- For compact composite sections,

$$M_n = M_p$$

where

M_p = moment corresponding to plastic stress distribution over the composite cross section, kip-in. (N-mm).

- For non-compact cross sections,

$$M_n = M_p - \left(M_p - M_y\right)\left(\frac{\lambda - \lambda_p}{\lambda_r - \lambda_p}\right) \tag{7.22}$$

where λ, λ_p, and λ_r are width-to-thickness ratios, as determined from Table 7.4 adapted from AISC Table I1.1b.

TABLE 7.4
Limiting With/Thickness Ratio in Composite Section Under Flexural

Element	Width/thickness	λp	λr	Mx. permitted
Flanges of rectangular HSS and box sections of uniform thickness	b/t	$2.26\ (E/fy)^{0.5}$	$3.0\ (E/fy)^{0.5}$	$5.0\ (E/fy)^{0.5}$
Webs of rectangular HSS and box sections of uniform thickness	h/t	$3.0\ (E/fy)^{0.5}$	$5.0\ (E/fy)^{0.5}$	$5.7\ (E/fy)^{0.5}$
Round HSS	D/t	0.9E/fy	0.31E/fy	0.31E/fy

Where, λp is from compact section to non compact section
λr from non compact section to slender element composite

M_y = yield moment corresponding to the yielding of the tension flange and first yield of the compression flange, kip-in. (N-mm). The capacity at first yield shall be calculated assuming a linear elastic stress distribution with the maximum concrete compressive stress limited to 0.7 f_c' and the maximum steel stress limited to F_y.

For slender-element composite sections, M_n shall be determined as the first yield moment. The compression flange stress shall be limited to the critical buckling stress, F_n, determined using the following equations:

1. For rectangular filled sections,

$$F_n = \frac{9E_s}{\lambda^2} \tag{7.23}$$

2. For round filled sections,

$$F_n = \frac{0.72F_y}{\left[\left(\frac{D}{t}\right)\frac{F_y}{E_s}\right]^{0.2}} \tag{7.24}$$

I2-10 or I2-11. The concrete stress distribution shall be linear elastic with the maximum compressive stress limited to 0.70 f_c'.

7.6 COMPOSITE COLUMN

The composite section of the column is very important as it can be used in a composite member structural system or in the case of repair, where a hollow pipe can be filled with concrete to increase its stiffness.

There are mainly three types of composite columns, as presented in Figure 7.12. The first type resembles an encased beam, where the steel member is inserted inside the reinforced concrete column, as shown in Figure 7.12a. In the other cases, either a rectangular or a circular tube is filled with concrete, as illustrated in Figure 7.12b and c, respectively.

For the calculation of column composite sections, it is mandatory that the total cross-sectional area of steel must be greater than 4% of the gross column area. If it is less than 4%, it will be designed as a reinforced concrete section.

To avoid local buckling, the minimum wall thickness of steel rectangular tubing filled with concrete shall be determined by applying the equation $b(F_y/3E_s)^{0.5}$ for each face of width, b. The minimum wall thickness for circular sections, with an outside diameter of D, shall be determined using the equation $D(F_y/3E_s)^{0.5}$.

7.6.1 Composite Column Design by ECP

As per ECP, the column design will be as follows:

For inelastic buckling $\lambda \leq 100$,

$$F_c = \left(0.58 - aF_{ym}\lambda^2\right)F_{ym} \tag{7.25}$$

For elastic buckling $\lambda > 100$,

$$F_c = 3.57\ E_m/\lambda^2 \tag{7.26}$$

$$F_{ym} = F_y + C_1 F_{yr}\left(A_r/A_s\right) + C_2 f_c\left(A_c/A_s\right) \tag{7.27}$$

$$E_m = E_s + C_3 E_c\left(A_c/A_s\right) \tag{7.28}$$

For concrete encased section,

$$C_1 = 0.7,\ C_2 = 0.48 \text{ and } C_3 = 0.20$$

For concrete filled pipe,

$$C_1 = 1.0,\ C_2 = 0.68 \text{ and } C_3 = 0.40$$

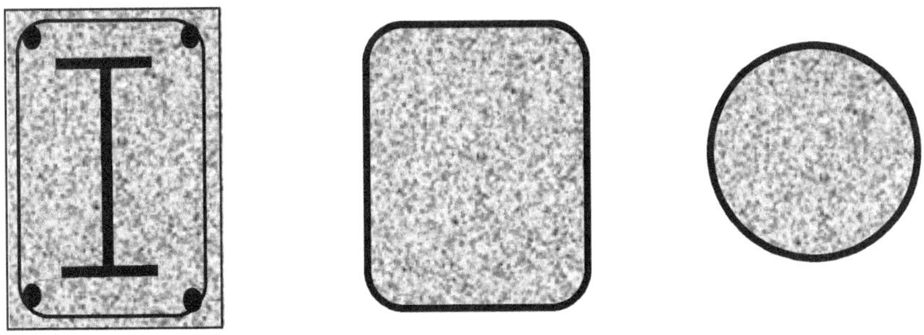

(c) Concrete encased I-Beam

(b) Concrete filled Rectangular tube

(a) Concrete filled pipe

FIGURE 7.12 Types of composite column.

where

f_c = concrete strength after 28 days
A_s = area of steel section, pipe, or tubing
A_r = area of longitudinal reinforcing bars
A_c = area of concrete excluding A_r and A_s
F_{yr} = yield stress of longitudinal reinforcing bars
F_{ym} = modified yield stress
r_m = radius of gyration of the steel shape, pipe or tubing.

7.6.2 Column Design by EC4

As per EC4, this clause applies to the columns and compression members in general, with steel grades S235 to S460 and normal weight concrete of strength classes *C20/25* to *C50/60.*

It also applies in the case of isolated columns and composite compression members in framed structures where the other structural members are either composite or steel.

(4) The steel contribution ratio δ should fulfil the following condition

$$0.2 \le \delta \le 0.9$$

The steel contribution ratio is calculated as follows:

$$\delta = \frac{A_{a\ f_{yd}}}{N_{pl,Rd}} \tag{7.29}$$

Considering that, $N_{pl,\,Rd}$ is the plastic resistance to compression of a composite cross section, it should be calculated by adding the plastic resistances of its components:

$$N_{pl,Rd} = A_a f_{yd} + 0.85\, A_c f_c + A_s f_{sd} \tag{7.30}$$

where

A_a is the cross-sectional area of the structural steel section, A_s is the steel reinforcement cross-sectional area, f_y is the design value of yield strength of structural steel, f_{sd} is the design value of yield strength of reinforcing steel, and fc is the cylinder compressive strength of concrete.

This equation applies for concrete-encased and partially concrete-encased steel sections. For concrete-filled sections, the coefficient 0.85 may be replaced by 1.0.

For concrete-filled tubes of circular cross section, account may be taken of the increase in strength of concrete caused by confinement, provided that the relative slenderness does not exceed 0.5 and $e/d < 0.1$, where e is the eccentricity of loading given by M_{Ed}/N_{Ed} and d is the external diameter of the column. The plastic resistance to compression may then be calculated from the following expression:

$$N_{pl,Rd} =_a A_a f_{yd} + A_c f_{cd}\left(1 +_c \frac{t}{d}\frac{f_y}{f_{ck}}\right) + A_s f_{sd} \tag{7.31}$$

where

t is the wall thickness of the steel tube.

For members with $e = 0$, the values $\eta_a = \eta_{ao}$ and $\eta_c = \eta_{co}$ are given by the following expressions:

$$\eta_{ao} = 0.25\left(3 + 2\overline{\lambda}\right) \text{ (but} \leq 1.0) \tag{7.32}$$

$$\eta_{co} = 4.9 - 18.5\,\overline{\lambda} + 17\overline{\lambda}^{2} \text{ (but} \geq 0) \tag{7.33}$$

In case of members in combined compression and bending with 0 <e/d≤ the values η_a and η_c should be determined from (6.36) and (6.37), where 11*ao* and 11*eo* are given by (6.34) and (6.35):

$$\eta_a = \eta_{ao} + (1 - \eta_{ao})\left(10\frac{e}{d}\right) \tag{7.34}$$

$$\eta_a = \eta_{co}\left(1 - \frac{10e}{d}\right) \tag{7.35}$$

For $e/d > 0.1$, $\eta_a = 1.0$ and $\eta_c = 0$.

The effects of local buckling may be neglected for a steel section fully encased in accordance with 6.7.5.1(2), and for other types of cross section, provided the maximum values of Table 6.3 are not exceeded.

a. For circular hollow section,

$$\text{Max}(d/t) = 90\left(235/f_y\right)^{0.5}$$

b. For rectangular hollow section,

$$\text{Max}(h/t) = 52\left(235/f_y\right)^{0.5}$$

c. Partially encased I-section:

$$\text{Max }\left(b/t_f\right) = 44\left(235/f_y\right)^{0.5}$$

where t is the thickness of the circular or rectangular hollow, d is the circular hollow section diameter, h is the rectangular hollow section long length, and B and t_f is the flange width and thickness for the I-section, respectively.

EC4 presents a simplified method of design. It is based on using members with a doubly symmetrical and uniform cross section over the member length, employing rolled, cold-formed, or welded steel sections. The simplified method is applicable if the structural steel component consists of two or more unconnected sections.

In this case, relative slenderness, $\bar{\lambda}$, must be less than or equal to 2.0:

$$\bar{\lambda} = \sqrt{\frac{N_{pl,Rd}}{N_{cr}}} \tag{7.39}$$

N_{cr} is elastic critical normal force for the relevant buckling mode, calculated with the effective flexural stiffness $(EI)_{eff}$:

$$(EI)_{eff} = E_a I_a + E_s I_s + 0.6 E_{c,eff} I_c \tag{7.40}$$

where

$$E_{c,eff} = E_{cm} \frac{1}{1+\left(N_{G,Ed}/N_{Ed}\right)\varphi_t} \tag{7.41}$$

I_a, I_s, and I_c are the second moment of area of the structural steel section, the reinforcement for bending plane, and the uncracked concrete section, respectively.

φ_t is the creep coefficient, N_{Ed} is the total design normal force, and $N_{G,Ed}$ the part of this normal force that is permanent.

On the other hand, for the case of a fully encased steel section the maximum thickness of the concrete cover will be as follows:

$$\text{Max. } c_z = 0.3h$$

$$\text{Max. } c_y = 0.4b$$

For simplification of members in axial compression, the design value of the normal force N_{Ed} should satisfy

$$\frac{N_{Ed}}{\chi N_{pl,Rd}} \leq 1.0 \tag{7.42}$$

χ is, as discussed in the previous chapter, the reduction factor according to the buckling mode in terms of the relevant relative slenderness $\bar{\lambda}$.

Considering that the relevant buckling curves for cross sections of composite columns are given in Table 7.5, where ρ_s is the reinforcement ratio A_s/A_c.

TABLE 7.5
Composite Column and Corresponding Buckling Curves as per EC3

Cross-section	Limits	Axis of buckling	Buckling curve	Member imperfection
concrete encased section 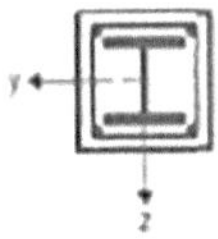		y-y	b	L/200
		z-z	c	L/150
partially concrete encased section 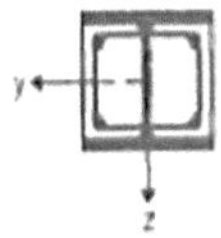		y-y	b	L/200
		z-z	c	L/150
circular and rectangular hollow steel section 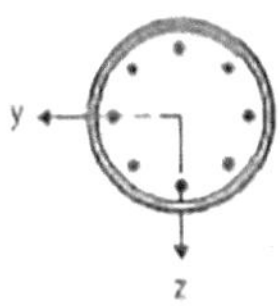	$\rho_s \le 3\%$	any	a	L/300
	$3\% < \rho_s \le 6\%$	any	b	L/200
circular hollow steel sections with additional I-section 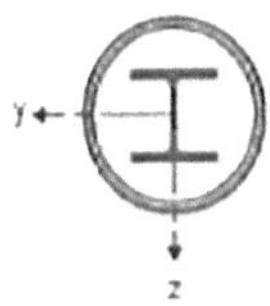		y-y	b	L/200
		z-z	b	L/200

7.6.3 Composite Column Design by AISC

7.6.3.1 Encased Composite Column

In general, the strength reduction factor, $\phi = 0.75$.

If $P_{no}/P_e \le 2.25$,

$$P_n = P_{no}\left(0.658^{\frac{P_{no}}{P_e}}\right) \tag{7.43}$$

If $P_{no}/P_e > 2.25$,

$$P_n = 0.877 P_e$$

P_e = elastic critical buckling load determined in accordance with Chapter C or Appendix 7, kips (N)

$$P_e = (\pi)^2 (EI)_{eff} / (Lc)^2 \tag{7.44}$$

$$(EI)_{eff} = E_s I_s + E_x I_{sr} + C_1 \; E_c I_c \tag{7.45}$$

$$C_1 = 0.25 + 3\left(A_s + A_{sr} / A_g\right) \le 0.7 \tag{7.46}$$

E_c = modulus of elasticity of concrete

$$= w_c^{1.5} \sqrt{f_c'}, \text{ksi} \left(0.043 w_c^{1.5} \sqrt{f_c'}, \text{MPa}\right) \tag{7.47}$$

f_c' = specified compressive strength of concrete, ksi (MPa)
w_c = weight of concrete per unit volume ($90 \le w_c \le 155$ lb/ft^3 or $1{,}500 \le w_c \le$ 2,500 kg/m^3)
E_s = modulus of elasticity of steel = 29,000 ksi (200,000 MPa)
I_c = moment of inertia of the concrete section about the elastic neutral axis of the composite section, in.^{4} (mm^4)
I_s = moment of inertia of steel shape about the elastic neutral axis of the composite section, in.^{4} (mm^4)
I_{sr} = moment of inertia of reinforcing bars about the elastic neutral axis of the composite section, in.^{4} (mm^4)
L_c = effective length of the member, in. (mm) = KL
K = effective length factor
L = laterally unbraced length of the member, in. (mm)
P_{no} = nominal axial compressive strength without consideration of length effects, kips (N)

$$= F_y A_s + F_{ysr} A_{sr} + 0.85 f_c' \, A_c \tag{7.48}$$

A_c = area of concrete, in.^{2} (mm^2)
A_s = cross-sectional area of the structural steel section, in.^{2} (mm^2)
F_y = specified minimum yield stress of the structural steel section, ksi (MPa)
F_{ysr} = specified minimum yield stress of reinforcing steel, ksi (MPa)

7.6.3.2 Tension Member

In case of tension member, the axial tension capacity will be as follows:

$$P_n = F_y A_s + F_{yrs} A_{sr} \tag{7.49}$$

In this case, the reduction strength factor, $\phi = 0.9$.

7.6.3.3 Filled Composite Column

For filled composite members, the following limitations shall be met: (a) The cross-sectional area of the structural steel section shall comprise at least 1% of the total composite cross section.

Minimum longitudinal reinforcement is not required. If longitudinal reinforcement is provided, internal transverse reinforcement is not required for strength; however, minimum internal transverse reinforcement shall be provided. Transverse reinforcement, where specified as ties or hoops, shall consist of a minimum of either a No. 3 (10 mm) bar spaced at a maximum of 12 in. (300 mm) on centre, or a No. 4 (13 mm) bar or larger spaced at a maximum of 16 in. (400 mm) on centre. Deformed wire or welded wire reinforcement of equivalent area is permitted.

A. Compressive strength for compact cross sections:
   $P_{no} = P_p$ as P_p is the plastic axial compressive strength

$$P_p = F_y A_s + C_2 f_c'\left(A_c + A_{Sr}\ E_s/E_c\right)$$

   $C_2 = 0.85$ for rectangular sections and 0.95 for round sections
B. For non-compact composite sections,

$$P_{no} = P_p - \frac{P_p - P_y}{\left(\lambda_r - \lambda_p\right)^2}\left(\lambda - \lambda_p\right)^2 \tag{7.50}$$

$$P_y = F_y A_s + 0.7 f_c'\left(A_c + A_{sr}\left(E_s/E_c\right)\right) \tag{7.51}$$

C. For slender composite sections,

$$P_y = F_n A_s + 0.7 f_c'\left(A_c + A_{sr}\left(E_s/E_c\right)\right)$$

The critical buckling stress for the structural steel section of filled composite members, F_n, is calculated from equations 7.23 and 7.24

The effective stiffness of the composite section, $(EI)_{eff}$, for all sections shall be calculated as follows:

$$(EI)_{eff} = E_s I_s + E_S I_{sr} + C_3 E_c I_c$$

$$C_3 = 0.45 + 3\left(A_s + A_{sr}/A_g\right) \le 0.9$$

7.7 COMPOSITE BEAM–COLUMN

7.7.1 Composite Beam–Column as per EC

In the case of a beam–column, as presented in previous chapters for frames, the composite section will follow the equations below, considering that lateral–torsional buckling does not affect this case.

The resistance of a cross section to combined compression and bending, and the corresponding interaction curve, may be calculated assuming rectangular stress blocks, as shown in Figure 7.13, taking into account the design shear force V_{Ed} in accordance with (3). The tensile strength of the concrete should be neglected.

The influence of transverse shear forces on the resistance to bending and normal force should be considered when determining the interaction curve, if the shear force

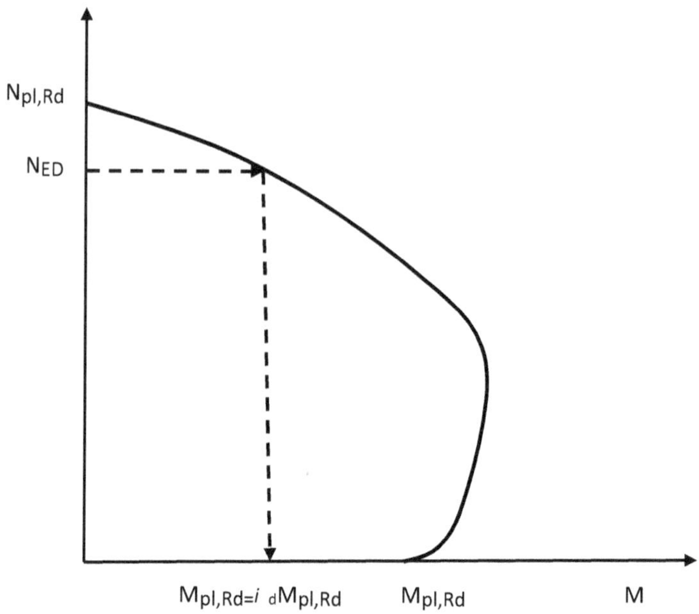

FIGURE 7.13 Interaction curve for combined compression and uniaxial bending.

$V_{a,Ed}$ on the steel section exceeds 50% of the design shear resistance $V_{pl,a,Rd}$ of the steel section, see 6.2.2.2.

Where $V_{a,Ed} > 0{,}5\ V_{pl,a,Rd}$, the influence of the transverse shear on the resistance in combined bending and compression should be taken into account by a reduced design steel strength $(1-p)$ .1~d in the shear area Av in accordance with 6.2.2.4(2) and Figure 6.18.

The shear force $V_{a,Ed}$ should not exceed the resistance to shear of the steel section determined according to 6.2.2. The resistance to shear $V_{c,Ed}$ of the reinforced concrete part should be verified in accordance with EN 1992-1-1, 6.2.

Unless a more accurate analysis is used, V_{Ed} may be distributed into $V_{a,Ed}$ acting on the structural steel and $V_{c,Ed}$ acting on the reinforced concrete section by

$$V_{a,Ed} = V_{Ed}\frac{M_{pl,a,Rd}}{M_{pl,Rd}} \qquad (7.52)$$

$$V_{c,Ed} = V_{Ed} - V_{a,Ed} \qquad (7.53)$$

where

$M_{pl,a,Rd}$ is the plastic resistance moment of the steel section and
$M_{pl,Rd}$ is the plastic resistance moment of the composite section.

For simplification, V_{ed} may be assumed to act on the structural steel section alone.

7.7.2 Composite Beam–Column as per AISC

In general, the interaction between flexure and axial forces in composite members shall account for stability. The calculation of compressive strength and the flexural strength shall be calculated as per the previous sections.

- In case of encased composite members and for filled composite members with compact composite sections, the interaction between axial force and flexure shall be based on the interaction equations as per Chapter 6
- For filled composite members with non-compact composite or slender-element composite sections, the interaction between axial force and flexure shall be based either on the interaction equations in AISC-19 of section H1.1, the method defined in Section I1.2d, or the following equations:

If $P_r/P_c \geq C_p$,

$$\frac{P_r}{P_c} + \frac{1-C_p}{C_m}\left(\frac{M_r}{M_c}\right) \leq 1.0 \tag{7.54}$$

If $P_r/P_c < C_p$,

$$\left(\frac{1-C_m}{C_p}\right)\left(\frac{P_r}{P_c}\right) + \frac{M_r}{M_c} \leq 1.0 \tag{7.55}$$

Mr = required flexural stresses, as per LRFD load combinations
M_c = design flexural strength determined= $\phi_b M_n$
P_r = axial load LRFD load combinations, kips (N)
P_c = design axial strength, = $\phi_c P_n$
ϕ_c = resistance factor for compression = 0.75
ϕ_b = resistance factor for flexure = 0.90
C_p and C_m from the following Table 7.6 which is adapted from Table I5.1 in AISC

TABLE 7.6*
The Coefficient C_p And C_m

		C_m	
Composite member type	**C_p**	**$C_{sr} \geq 0.5$**	**$C_{sr} < 0.5$**
Rectangular	$C_p=0.17/(C_{sr})^{0.4}$	$C_p = 1.06/(C_{sr})^{0.11} \geq 1.0$	$C_p = 0.90/(C_{sr})^{0.36} \leq 1.67$
Round HSS	$C_p=0.27/(C_{sr})^{0.4}$	$C_p = 1.10/(C_{sr})^{0.08} \geq 1.0$	$C_p = 0.95/(C_{sr})^{0.32} \leq 1.67$

* This table adapted from AISC—22 Table I5.1
** Csr= As fy+Asr Fyr/Acfc'

7.7.2.1 Beam–Column in AISC

$$f_{ca}/F_c + \left(f_{bx}/0.72F_y\right)A_1 + \left(f_{by}/0.72\ F_y\right)A_2 \le 1.0 \qquad (7.54)$$

where

f_{ca} = actual comp. Stress due to axial force computed on steel section only.
F_c = as before in equations 10.4 and 10.5

$$A_1 = C_{mx}/\left(1 - f_{ca}/F_{emx}\right) \ge 1.0 \qquad (7.55)$$

$$A_2 = C_{my}/\left(1 - f_{ca}/F_{emy}\right) \ge 1.0 \qquad (7.56)$$

C_{mx}, C_{my} = moment modification factor.

$$F_{emx} = 3.57\ E_m/\left(l_x\right)^2 \qquad (7.57)$$

f_{bx}, f_{by} = bending stresses about the x- and y-axes.

Example

This example illustrates the design of a 6 m long composite beam subjected to a uniformly distributed load (UDL) at 4 m centres. The composite slab is 130 mm deep, featuring 1.0 mm gauge ComFlor 60 (Corus) running perpendicular to the steel beam. The main beam span length is 6.5 m. The design checks include the moment resistance of the composite beam, the number of shear connectors, vertical shear, and transverse reinforcement.

For the answer in this Excel sheet, contact us at www.elreedyman.com to receive the Excel sheet.

ELREEDYMAN www.elreedyman.com	Client :-		Page no.	
	Project :-		REV. No :-	
			Calcs By :-	
	Title :-		Check. By :-	
			Approve By :-	
		Calculation sheet	Date:	

Beam Calculation

INPUT DATA

MEMBER SIZE:	UKB254x102x22	
YIELD STRESS OF STEEL (Fy):	275	Mpa
Ultimate strength	410	Mpa
STEEL REINFORCEMENT YIELD STRENGTH< fyk	500	Mpa
ELASTIC MODULUS (E):	210,000	Mpa
BEAM SPAN	6.50	m
BEAM SPACING, S	4.00	m
TOTAL SLAB DEPTH,h	130.00	mm
Depth of concrete above profile,hc	70.00	mm
Deck profile heoght, hp	60.00	mm
SHEAR CONNECTOR		
diamater	19.00	mm
overall height before welding,hsc	100.00	mm
height after welding	95.00	
fys	450.00	MPa
G	80770.00	MPa
Concrete		
Density wet	26.00	kN/m3
Density dry	25.00	kN/m3
cylider strength, fck	25.00	Mpa
Concrete design strength, fcd=	14.17	MPa
Secant modulus of elasticity	31.00	kN/mm2
Partial Factors		
γ_{Mo}=	1.00	
γ_c=	1.50	
γ_s=	1.15	
γ_v=	1.25	
γ_{vs}=	1.25	

Actions

Concrete weight (wet)=	2.522	kN/m^2		
Concrete weight (Dry)=	2.425	kN/m^2		
steel deck			0.11	kN/m^2
STEEL BEAM			0.20	kN/m^2
CEILING AND SERVICES			0.15	kN/m^2
FLOOR LOAD			2.50	kN/m^2
Construction loading inside &outside the working area			0.75	kN/m^2

Permenant actions

Construction stage=	0.31	kN/m^2
Composite stage=	2.885	kN/m^2

variable Action

Construction stage=	3.272	kN/m^2
Composite stage=	2.5	kN/m^2

Ultimate limit state

Constructio stage total distributed load=	5.30	kN/m^2
Construction Total Load=	137.67	kN
Composite stage total distributed load=	7.35	kN/m^2
Composite total load=	191	kN

DESIGN LOADING

ALLOWABLE STRESS DESIGN

BENDING MOMENT CONSTRUCTION (MyED):	111.9	kN-m
BENDING MOMENT COMPOSITE (My,ED):	155	kN-m
VED	96	kN

MEMBER PROPERTIES

GENERAL

A =	28.00	cm^2	
d =	254.00	mm	
tw =	5.70	mm	
bf =	101.60	mm	
tf =	6.80	mm	
h = T =	254	mm	
I_t =		cm^4	torsion constant
hc =	254.00	mm	
I_w =		cm^6	Warping constant

Y - AXIS

Iy =	880.7565	cm^4
Sy =	13.18	cm^3
ry =	7.60	cm
Wpl,y =	259.00	cm^3

Z - AXIS

Iz =	232.8023	cm^4
Sz =	39.42	cm^3
rz =	2.964569	cm
Wpl,z=	0.00	cm^3

ε=	0.92	
cf/(tf ε)	5.61 <9	The flange is class 1
cw/(twε)	40.80 <62	The web is class 1

Compression Resistance of Concrete Slab

At mid-span the effective width of the compression flange of the composite beam is determined from:

$$b_{eff} = b_o + \sum b_{e1}$$

b_{e1}= 0.81 m

Assume a single line of shear studs, therefore,bo=0m

b_{eff}= 1.625 m

Compression resisrtance of concrete slabis determined from:

$$N_{c,f} = f_{cd} b_{eff} h_c$$

Nc, f= 1611.46 kN

Tensile resisrtance of steel section:

$$N_{pl,a} = f_d A_a = \frac{f_y A_a}{\gamma_{MO}}$$

Npl,a= 770 kN

Locationof natural axis

Since Npl,a < Nc,f the plastic neutral axis lies in the concrete flange.

Design bending resistance with full shear connection

$$M_{pl,Rd} = N_{pl,a}\left(\frac{h_a}{2} + h_s - \left[\frac{N_{pl,a}}{N_{c,f}}\frac{h_c}{2}\right]\right)$$

Mpl,Rd= 185.01 kN.m

U.C=My,Ed/Mpl,Rd= 0.84 < 1.00

Shears connector resistance

$$P_{Rd}=0.29\,\alpha d^2 (f_{ck} E_{cm})^{0.5} / \gamma_v$$

P_{Rd}= 73.73 kN

$P_{Rd}=0.8f_u\pi d^2/4\,\gamma_v$

P_{Rd}= 81.61

So P_{Rd}= 73.73 kN

Influence of deck shape

Deck crosses the beam (i.e. ribs transverse to the beam)

One stud per trough,

nr = 1.00

Reduction factor

$$k_t = \frac{0.7}{\sqrt{n_r}}\frac{b_o}{h_p}\left(\frac{h_{sc}}{h_p} - 1\right) \leq 1.0$$

For trapezoidal profiles, b_0 is the average width of a trough, taken here as (120 + 170)/ 2 = 145.00 mm

Kt= 1.13

Therefore, as kt = 1.0 no reduction in shear connector resistance is

required. Therefore,

P_{Rd} = 73.73 kN

Number of shear studs in half span

Use one shear connector per trough, therefore,

Stud spacing along beam = 300 mm

Centre line to centre line span of 3 m should be reduced to allow for the primary beam width or the column width (assume 254 mm).

n= 9.68

Assume n= 9 stud shear connectors per half span

Degree of shear connection

Total resistance of 9 shear connectors

Rq= 663.57 kN

Rq/Npl.a= 0.86 η= 0.86

As this is less than 1.0, this beam has partial shear connection. Therefore, the minimum shear connection requirements must be checked, and the moment resistance reassessed.

Minimum shear connection requirements

The minimum shear connection requirement is calculated from:

(for Le < 25m):

$$\eta \geq 1 - \left(\frac{355}{f_y}\right)(0.75 - 0.03\,L_e)$$

For a simply supported beam, Le is equal to the span.

η≥ 0.28 < 0.4

Therefore the degree of shear connection must be at least 0.4. As shown above, there are a sufficient number of shear connectors to achieve this.

Design bending moment resistance with partial shear connection

The design bending moment can conservatively be calculated using:

$$M_{Rd} = M_{pl,a,Rd} + (M_{pl,Rd} - M_{pl,a,Rd})\eta$$

$M_{pl,a,Rd}$ is the plastic moment resistance of the steel section:

$M_{pl,a,Rd}$= 71.23 kN.m

So the design bending moment resistance is:

M_{Rd}= 169.29 kN.m

$M_{y,Ed}/M_{Rd}$= 0.92 < 1

Therefore, the design bending resistance of the composite beam with partial shear connection is adequate.

Resistance to vertical shear

Vertical shear resistance of the composite beam is:

$$V_{pl,Rd} = V_{pl,a,Rd} = \frac{A_v(f_y/\sqrt{3})}{\gamma_{M0}}$$

For rolled I and H sections loaded parallel to the web:

A_v=A-2bt_f+t_f(t_w+2r) ≥ $\eta h_w t_w$

Av= 1560.36 mm2 $\eta h_w t_w$= 1447.8

VPl.Rd= 248.03 kN

$V_{ed}/V_{Pl,Rd}$= 0.39 <1.0

Design of the transverse reinforcement

For simplicity, neglect the contribution of the decking and check the resistance of the reinforced concrete flange to splitting.
The area of reinforcement (Asf) can be determined using the following equation:

$$\frac{A_{sf} f_{yd}}{s_f} \geq V_{Ed} \frac{h_f}{\cot\theta_t}$$

where:
h_f is the depth of concrete above the metal decking, therefore,
h_f = hc = 70 mm
S_f is the spacing of the transverse reinforcement
f_{yd}= 435 N/mm²
26.3° ≤ θ_f ≤ 45° for compression flanges

The longitudinal shear stress is the stress transferred from the steel beam to the concrete. This is determined from the minimum resistance of the steel, concrete and shear connectors. In this example, with partial shear connection, that maximum force that can be transferred is limited by the resistance of the shear connectors over half of the span,
This force must be transferred over each half-span. As there are two shear planes (one on either side of the beam, running parallel to it), the longitudinal shear stress is:

$$V_{Ed} = \frac{R_q}{2h_f s}$$ 1.2 N/mm^2

For minimum area of transverse reinforcement assume θ = 26.5°

$$\frac{A_{sf} f_{yd}}{s_f} \geq V_{Ed} \frac{h_f}{\cot\theta_t}$$ 924.33 mm^2/m

Crushing of the concrete compression strut

The model for resisting the longitudinal shear assumes compression struts form in the concrete slab

Verify that:

$$V_{Ed} \leq V f_{cd} \sin\theta_f \cos\theta_f$$

where

V =0.6(1-f_{ck}/250)= 0.6

$V f_{cd} \sin\theta_f \cos\theta_f =$ 1.69

V_{Ed}=	1.18 N/mm^2	<	1.69	OK safe

REFERENCES

AISC (2022). *An American National Standard Specification for Structural Steel Buildings*, ANSI/AISC 360-22, American Institute of Steel Construction, Chicago, IL.

BSI (2004). *Eurocode 2: Design of Concrete Structures. General Rules and Rules for Buildings* (Incorporating Corrigendum January 2008 and November 2010), British Standards Institution, London.

CEN (2005). *Design of Composite Steel and Concrete Structures - Part 1.1: General Rules and Rules for Buildings*, EN 1994-1-1. 2005.

Fisher, J.W. (1970). Design of Composite Beams with Formed Metal Deck, *AISC Engineering Journal*, Vol. 7, No. 3, pp. 88–96, July 1970.

Grant, J.A., Fisher, J.W., and Slutter, R.O. (1977). Composite Beams with Formed Steel Deck, *AISC Engineering Journal*, Vol. 14, No. 1, pp. 24–43, 1977.

McGarraugh, J.B. and Baldwin, J.W. (1971). Lightweight Concrete-on-steel Composite Beams, *AISC Engineering Journal*, Vol. 8, No. 3, pp. 90–98, July 1971.

Ollgaard, J.G., Slutter, R.G. and Fisher, J.W. (1971). Shear Strength of Stud Shear Connections in Lightweight and Normal Weight Concrete. *AISC Engineering Journal*, Vol. 8, No. 2, pp. 55–64, April 1971.

SCI (2010). *NCCI: Modified Limitation on Partial Shear Connection in Beams for Buildings*, PN002a-GB, www.steel-ncci.co.uk.

8 Connections Design

8.1 INTRODUCTION

This chapter discusses connection design using the European code and AISC approach. Connections can be welded or bolted, based on the design and fabrication procedure, which may be constrained by whether the connection will be fabricated onsite or in the workshop. Consequently, various factors influence the decision regarding the type of connection. According to the European code, joint design follows EN 1993-1-8, which refers to joints as a general term rather than specific connections.

The following rules should be adhered to when selecting a welded connection. Welding in the field is avoided if possible due to specific requirements for welding conditions. Field welding should not be performed during rain, snow, or at temperatures below 0°F, and in certain ambient temperatures, preheating of the material to be welded is required. The AWS Code D1.1 (2004b) specifies minimum preheat temperatures, which are designed to prevent cracking.

Connections shall conform to the following requirements:

a. Shop connections may be either bolted or welded. Field connections shall be bolted wherever possible. Connections may be field welded only when conditions render a bolted connection unsuitable.
b. Bolted connections for primary members shall utilize high-strength bolts in accordance with ASTM A-325-N, bearing-type connections with threads included in the shear plane. Slip-critical-type connections shall be employed for connections subjected to vibration or repeated stress reversal.
c. Standard connections shall be designed by the fabricator in accordance with the project construction specifications and the loads specified in the drawings. Moment connections and special connections shall be designed by the engineer and depicted in the engineering drawings.
d. Moment connections shall preferably be of the bolted end plate type.

In general, the differences between bolting and welding that influence selection are summarized as follows:

- Bolting is generally a faster operation than welding
- Bolting does not have the temperature and weather condition requirements that are associated with welding
- Unexpected weather changes may delay welding operations

DOI: 10.1201/9781003307303-8

8.2 BOLTS

Four basic types of bolts are commonly in use. They are designated by ASTM as A307, A325, A490, and A449. A307 bolts are referred to as unfinished or ordinary bolts and are made from low carbon steel. Two grades (A and B) are available, and their diameters range from 6 to 100 mm. They are primarily used for low-stress connections and for secondary members. A325 and A490 bolts are classified as high-strength bolts. A325 bolts are made from heat-treated medium carbon steel.

There are three types of A325 and A490 bolts available on the market:

1. Type 1: bolts made of medium carbon steel
2. Type 2: bolts made of low carbon steel
3. Type 3: bolts having atmospheric corrosion resistance and weathering characteristics comparable to A242 and A588 steel.

A490 bolts, made from quenched and tempered alloy steel, possess higher strength than A325 bolts and also come in the same three types (types 1 to 3).

Both A325 and A490 bolts are available in diameters ranging from 12 to 38 mm in 1/8 in. increments and are used for general construction purposes. A449 bolts are available in diameters up to 75 mm, making them suitable for applications where diameters exceeding 38 mm are required. They are also used for anchor bolts and threaded rods.

High-strength bolts can be tightened to two conditions of tightness: snug-tight and fully tight.

Snug-tight conditions can be achieved through a few impacts of an impact wrench or the full effort of a worker using a standard spud wrench.

In European codes, yield strength f_{yb} and ultimate tensile strength f_{ub} for bolt classes 4.6, 5.6, 5.8, 6.8, 8.8 and 0.9 are given in Table 8.1. These values should be adopted as characteristic values in design calculations.

Accordingly, only bolt assemblies of classes 8.8 and 10.9 that conform to the requirements specified in 1.2.4 Reference Standards: Group 4 for High Strength Structural Bolting, which involves preloading with controlled tightening in accordance with the requirements in 1.2.7 Reference Standards: Group 7, may be used as preloaded bolts.

Snug-tight conditions must be clearly identified on the design drawing and are permitted only if the bolts are not subjected to tension loads; loosening or fatigue due to vibration or load fluctuations must not be design considerations. Bolts used in slip-critical conditions (i.e. conditions where the integrity of the connected parts

TABLE 8.1
Nominal Values for Bolts Yield, f_{yb}, and Ultimate Strength, f_{ub}

Bolt class	4.6	4.8	5.6	5.8	6.8	8.8	10.9
f_{yb} (MPa)	240	320	300	400	480	640	900
f_{ub} (MPa)	400	400	500	500	600	800	1,000

depends on the frictional force developed between the interfaces of the joint) and in situations where the bolts are subjected to direct tension must be fully tightened to develop a pre-tension force equal to approximately 70% of the minimum tensile stress F_u of the material from which the bolts are made.

Different bolted connections and those bolted with welding are presented in Figures 8.1–8.10.

The connection with the seated angle or plate, as in Figures 5.8 and 8.8 transfers a beam reaction force R_{Ed} to the supporting member through the angle seat. The top cleat is solely for lateral restraint and may be bolted either to the top of the web or to the top flange. The angle seat may be bolted or fillet welded to the supporting

FIGURE 8.1 Shear connection.

FIGURE 8.2 Single plate connection.

FIGURE 8.3 Double angle connection.

FIGURE 8.4 Welded and bolted angle.

FIGURE 8.5 Seated connection.

FIGURE 8.6 Special bolted connection.

FIGURE 8.7 Moment connection.

FIGURE 8.8 Seated plate connection.

member. The joint has very little moment capacity and is classified as a nominally pinned joint. It may be used for simple construction.

The beam reaction is transferred by bearing, shear, and bending of the horizontal leg of the angle, as well as by vertical shear through the connectors and horizontal forces in the connectors and between the vertical leg and the supporting member.

FIGURE 8.9 Extended end plate connection.

FIGURE 8.10 Fabrication of bolted connection.

In this joint, the beam is designed for zero end moment, while the supporting member is designed for the eccentric beam reaction. The beam web may need to be stiffened to resist shear and bearing. Although this joint is easily designed, its use is often discouraged due to the erection difficulties associated with the close depth tolerances required at the top and bottom of the beam.

The stresses on the bolts can be demonstrated in Figure 8.11. This can be accomplished by using the turn-of-the-nut method, the calibrated wrench method, or by employing alternate design fasteners or direct tension indicators.

In the case of bolted connections, the following tools can be used as shown in Figures 8.12 and 8.13. Common tools used by ironworkers include spud wrenches, pins, and correction bars of various sizes, as shown in Figure 8.12. Impact wrenches will be needed for certain installations, as shown in Figure 8.13.

The marking on the bolts will conform to ASME standards as well as Eurocode, as presented in Figure 8.14. The components of the bolts are depicted in the schematic drawings in Figure 8.15.

From ECS2008 Table 8.2 presents the sizes of the bolts along with their areas, A_s, and the area of stresses with the required torque (M_a) to achieve the required load. The shearing force resistance, P_s, is presented in Case I for primary stresses due to dead load, live load, or superimposed load, as well as dynamic effects and centrifugal force. In Case II, the primary and additional stresses from Case I, along with earthquake load, wind load, braking force, and lateral shock, are addressed.

The values in the above table can be applied for grade 8.8 by reducing the values by 30%.

The length of the bolts depends on the thickness of the connected parts and the required bolt end protrusion beyond the nut face, as specified in EN1090-2[2008].

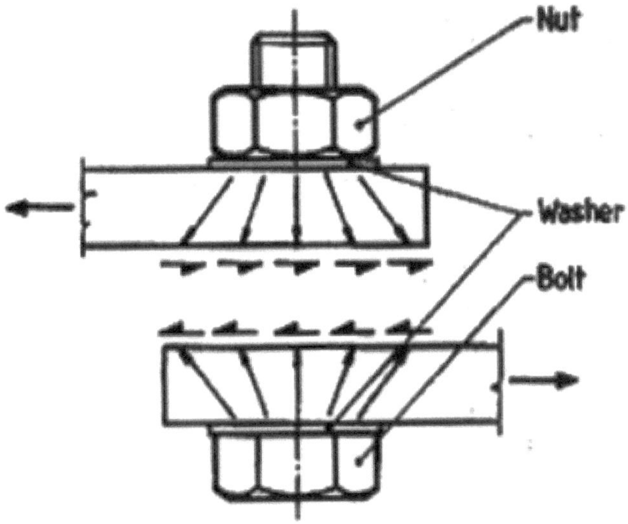

FIGURE 8.11 High-strength bolt stresses.

FIGURE 8.12 Tools for erecting connection.

The above table is match with EN1993.1.8 (2005) concerning the design slip resistance of preloaded bolts, which is covered by this equation:

$$F_s, R_d = k_s \cdot \eta \cdot \mu \cdot F_{p,c} / \gamma_{M3} \tag{8.1}$$

where $F_{p,c}$ is the preloading force, η is the number of friction surfaces, k_s equals 1.0 for normal holes, 0.85 for oversize holes or short slotted holes with the axis of the slot perpendicular to the force, 0.7 for long slotted holes with the axis of the slot perpendicular to the force, 0.76 for short slotted holes with the axis of the slot parallel to the force, and 0.63 for long slotted holes with the axis of the slot parallel to the force.

FIGURE 8.13 Hydraulic winch.

FIGURE 8.14 Photo showing the number of bolts.

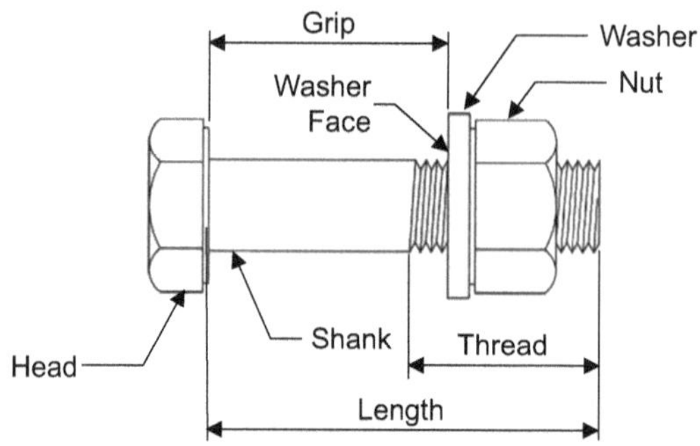

FIGURE 8.15 Bolt component.

TABLE 8.2
Properties and Strength of High-Strength Bolts (Grade 10.9)

					Permissible Friction Load of One Bolt per One Friction Surface (p_s) Tonnes			
					St. 37 & 42–44 ($\mu = 0.4$)		St. 50–55 ($\mu = 0.5$)	
D	A, mm²	A_s, mm²	T	M_a	I	II	I	II
M12	113	84	5.29	12	1.69	2.01	2.11	2.52
M16	201	157	9.89	31	3.16	3.37	3.95	4.71
M20	314	245	15.43	62	4.93	5.90	6.17	7.36
M22	380	303	19.08	84	6.10	7.27	7.63	9.10
M24	452	353	22.23	107	7.11	8.45	8.89	10.60
M27	573	459	28.91	157	9.25	11.03	11.56	13.78
M30	706	561	35.34	213	11.30	13.48	14.13	16.86
M36	1,018	817	51.47	372	16.47	19.64	20.58	24.55

μ is the slip factor with the following values: 0.5 for category A friction surfaces (blasted surfaces to remove loose rust, not pitted) and 0.4 for category B friction surfaces (spray-metallized with an Al- or Zn-based product with alkali-zinc silicate paint, thickness 50–80 μm). For category C, the value is 0.3, which applies to friction surfaces cleaned by wire-brushing or flame cutting, with loose rust removed; the value is 0.2 for category D friction surfaces (untreated as-rolled surfaces).

Considering that γ_{M3} is equal to 1.25 for category C slip-resistant connections and γ_{M3} is equal to 1.10 for category B slip-resistant connections.

8.2.1 Bolts Detailing

Connection detailing is very critical and important as it should be reflected in the design requirements.

The steel detailing guide is shown in Figures 8.16 and 8.17.

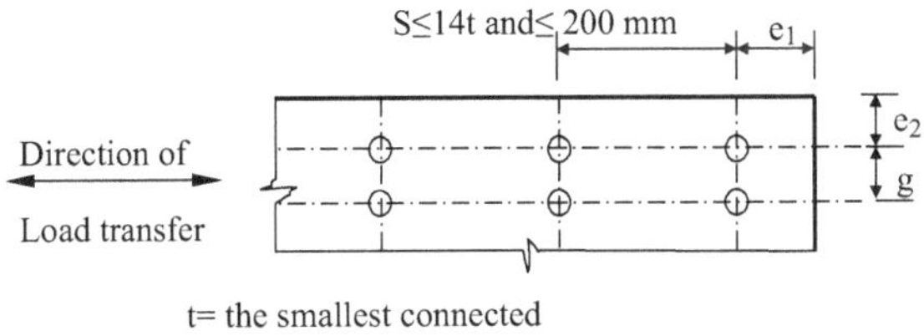

FIGURE 8.16 Spacing in tension or compression member.

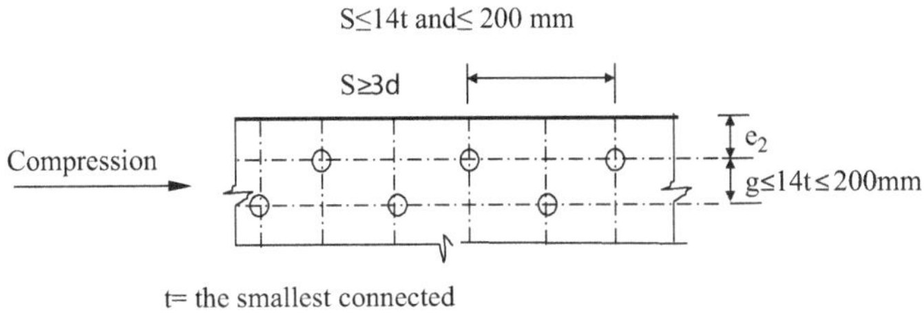

FIGURE 8.17 Staggered spacing in compression.

Where, as per EN1998.1.8 the minimum values of e_1 and e_2 is 1.2*d* as in Figures 8.16 and 8.17. In most practical cases, detailing e_1 and e_2 is 1.5*d*.

8.2.2 Bolts Design

8.2.2.1 Bolts Shear Resistance

The resistance F_v of a bolt in shear depends on the shear strength of the bolt (with tensile strength f_{ub}) and the area *A* of the bolt in a particular shear plane (either the gross area or the tensile stress area through the threads A_s, as appropriate). It can be expressed in the following form:

$$F_v = \alpha_v f_{ub} A / \gamma_{M2} \tag{8.2}$$

in which $\alpha_v = 0.6$ for classes 4.6, 5.6, and 8.8, except for Grade 10.9, 4.8, and 5.8, 6.8 bolts, when the shear plane passes through the threaded portion of the bolt, in which case $\alpha_v = 0.5$. The factor $\alpha_v = 0.6$ is close to the theoretical yield value of $\tau_y/f_y \approx 0.577$ and the experimental value of 0.62 reported by Kulak et al. (1987).

Here, $\gamma_{M2} = 1.25$ is the partial factor for the connector resistance (1.50 for Grade 4.6 bolts). It is common and conservative to determine the area A_n at a shear plane by substituting the tensile stress area A_s for the shank area *A* of the bolt.

In cases where bolts are used in long joints with a distance L_j between the end bolts in the joint, the shear resistance of all bolts will be reduced by the following factor in equation 8.2 considering that this factor should be between 0.75 and 1.0.

$$b_{Lf} = 1 - \left[\left(L_j - 15d \right) / 200d \right] \tag{8.3}$$

8.2.2.2 Bolts Tension Resistance

The resistance of a bolt in tension depends on the tensile strength f_{ub} of the bolt and the minimum cross-sectional area of the threaded length of the bolt. The design force $F_{t,Ed}$ is limited by EC3-1-8 to

$$F_{t,Ed} \leq F_{t,Rd} = 0.9 f_{ub} A_s / \gamma_{M2}, \tag{8.4}$$

where $\gamma_{M2} = 1.25$ (1.50 for Grade 4.6 bolts) and A_s is the tensile stress area of the bolt. The use of $0.9f_{ub}$ for the limit state of bolt fracture in addition to the partial factor γ_{M2} for the connector resistance reflects the reduced ductility at tensile fracture compared with shear failure.

If any of the connecting plates is sufficiently flexible, then additional prying forces may be induced in the bolts. As per Trahair et al. (2008), EC3-1-8 does not provide specific guidance on how to determine the prying force in the bolt, even in the T-stub component-based method of design. Under certain circumstances that minimize plate flexure in a joint component, such that the additional tensile forces caused by prying are small, it has been suggested that the prying forces need not be calculated, provided that $F_{t,Rd}$ is determined from

$$F_{t,Rd} = 0.72 f_{ub} A_s / \gamma_{M2}, \tag{8.5}$$

which results from the insertion of a reduction factor of 0.8 into equation 8.3.

8.2.2.3 Bearing Resistance

Apply this equation to calculate bearing resistance:

$$F_{b,Rd} = k_1 a_b f_u d \; t / g_{M2} \tag{8.6}$$

where a_b is the smallest of a_d, f_{ub}/f_u, or 1.0.

A. In the direction of load transfer,
   $a_d = e_1/3d$ for end bolts
   $= (S/3d) - 0.25$ for inner bolts
B. Perpendicular to the direction of load transfer
   - For edge bolts: k_1 is the smallest of $2.8\ (e_2/d) - 1.7$ or 2.5
   - For inner bolts: k_1 is the smallest of $1.4\ (S/d) - 1.7$ or 2.5

If e_2 is $1.5d$ or more, then in this case $k_1 = 2.5$, and a_d is equal to 0.5 for edge bolts and 0.75 for inner bolts, so to be conservative it is taken as 0.5.

Considering that oversized holes reduce bearing resistance to 0.8 times the resistance for bolts in normal holes. In the case of slotted holes, where the longitudinal axis of the slotted hole is perpendicular to the direction of force transfer, the bearing resistance for bolts is 0.6 times that of bolts in round, normal holes.

8.2.2.4 Bearing Bolts in Shear and Tension

According to Kulak et al. (1987), test results for bearing bolts in shear and tension suggest a circular interaction relationship for the strength limit state. EC3-1-8 employs a more conservative interaction between shear and tension, which can be expressed in the form of equation 8.7

$$\frac{F_{v,Ed}}{F_{v,Rd}} + \frac{F_{t,Ed}}{1.4F_{t,Rd}} \leq 1 \tag{8.7}$$

where $F_{v,Rd}$ is the shear resistance when there is no tension and $F_{t,Rd}$ is the tension resistance when there is no shear.

8.2.2.5 Angles Connected by One Leg

A single angle in tension connected by a single row of bolts in one leg (see Figure 3.9) may be treated as concentrically loaded over an effective net section, for which the design ultimate resistance should be determined as follows:

In case of

- 1 bolt:

$$N_{u,Rd} = 2.0(e_2 - 0.5d)t\, f_u / g_{M2}$$

- 2 bolts:

$$N_{u,Rd} = \beta_2 A_{net} f_u / \gamma_{M2} \qquad (\beta_2 = 0.4 \text{ for } p_1 \leq 2.5 \text{ and} = 0.7 \text{ for } p_1 \geq 5.0)$$

- 3 bolts or more:

$$N_{u,Rd} = \beta_3 A_{net} f_u / g_{M2} \qquad (\beta_3 = 0.5 \text{ for } p_1 \leq 2.5 \text{ and} = 0.7 \text{ for } p_1 \geq 5.0)$$

β_2 and β_3 are reduction factors dependent on the pitch p_1 as described above. For intermediate values of p_1, the value of β may be determined by linear interpolation. A_{net} is the net area of the angle.

For an unequal-leg angle connected by its smaller leg, A_{net} should be taken as equal to the net section area of an equivalent equal-leg angle of size equal to that of the smaller leg (Figure 8.18).

8.2.2.6 Design of Block Tearing

Block tearing consists of failure in shear at the row of bolts along the shear face of the hole group, accompanied by tensile rupture along the line of bolt holes on the tension face of the bolt group.

For a symmetric bolt group subject to concentric loading the design block tearing resistance, $V_{eff,1Rd}$, is given by

$$V_{eff,1,Rd} = f_u A_{nt} / \gamma_{M2} + \left(1/(3)^{0.5}\right) f_y A_{nv} / \gamma_{M0} \tag{8.8}$$

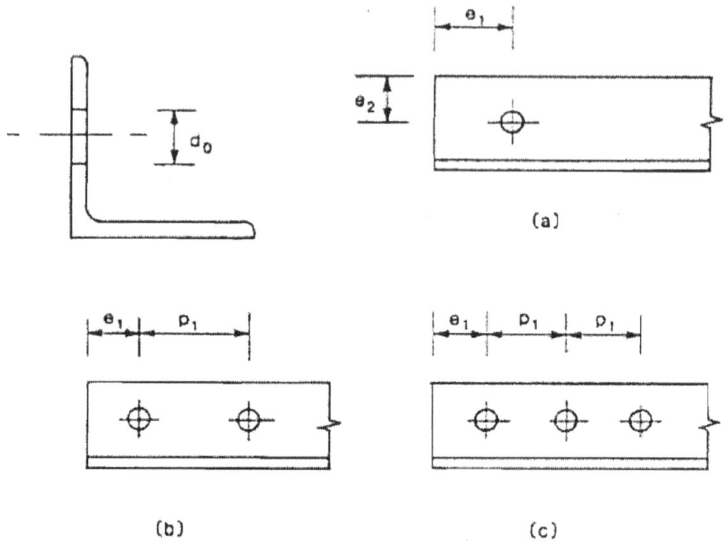

FIGURE 8.18 Angles connected by one leg.

where

A_{nt} is the net area subjected to tension;
A_{nv} is the net area subjected to shear.
γ_{M0} is equal to 1.0, which is the partial safety factor for steel
γ_{M2} is 1.1, which is the partial safety factor for fracture.

For a bolt group subject to eccentric loading, the design block shear tearing resistance, $V_{eff,2,Rd}$, is given by

$$V_{eff,l,Rd} = 0.5 f_u A_{nt} / \gamma_{M2} + \left(1/(3)^{0.5}\right) f_y A_{nv} / \gamma_{M0} \tag{8.9}$$

8.2.2.7 Splices of Members

Members are frequently spliced using splice plates in the flanges and the web, which are bolted or welded to the flanges and the web of the initial cross section. Figure 8.19 presents a beam splice. Beam splices may be full or partial strength, i.e. the splice may be stronger than, or not as strong as, the initial cross section. In full strength splices, the strength of the splice plates and the connecting media is at least equal to the strength of the corresponding part of the section. This means that the product of the area and the yield strength is higher than the corresponding value for the initial cross section. Such a design often leads to excessively thick and long splice plates, so in practice partial strength splices may be used, which are capable of transferring the applied forces and moments at the location of the splice. In addition, the splices should have some minimum dimensions, e.g. the splice plates should have at least half the thickness of the connected parts.

FIGURE 8.19 Beam splice.

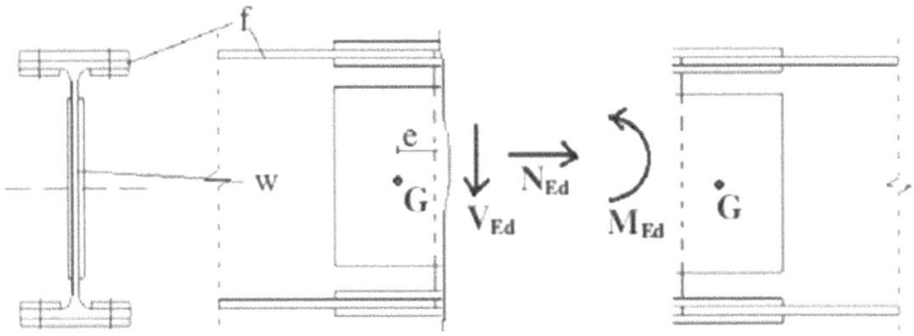

FIGURE 8.20 Forces and moments on beam splice.

As per Linder (2008), for splices of compression columns, part of the forces may be considered to be transferred through contact, so only flange splice plates may be used, while web splice plates may be avoided [5.28].

In beam splices, the splice plates transfer forces and moments that are resisted by the part of the cross section they join. Accordingly, the plates that join the flanges must transfer the partial forces that are resisted by the flange of the cross section, while the plates joining the web transfer the partial forces and moments resisted by the web. The partial forces may be determined by considering the elastic behaviour of the cross section.

Figure 8.20 illustrates the partial forces and moments resisted by the flanges and the web of an *I*-cross section subjected to axial forces, shear forces, and bending moments.

As per Vayas et al. (2019), the design of the different splice connections is as follows:

A. Design axial force N_{Ed}

The partial forces transferred by the flanges and the web are as follows:

$$\text{Flanges}: N_{f,N} = A_f / A \cdot N_{Ed} \quad (8.10)$$

$$\text{Web}: N_{w,N} = A_w / A \cdot N_{Ed} \quad (8.11)$$

where

A_f is the cross-sectional area of one flange
A_w is the cross-sectional area of the web
A is the area of the entire cross section

If N_{Ed} is a compression force, part of it is transferred through contact.

B. Design shear force V_{Ed}

The entire shear force is transferred by the web so it is

$$\text{Web}: V_w = V_{Ed}$$

C. Design bending moment M_{Ed}

Considering an elastic stress distribution due to the applied moment, it can be seen that the flanges are transferring axial forces, while the web transfers an axial force and a bending moment. The partial forces are determined as follows:

$$\text{Flanges}: N_{f,M} = S_f / I \cdot M_{Ed} \quad (8.12)$$

$$\text{Web}: N_{w,M} = S_w / I \cdot M_{Ed} \quad (8.13)$$

$$M_{w,M} = I_w / I \cdot M_{Ed} \quad (8.14)$$

where

S_f is the first moment of area of the relevant flange with respect to the cross-sectional centroid
S_w is the first moment of area of the web with respect to the cross-sectional centroid
I_w is the second moment of area of the web with respect to its own centroid
I is the second moment of area of the entire cross section.

8.3 WELDING

Welded connections are connections in which the components are primarily joined together by welds. Welds can be classified according to

- Types of Welds: groove, fillet, plug, and slot welds.
- Positions of the Welds: horizontal, vertical, overhead, and flat welds.
- Types of Joints: butt, lap, corner, edge, and tee.

Although fillet welds are generally weaker than groove welds, they are used more often because they allow for larger tolerances during erection compared to groove welds. Plug and slot welds are expensive to make and do not provide much reliability in transmitting tensile forces perpendicular to the surfaces. Furthermore, the quality control of such welds is difficult because inspecting them can be quite arduous. As a result, plug and slot welds are typically used only for stitching different parts of the members together.

8.3.1 Welding Symbols

A shorthand notation that provides important information on the location, size, length, etc., for the various types of welds has been developed by the American Welding Society to facilitate the detailing of welds. The welding system notation and arrow configuration in Figure 8.21a and b respectively.

8.3.2 Strength of Welds

In LRFD, the design strength of welds is taken as the smaller of the design strength of the base material and the design strength of the weld electrode. These allowable stresses and design strengths are summarized in Table 8.3. When a design uses ASD, the computed stress in the weld shall not exceed its allowable value. When a design uses LRFD, the design strength of welds should exceed the required strength obtained by dividing the load to be transmitted by the effective area of the welds. This table follows AISC; however, the European code aligns with the principles of LRFD.

The general limitations to the welding will be as follows:

- $S \leq \text{min of } t_1 \text{ or } t_2$
- min of $t_1$or t_2 < 20 mm $S \geq 5$ mm

  20 – 30 mm $S \geq 6$ mm

  30 – 50 mm $S \geq 8$ mm

  50 – 100 mm $S \geq 10$ mm

- For building $S \geq 4$ mm
- L_{min} = largest of (4S or 5 cm)
- L_{max} = 70S
- No limit for "L" for beam–column connections or flange-to-web welds in welded built-up girders.

For the European code, as in EN1993.1.8, the design resistance of fillet welds is as follows:

$$\left[\sigma_+^2 + 3\left(\tau_+^2 + \tau_{\parallel}^2\right)\right]^{0.5} \leq f_U \big/ \left(\beta_w \gamma_{M2}\right) \text{ and } \sigma_+ \leq 0.9 f_u / \gamma_{M2} \tag{8.15}$$

where σ_+ is the normal stress perpendicular to the throat, $\sigma_{\parallel}$ is the normal stress parallel to the axis of the weld, τ_+ is the shear stress in the plane of the throat and perpendicular to the axis of the weld, $\tau_{\parallel}$ is the shear stress in the plane of the throat

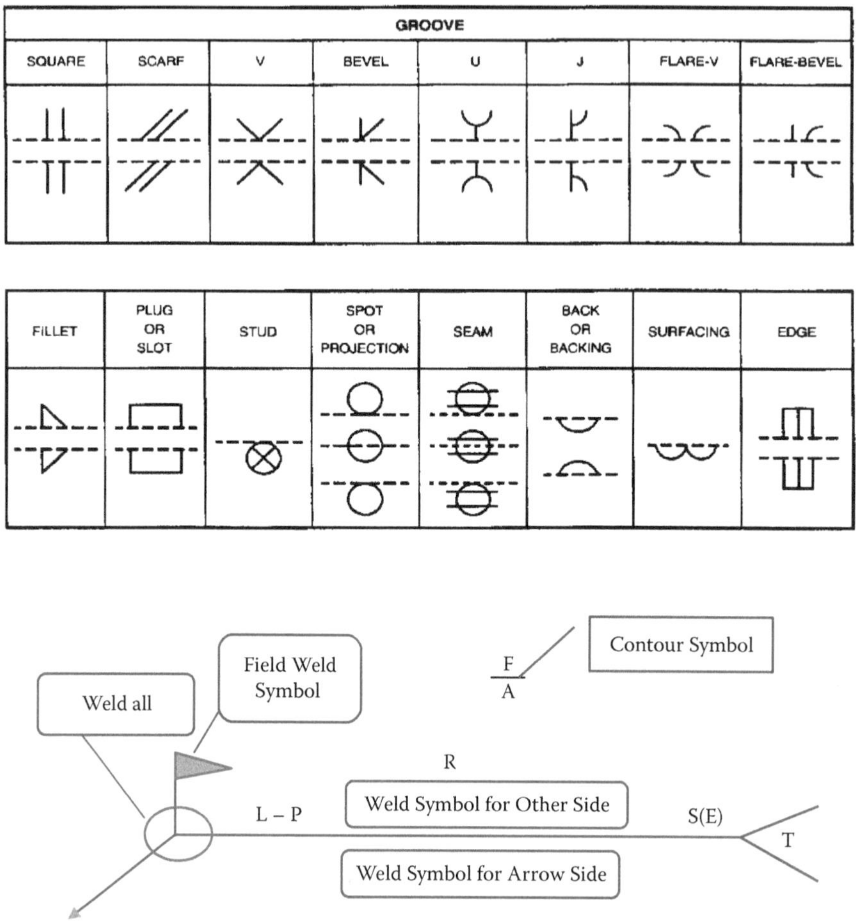

FIGURE 8.21 Welding symbol.

TABLE 8.3
Strength of Welds

Types of Weld and Stress	Material	ASD Allowable stress	LRFD ϕF_{BM} or ϕF_W	Required Weld Strength Level
		Full penetration groove weld		
Tension normal to effective Area	Base	Same as base metal	$0{:}90F_y$	"Matching" weld must be used
Compression normal to effective area	Base	Same as base metal	$0{:}90F_y$	Weld metal with a strength level equal to or less than "matching" must be used
Tension of compression parallel to the axis of weld	Base	Same as base metal	$0{:}90F_y$	
Shear on the effective area	Base weld electrode	0.3 × nominal tensile strength of weld metal	$0.9(0.6F_y)$ 0.8(0.6FEXX)	
		Partial penetration groove welds		
Compression normal to the effective area	Base	Same as base metal	$0{:}90F_y$	Weld metal with a strength level equal to or less than "matching" must be used
Tension or compression parallel to axis of weld *d* Shear parallel to axis of weld	Base weld electrode	0.30 × nominal tensile strength of weld metal	$0{:}75[{:}60FEXX]$	
Tension normal to the effective area	Base weld electrode	0.30 × nominal tensile strength of weld metal ≤0.18 × yield stress of base metal	$0.90F_y$ 0.8[0.6FEXX]	
		Fillet welds		
Stress on effective area	Base weld electrode	0.30 × nominal tensile strength of weld metal	$0{:}75[{:}60FEXX]$ $0.9F_y$	Weld metal with a strength level equal to or less than "matching" must be used
Tension or compression parallel to axis of weld	Base	0.30 × nominal tensile strength of weld metal	$0.9F_y$	

Notes:

- See AWS D1.1 for "matching" weld material.
- Weld metal one strength level stronger than "matching" weld metal will be permitted.
- Fillet welds partial penetration groove welds joining component elements of built-up members such as flange-to-web connections may be designed without regard to the tensile or compressive stress in these elements parallel to the axis of the welds.

parallel to the axis of the weld, f_u is the nominal ultimate strength of the weaker part being joined, and β_w is the appropriate correlation factor obtained from Table 8.4, depending on the steel grade.

The correlation factor presented in Table 8.4 varies for different grades, considering that it is applied for different standards: EN10025, EN10210 and EN10219.

TABLE 8.4
Correlation Factor β_w for Fillet Welds

Steel Grade	Correlation Factor β_w
S235	0.8
S275	0.85
S335	0.9
S420	1.0
S460	1.0

The code also provides a simple formula, independent of the orientation of the weld throat plane to the applied force, stating that the design resistance per unit length should be determined from:

The design shear strength f_{vwd} of the weld will be calculated using this formula:

$$f_{vwd} = 0.577\, f_u / (\beta_w \gamma_{M2}) \tag{8.16}$$

where f_u is the nominal ultimate tensile strength of the weaker part being joined and β_w is the appropriate correlation factor taken from Table 8.4.

In the case of a butt weld, there is no check required, as the weld is made with a suitable consumable that will produce all-weld tensile specimens having both a minimum yield strength and a minimum tensile strength not less than those specified for the parent metal. Therefore, if the steel section is safe, then the weld will be safe.

Example 8.1: Welded Connection

In Figure 8.22, an unequal angle of $80 \times 200 \times 120$ is connected to the column by welding, and a force of $P = 6$ tonnes is applied to this angle.

Calculation

Straining action:

$Q = 60$ kN.m $= 60 \times 45 = 2{,}700$ kN.m

SECTION 1-1 (take $S_w = 10$ mm)

$q = 60/(2 \times 120 \times 10) = 0.025$ kN/mm^2

$q(M_t) = 2{,}700 \times 60/(2 \times (10 \times (120)^3/12)) = 0.0563$ kN/mm^2

$f_{vwd} = 355/(3)^{0.5}/0.9/1.25 = 182.2$ N/mm^2

$q(t) = [(0.025)^2 + (0.0563)^2]^{0.5} = 0.061 < 0.18$ kN/mm^2 O.K

SECTION 2-2

$I_x = 2 \times (120)^3 \times 10/12 = 2{,}880{,}000$ mm^4

$F_t = 2{,}700 \times (120/2)/2{,}880{,}000 = 0.05625$ kN/mm^2

$F_{eq} = [(0.05625)^2 + 3(0.025)^2]^{0.5} = 0.07 < 0.18 \times 1.1$ OK

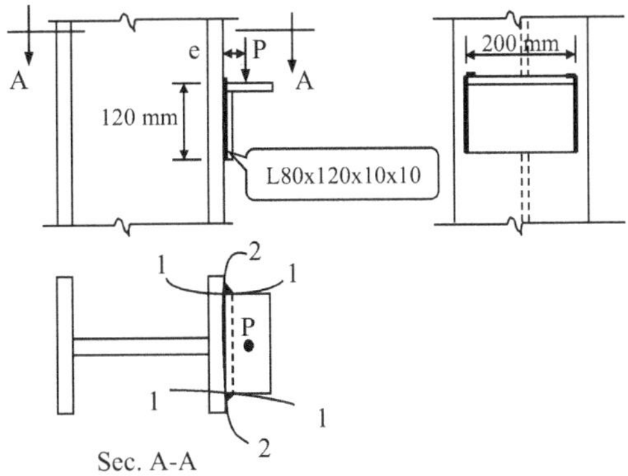

FIGURE 8.22 Welded connection.

8.3.2.1 Welding in Existing Structures

Welding to an existing structure, in cases of performing an extension or during retrofit projects requires careful consideration of numerous factors:

- Determine weldability by identifying the steel grade to establish a welding procedure
- Select and Design the Weld: Fillet welds are preferred, and over-welding should be avoided
- Surface Preparation: Remove contaminants such as paint, oil, and grease
- Loads during Retrofit: An engineer should determine the extent to which a member will be permitted to carry loads while heating, welding, or cutting
- Fire Hazards: Follow all governing fire codes, regulations, and safety rules to avoid fires

Example 8.2: Bracing Connection Design

The normal factored action force is $P_x = 80$ kN, and the bracing consists of two equal angles of 100 × 100 × 10 mm.

Solution:

This calculation was performed using an Excel sheet. You can contact me or download it from the website www.elreedyma.com

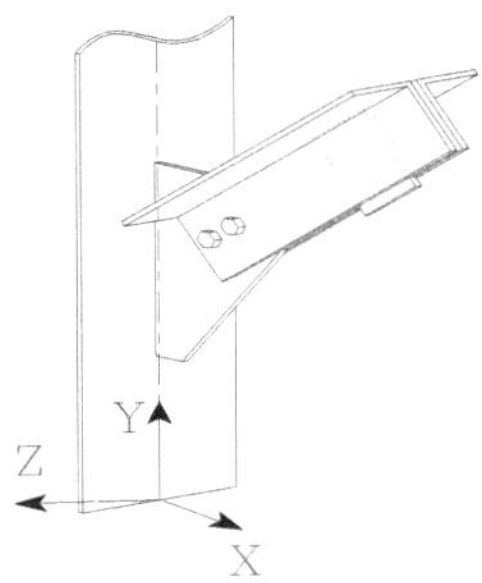

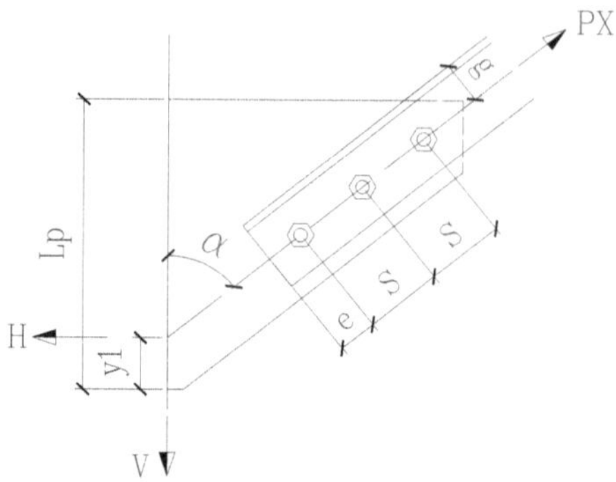

Section properties:

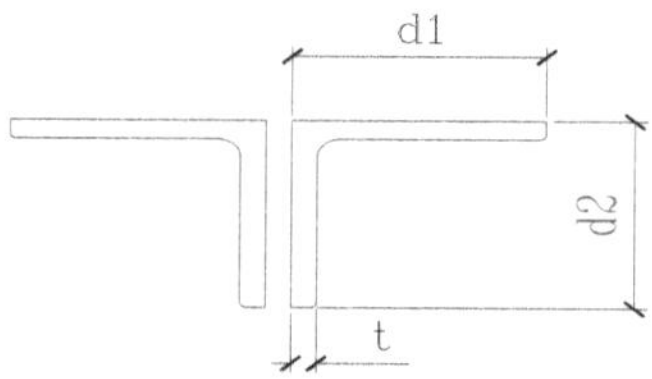

$d_1 = 100\,mm$
$d_2 = 100\,mm$
$t = 10\,mm$
$Ag = 1{,}920\,mm^2$
The steel structure member and the connection plates are of Grade S275.
$F_u = 430\ N/mm^2$ ultimate strength
$F_y = 275\ N/mm^2$ yield strength

- Connection bolts

  Choose bolt grade ASTM A325 Grade 5.6.

$d = 20\ mm$	bolt diameter
$d_h = 22\ mm$	hole diameter
A_b gross $= 314\,mm^2$	bolt gross area
A_b net $= 245\,mm^2$	bolt net area
$f_{vb} = 300\ N/mm^2$	
$f_{ub} = 500\ N/mm^2$	

- Geometry

$e = 35\,mm$	edge distance
$S = 70\,mm$	spacing between bolts
$g = 55\,mm$	gauge of bolts
$N_b = 2$	number of bolts
$t_p = 12\,mm$	gusset plate thickness
$L_p = 200\,mm$	gusset plate length
$Y_1 = 0$	
$L_1 = 120\,mm$	buckling length
$K = 0.5$	effective length factor
$\alpha = 30°$	
$S_w = 8\,mm$	weld size for gusset plate

- Check stress for bolts
  Bolt shear strength

  $Fv = \alpha vfub\ A/\gamma_{M2}$
  $F_v = 0.6 \times 500 \times 245/1.25/1{,}000 = 58.8$ kN
  F_s(act) actual shear force per bolt $= P_x/N_b$
  F_s(act) $= 40$ kN $< Fv$ *Ok safe*

- Check for brace
  1. Stress on effective are
     A_{net}, net area for brace section $= 2(A_g - txd_h)$
     $A_{net} = 3{,}400\ \text{mm}^2$
     A_e = effective area $= 0.75 \times A_{net} = 2{,}550\ \text{mm}^2$
     $N_{pl,Rd} = Afy/\gamma_{M0} = 3{,}840 \times 275/1.0/1{,}000 = 1{,}056$ kN.
     $N_{u,Rd} = \beta_2\ Anet\ fu/\gamma_{M2} = 0.7 \times 3{,}400 \times 430/1.1 = 930.4$ kN.
     B_2 is 0.7 as $s/d - 3.5$
     $N_{t,Rd} = 930.4$ kN (the lesser of $N_{pl,Rd}$ and $N_{u,Rd}$) $> P_x$ *OK safe*

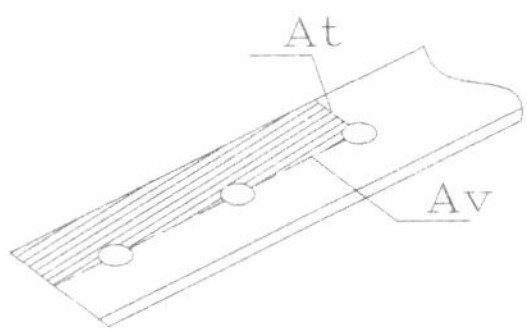

  2. Block shear capacity
     Shear area $A_{nv} = 2 \times t \times [(e + (N_b - 1) \times s - (N_b - 0.5) \times d_h]$
     $A_{nv} = 1{,}440\ \text{mm}^2$
     Tension area $A_{nt} = 2 \times t \times [d_1 - g - 0.5 \times d_h]$
     $A_{nt} = 680\ \text{mm}^2$
     $V_{eff,1,Rd} = f_u\ A_{nt}/\gamma_{M2} + (f_y/(3)^{0.5})A_{nv}/\gamma_{M0} = 430 \times 680/1.1/1{,}000 + 275 \times 1{,}440/1.732/1{,}000 = 265.82\ \text{kN} + 228.64\ \text{kN} = 494.46\ \text{kN} > P_x$ OK safe

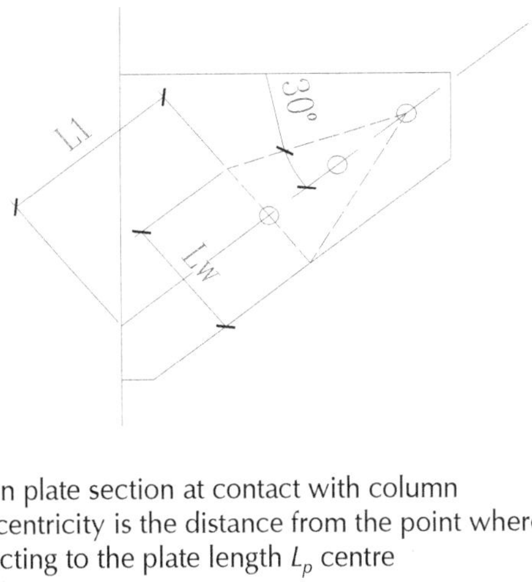

  3. Stress on plate section at contact with column
     E_{cc} = Eccentricity is the distance from the point where the axial force is acting to the plate length L_p centre
     $H = P_x \sin \alpha$
     $V = P_x \cos \alpha$
     $M = E_{cc} \times H$

4. Stress on plate section at contact with column
   E_{cc}, Distance from the point where the axial force is acting to the plate length L_p centre
   $H = P_x \sin\alpha$
   $V = P_x \cos\alpha$
   $M = E_{cc} \times H$

   $$f_t(\text{act}), \text{ actual tensile stress } = H/(L_p \times t_p) + M/(t_p \times L_p^2/6)$$

   $f_t(\text{act}) = 67\ \text{N/mm}^2 > f_y$ Ok safe

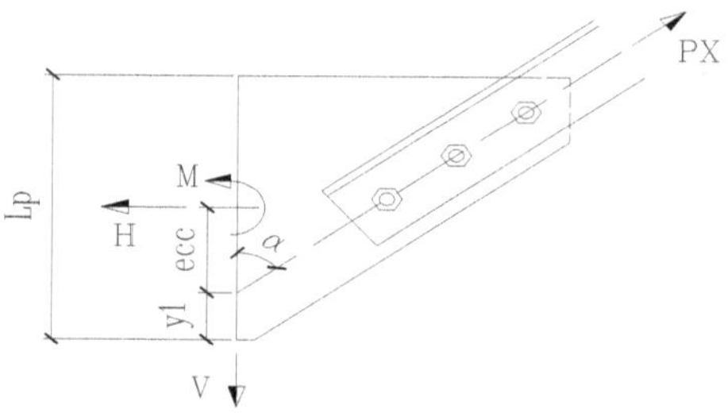

5. Block shear capacity
   Shear area $An_v = t_p \times [(e + (N_b - 1) \times s - (N_b - 0.5) \times d_h]$
   $A_v = 864\ \text{mm}^2$
   Tension area $A_{nt} = t_p \times [d_1 - g - 0.5 \times d_h]$
   $A_{nt} = 408\ \text{mm}^2$
   $V_{eff,1,Rd} = f_u\ A_{nt}/\gamma_{M2} + (f_y/(3)^{0.5})A_{nv}/\gamma_{M0} = 430 \times 408/1.1/1{,}000 + 275 \times 864/1.732/1{,}000 = 159.5\ \text{kN} + 137.2\text{kN} = 296.7\ \text{kN} > P_x$ ok safe

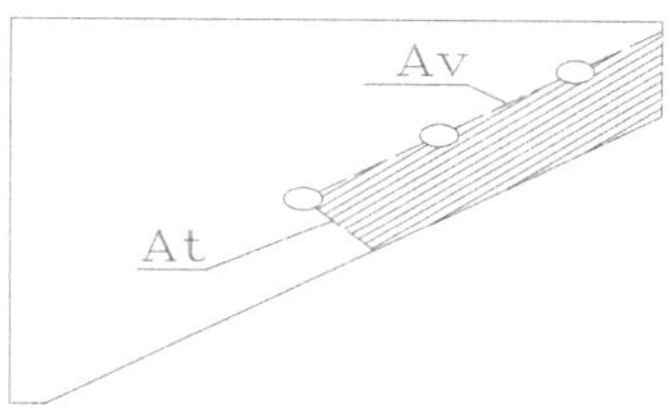

- Check for welds
  We will use Electrode E70.

$F_{vwd} = f_y\ (3)^{0.5}/0.9/1.25 = 182\ \text{N/mm}^2$
Weld stress components:
$f_1 = H/(2 \times Lp \times sw) + M/2 \times (sw \times Lp^2/6)$
$f_1 = 50\ \text{N/mm}^2$
$f_2 = V/(2 \times L_p \times s_w)$
$f_2 = 21.7\ \text{N/mm}^2$

$$f_r = \sqrt{f_1^2 + f_2^2}$$

$f_r = 54.5\ \text{N/mm}^2$
$f_{vwd} > f_r$ OK SAFE

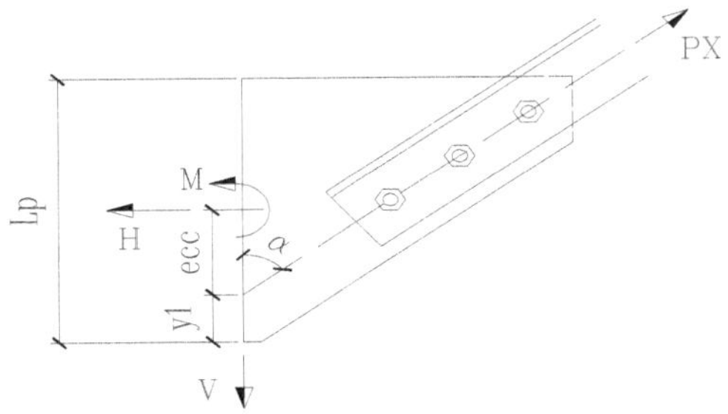

8.4 BEAM TO COLUMN JOINT

The joint between I-shaped beams and columns, where the beam flanges and the web are welded to the column flange. The individual components for a joint subjected to a bending moment are the column web in tension, compression, and shear, and the beam flange and web in compression.

The compression from the beam flange transfers to the column web, as illustrated in the following figure.

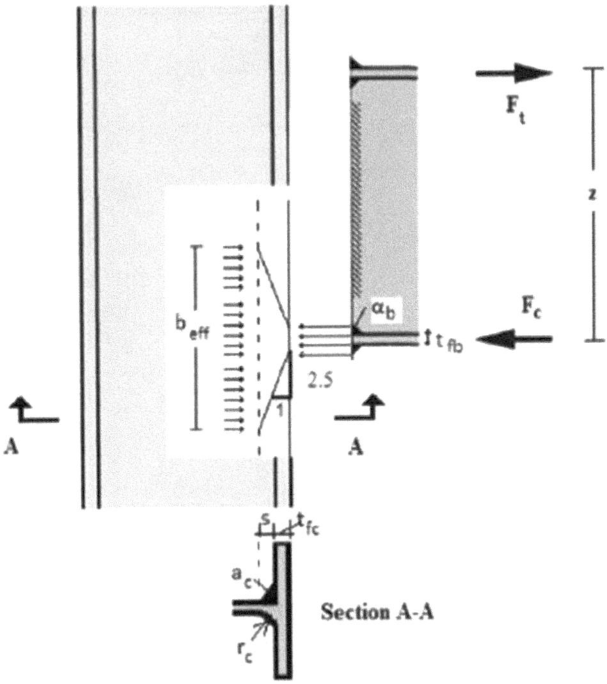

The design resistances of basic components for unstiffened welded beam-to-column connections are calculated based on the following:

1. Column web in shear

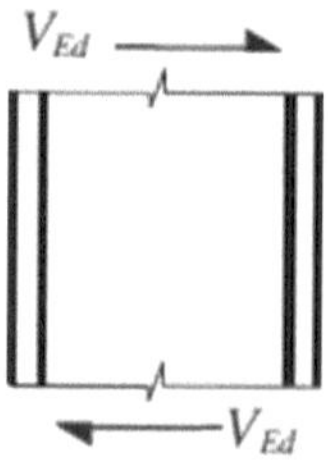

$$V_{wp,Rd} = 0.9 f_{y,wc} A_{vc} / \left((3)^{0.5} g_{MO} \right) \tag{8.17}$$

A_{vc} is the column shear area, assuming no shear buckling.

2. Column web in compression

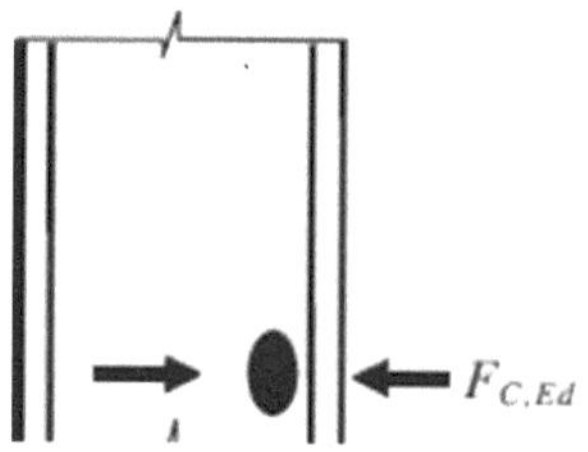

$$F_{c,wc,Rd} = \omega k_{wc} b_{eff,c,wc} t_{wc} f_{y,wc} / \gamma_{M0},$$

but

$$F_{c,wc,Rd} \le \omega k_{wc} \rho b_{eff,c,wc} t_{wc} f_{y,wc} / \gamma_{M1} \tag{8.18}$$

where

$$b_{eff,c,wc} = t_{fb} + 2\sqrt{2} a_b + 5 \left(t_{fc} + s \right) \tag{8.19}$$

$s = r_c$ for rolled section column and $= (2)^{0.5}\ \alpha_c$ for welded section.
ω is the reduction factor to consider interaction with shear in the column, as shown in the following table 8.5 in clause 6.2.6.2

Approximation values for the transformation parameter, β, are as follows:

A. There is a single beam connected to the column, $\beta = 1.0$.
B. Two beams in opposite directions are connected to the column.

- $M_{bl,ED} = M_{br,ED}$ $\beta = 0$
- $M_{br,ED} / M_{bl,ED} > 0$ $\beta = 1$
- $M_{br,ED} / M_{bl,ED} < 0$ $\beta = 2$
- $M_{br,ED} + M_{bl,ED} = 0$ $\beta = 2$

TABLE 8.5
Reduction Factor ω for Interaction with Shear

Transformation Parameter, β	Reduction Factor, ω
$0 \leq \beta \leq 0.5$	$\omega = 1$
$0.5 \leq \beta \leq 1$	$\omega = \omega_1 + 2(1-\beta)(1-\omega_1)$
$\beta = 1$	$\omega = \omega_1$
$1 < \beta < 2$	$\omega = \omega_1 + (\beta - 1)(\omega_2 - \omega)$
$\beta = 2$	$\omega = \omega_2$

$\omega_1 = 1/[1+1.3(b_{eff,wc}\, t_{wc}/A_{vc})^2]^{0.5}$.
$\omega_2 = 1/[1+5.2(b_{eff,wc}\, t_{wc}/A_{vc})^2]^{0.5}$.
A_{vc} is the shear area of the column.
β is the transformation parameter.

The approximate value of the transformation parameter, β:

1. End connection, $\beta = 1$

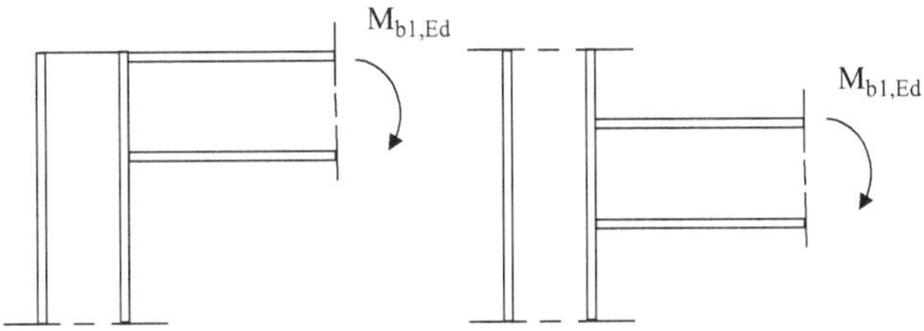

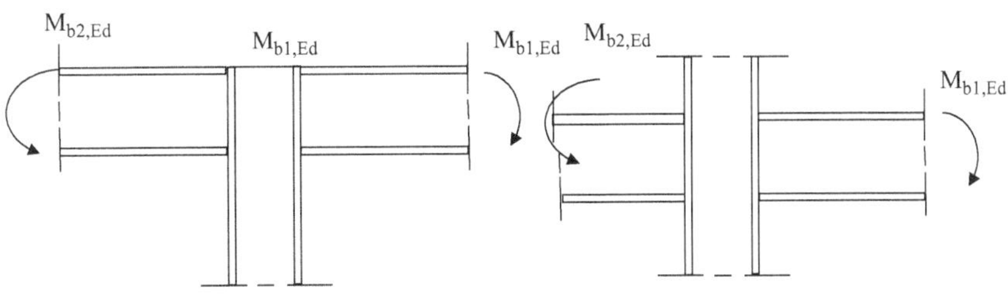

2. Connection for external column:

- $M_{b1,Ed} = M_{b2,Ed}$ $\quad \beta = 0$
- $M_{b1,Ed}/M_{b2,Ed} > 0$ $\quad \beta = 1$
- $M_{b1,Ed}/M_{b2,Ed} < 0$ $\quad \beta = 2$
- $M_{b1,Ed} + M_{b2,Ed} = 0$ $\quad \beta = 2$

where

ρ is the reduction factor to consider compression buckling in the column web from [5.2]

ρ is the plate buckling reduction factor

$$\rho = 1.0 \text{ if } \lambda_p \leq 0.72$$

$$\rho = (\lambda_p - 0.2)/(\lambda_p)^2 \tag{8.20}$$

where l_p is the plate slenderness calculated from the following equation:

$$\lambda_p = 0.932(b_{eff} d_{wc} f_{ywc}/E(t_{wc})^2 \tag{8.21}$$

- In case of rolled I or H column sections, $d_{wc} = h_c - 2(t_{fc} + r_c)$
- In case of welded I or H column sections, $d_{wc} = h_c - 2(t_{fc} + 1.414\ a_c)$

k_{wc} reduction factor to consider interaction with compression stresses along the column due to axial force and bending. As per EN1.8(2005) clause 6.2.6, this is 1.0, and no reduction is necessary. It can therefore be omitted in preliminary calculations when the longitudinal stress is unknown and checked later.

$M_{b,Rd}$ Design bending resistance of the beam, possibly reduced due to shear. h is the height of the beam.

3. Column web in tension

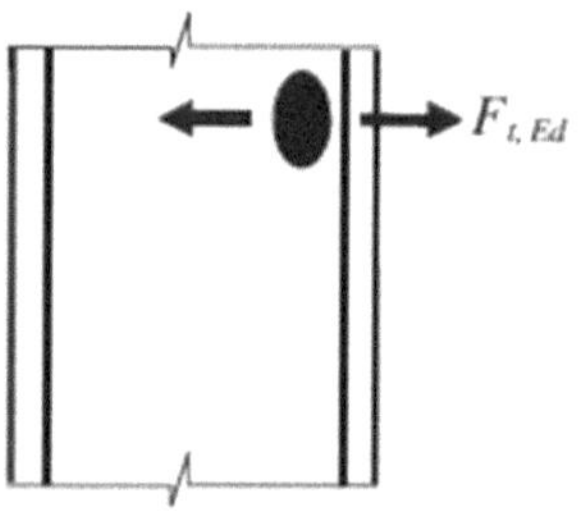

4. Beam flange and web in compression

$$F_{c,fb,Rd} = M_{b,Rd}/h - t\ fb \tag{8.22}$$

8.5 BASE CONNECTION

8.5.1 Base Plate Design

In the case of anchor bolts, the mechanical properties of these bolts shall be in accordance with EN ISO 898-1 or fabricated from hot-rolled steel conforming to EN 10025-2 to EN 10025-4. If specified, reinforcing steels may be used. In this case, they shall conform to EN 10080, and the steel grade shall be specified.

In steel-to-concrete joints, the flange of an equivalent I-stub in compression may be used to model the design resistances for the combination of the following basic components:

the steel base plate in bending under the bearing pressure on the foundation;
the concrete and/or grout joint material in bearing.

The total effective length l_{eff} and the total effective width b_{eff} of an equivalent I-stub should be such that the design compression resistance of the I-stub is equivalent to that of the basic joint component it represents.

As per clause 6.2.5, the design compression resistance of a T-stub flange $F_{c,Rd}$ should be determined as follows:

$$F_{c,Rd} = f_{jd} b_{eff} l_{eff} \ldots \tag{8.23}$$

where

b_{eff} is the effective width of the I-stub flange, see 6.2.5(5) and 6.2.5(6)
l_{eff} is the effective length of the I-stub flange, see 6.2.5(5) and 6.2.5(6)
f_{jd} is the bearing strength of the joint, calculated as follows:

$$f_{jd} = \beta_j F_{Rdu} / (b_{eff} l_{eff}) \tag{8.24}$$

β_j is the foundation joint material coefficient, which may be taken as equal to 0.67, provided that the characteristic strength of the grout is not less than 0.2 times the characteristic strength of the concrete foundation and the thickness of the grout is not greater than 0.2 times the smallest width of the steel base plate. In cases where the thickness of the grout is more than 50 mm, the characteristic strength of the grout should be at least the same as that of the concrete foundation.

F_{Rdu} is the concentrated design resistance force given in EN 1992, where A_{co} is to be taken as $(b_{eff} l_{eff})$.

The forces transferred through an I-stub should be assumed to spread uniformly, as shown in Figure 6.4a and b. The pressure on the resulting bearing area should not exceed the design bearing strength f_{jd}, and the additional bearing width, c, should not exceed the value in the following equation:

$$C = t[f_y / (3 f_{jd} \gamma_{M0})]^{0.5} \tag{8.25}$$

where t and f_y are the thickness and yield strength of the T-stub flange, respectively.

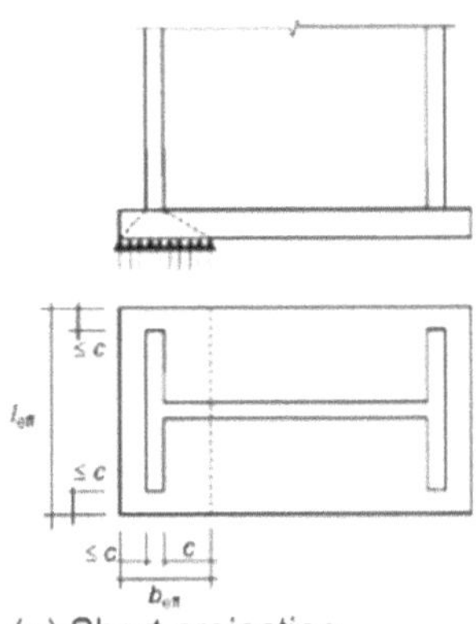

(a) Short projection

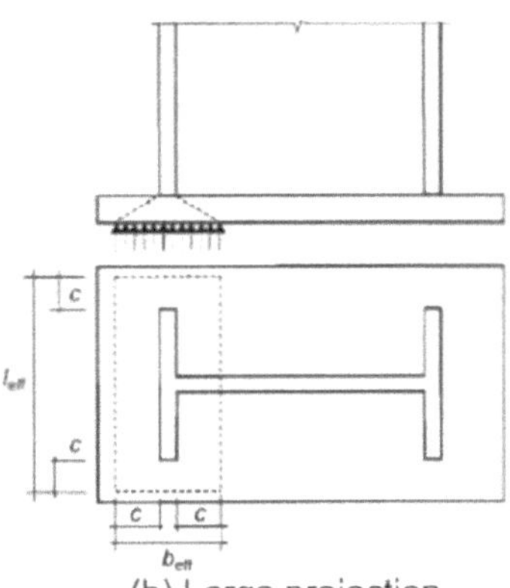

(b) Large projection

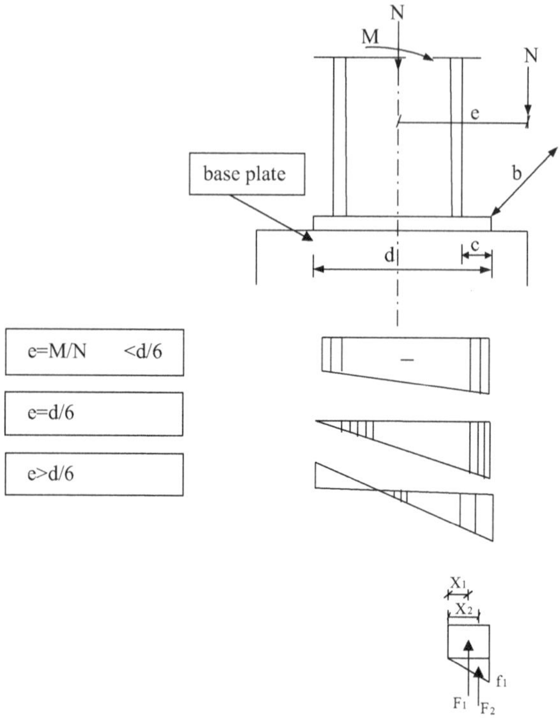

FIGURE 8.23 Base plate stresses.

8.5.1.1 Stresses in Fixed Support

In general, the stresses on the base plate and the anchor bolt in the case of fixed support are presented in Figure 8.23.

The stresses on the base plate are shown in Figure 8.23, where *d* and *b* are the length and width of the base plate, respectively.

$$f_{1,2} = (-N / A) + \left[6\left(Ne / bd^2\right)\right] \tag{8.26}$$

$$M = F_1.X_1 + F_2.\,X_2 \tag{8.27}$$

Example 8.3: Base Plate for Fixed Connection

Figure 8.24 presents the case study for fixed support.

Design a fixed base connection to support a vertical total action $N = 500$ kN, a bending moment $M = 30$ m. kN and a horizontal action force $H = 50$ kN. The column section is HEB 300 with $F_y = 275$ MPa and $t_f = 19$ mm.

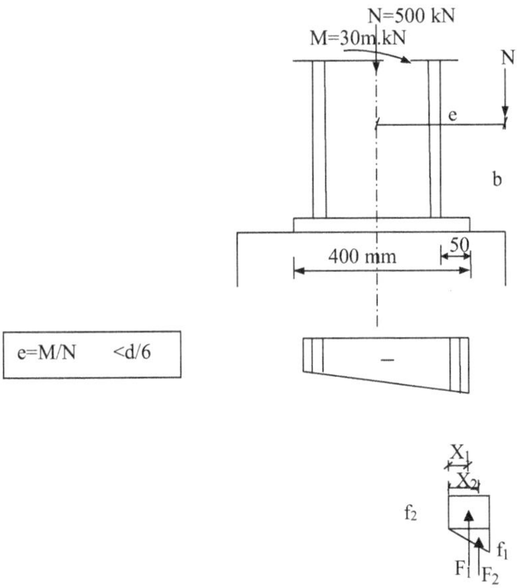

FIGURE 8.24 Base plate stresses.

Solution:

$$f_a = 6M/\left(b.t^2\right)$$

Get base plate thickness (t)

The base plate dimensions are obtained from the following equations:

$D = 0.95d + 100$ in mm
$B = 0.8b + 100$ in mm

Assume the base plate is 400×400 mm.

$e = M/N = 30/500 = 0.06$ m $= 60$ mm $< 400/6$

So, all the base plate under compression as the resultant inside the core

Calculate stresses:

f_c' for the concrete is 30 MPa
$l_{eff} = 400$ mm
$C = 50$ mm
$b_{eff} = 2 \times 50 + 19 = 119$ mm
$f_{jd} = \beta_j\, F_{Rdu}/(b_{eff}\, l_{eff})$
$f_{jd} = 0.67 \times 30 = 20.1$ N/mm^2
$F_{c,Rd} = f_{jd}\, b_{eff}\, l_{eff}$
$= 20.1 \times 400 \times 119/1{,}000 = 956.76$ kN
$C = t\ [f_y/(3 f_{jd}\, \gamma_{M0})]^{0.5}$
$= 19[275/(3 \times 20.1 \times 1.0)^{0.5} = 147.5$ mm > 50 mm OK
$f_1 = -N/A + M.y/I$
$f_1 = (500{,}000\ \ N/(400 \times 400)) + 30 \times 1{,}000{,}000 \times 200/(400 \times (400)^3)/12 = 5.94$ N/mm^2

$f_1 < 30$ MPa, which is (f_c') the concrete characteristic strength.

$f_2 = (500{,}000\ \text{N}/(400 \times 400)) + 30 \times 1{,}000{,}000 \times 150/(400 \times (400)^3)/12 = 5.23$ N/mm^2

Calculate the moment as in Figure 8.24, where $X_1 = 25$ mm and $X_2 = 2/3 \times 50 = 33.3$ mm

$M_1 = 5.23 \times 50 \times 25 + ½\,(5.94 - 5.23) \times 50 \times 33.3 = 7{,}128.58$

$t = (6M/f_y/1.0)^{0.5} = 12.5$ mm; choose thickness 14 mm.

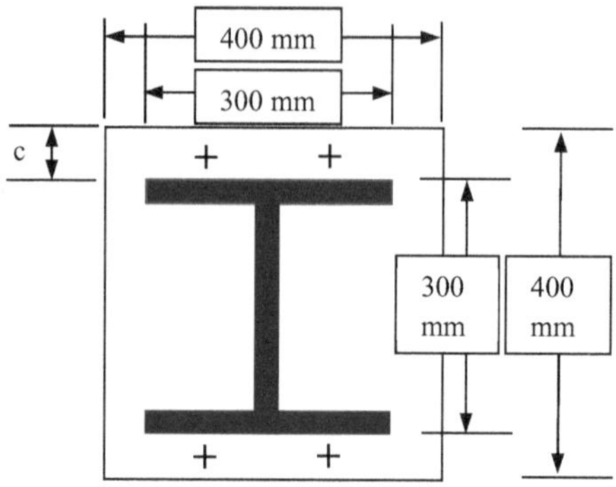

Example 8.4: Base Plate for Hinge Connection

This example involves designing a base plate for a hinge connection with a normal action force $N = 450$ kN and a shear action force $Q = 40$ KN. The steel section is HEB 300, and the concrete characteristic strength is 35 MPa, while the steel $f_y = 235$ MPa (Figure 8.25).

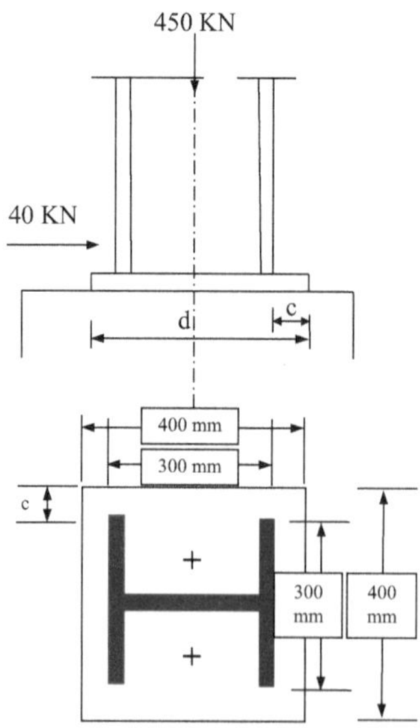

FIGURE 8.25 Base plate design.

Solution:

$f_c = 450 \times 1{,}000/385 \times 340 = 3.44$ MPa $< f_c'$ OK
$M = 3.44\ (50)^2/2 = 4{,}300$ mm.N/mm
$t = (6M/(f_y/1.0))^{0.5}$
$t = 10.5$ mm choose 12 mm

In case of welding
$F_{vwd} = f_y\ (3)^{0.5}/0.9/1.25 = 120$ N/mm²
$S_w = (40 \times 1{,}000/2 \times 120 \times 300) = 0.5$ mm; take minimum 5 mm

Bolts
Choose 2 Bolt Grade 5.6

$D = 20$ mm	bolt diameter
$d_h = 22$ mm	hole diameter
A_b gross $= 314$ mm²	bolt gross area
A_b net $= 245$ mm²	bolt net area
$f_{vb} = 300$ N/mm²	
$f_{ub} = 500$ N/mm²	

$F_v = (2)\ 0.6 \times 500 \times 245/1.25/1{,}000 = 117.6 > 40$ kN Ok safe

There is a common question from young engineers regarding how to distinguish between hinge connections and fixed connections. The plan view for the above two examples, with the position of the bolts, defines the type of connection easily.

8.5.2 Anchor Bolt Design

The design resistance of the anchor bolts should be taken as the smaller of the design tension resistance of the anchor bolt (see 3.6) and the bond resistance of the concrete on the anchor bolt according to EN 1992-1-1.

Definitions for the different types of anchor bolts:

Cast-in Anchor: A headed bolt, anchor rod, or hooked bolt that is installed before the concrete is placed.

Post-Installed Anchor: An anchor installed in hardened concrete. Expansion, undercut, and adhesive anchors are examples of post-installed anchors.

Pre-tensioned Anchor Bolt: An anchor bolt that is designed to be tensioned to a predefined force to prevent premature failure of the anchor bolt due to fatigue.

Shear Lug: A pipe or plate section welded to the bottom of a base plate that is used to resist base shear forces.

The cast-in anchor bolt is the most traditional type used in industrial projects; however, the other types are very specific for certain facilities and based on the fabricator's or manufacturer's requirements.

One of the following methods should be used to secure anchor bolts into the foundation:

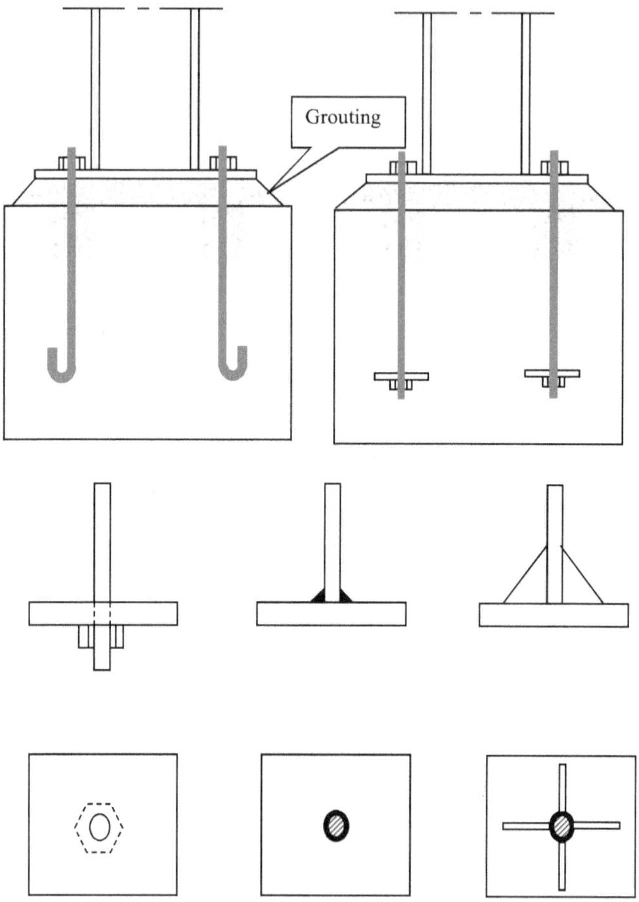

FIGURE 8.26 Anchor bolt detail.

- a hook, as shown in Figure 8.26
- a washer plate Presents in many shapes, as shown in Figure 8.26
- some other appropriate load distributing member embedded in some concrete,
- some other fixing that has been adequately tested and approved.

In general, when the bolts are provided with a hook, the anchorage length should be sufficient to prevent bond failure before the yielding of the bolt. The anchorage length should be calculated in accordance with EN 1992-1-1. This type of anchorage should not be used for bolts with a yield strength f_{yb} higher than 300 N/mm^2

8.5.3 Anchor Bolts, Nuts and Washers

8.5.3.1 Anchor Bolts

For most ordinary structures, ASTM A307 headed bolts, ASTM A36/A36M rods, or ASTM F1554 grade 36 rods should be specified for ordinary strength anchor bolts. Applications requiring high-strength materials should use ASTM A193/A193M grade B7 or ASTM F1554 grade 105 anchor bolts and rods.

Anchor bolt materials and properties are shown in Table 8.6, while the anchor bolt nut is presented in Table 8.7.

As per EN1993.1.8, the following materials may be used for anchor bolts:

- Steel grades conforming to 1.2.1 Reference Standards: Group 1;
- Steel grades conforming to 1.2.4 Reference Standards: Group 4;
- Steel grades used for reinforcing bars conforming to EN 10080;

provided that the nominal yield strength does not exceed 640 N/mm^2 when the anchor bolts are required to act in shear and not more than 900 N/mm^2 otherwise.

8.5.3.2 Washers

Because base plates typically have holes larger than oversized holes to allow for tolerances in the location of the anchor rod, washers are usually made from ASTM A36 steel plate. They may be round, square, or rectangular and generally have holes 2 mm larger than the anchor rod diameter. The thickness must be appropriate for the forces to be transferred.

TABLE 8.6
Anchor Bolt Materials and Properties

Anchor Material Type		F_y ksi (Mpa)	f_{ut} ksi (Mpa)	Gavlanize?	Ductile?
A307		Not clearly defined	60(410)	Yes	Yes
A 36 or f1554 grade 36		36(250)	58(400)	Yes	Yes
F1554 grade 105		105 (720)	125(860)	Yes	Yes
A 193 trade & 760	do ≤ 64 mm	105 (720)	125(860)	No	Yes
assessed on bolt	64 > do ≤ 102 mm	95 (660)	115(790)	No	Yes
diameter (d_b)	do > 102 mm	75(515)	100(690)	No	Yes

TABLE 8.7
Anchor Bolt Nut Specifications

Anchor Material Specification	Nut Material Specification
ASTM A307, GRADE A	ASTM A563, GR.A
ASTM A36/A36M	ASTM A563, GR.A
ASTM F1554, GR.36	ASTM A563, GR.A
ASTM A193/A193M, GR.B7	ASTM A563,DH HEAVY-HEX OR ASTM A194/A194M
ASTM F1554 GR.105	ASTM A563,DH HEAVY-HEX OR ASTM A194/A194M

Washers are not necessary for non-preloaded bolts in normal holes, unless for inclined connection surfaces at inclination angles larger than 20 or 30 degrees for bolts larger or correspondingly smaller than 20 mm (taper washers). For preloaded 8.8 or 10.9 bolts, one or two washers are needed for better load distribution in the connection elements. Washers are also required for oversized round holes, long slotted holes, and single lap joints with only one bolt or one row of bolts.

8.5.3.3 Sleeves

The sleeves are very important tools to solve small movement issues of anchor bolts during the installation of machine skids after pouring concrete. Therefore, sleeves are recommended for use when minor movement of the anchor bolt is desired after the bolt is set in concrete. There are two types of sleeves commonly used with anchors: a partial sleeve, which is primarily used for alignment requirements, and a full sleeve, which is used for both alignment and pre-tensioning.

It is worth mentioning that sleeves do not affect the tensile capacity of a headed anchor because the tension in the anchor is transferred to the concrete through the head and does not rely on the bond between the anchor and surrounding concrete. Sleeved anchors can only resist shear forces when the sleeve is filled with grout.

A partial length sleeve is used where precise alignment of anchors is required during equipment installation or where bolt groups or patterns involve six or more anchor bolts with interdependent dimensional requirements. In this situation, the sleeve should be filled with grout after installation is complete. Partial length sleeves should not be used for base plates on structural steel columns that have oversized holes.

Full-length anchor bolt sleeves are used when anchor bolts will be pre-tensioned to maintain the bolt under continuous tensile stresses during load reversal.

In cases where pre-tensioning of anchor bolts is required, the bolt must be unbonded over a well-defined length that is stretched. When sleeves are used for pre-tensioned bolts, the tops of these sleeves should be sealed, and the sleeves should be filled with elastomeric material to prevent grout or water from entering.

In some cases, for vessels with significant height and exposure to wind effects, the forces on the anchor bolt will be high. Thus, the friction between the anchor bolt and concrete may not solely manage the force on the anchor bolt, so it is recommended to attach a plate connected to the lower point of the anchor bolt, as shown in Figure 8.26.

8.5.4 Coatings and Corrosion

Corrosion is the main factor affecting anchor bolt materials. It is important to define the elements that will be in contact with the anchor bolts and the environmental conditions surrounding them, such as the water table, humidity, saltwater, and caustic gases.

Galvanizing is a common option for ASTM A307 bolts and ASTM A36/A36M and ASTM F1554 threaded rods. Hot-dip galvanizing is not recommended for ASTM A193/A193M grade B7 anchor bolts due to the possibility of hydrogen embrittlement that can occur from the hot-dip galvanizing process.

In practical design, the anchor should consider at least 3mm as a corrosion allowance in the anchor bolt diameter if it is located in highly corrosive environments, such as buildings near coastal areas or where they are periodically washed down, as in the case of sphere tank legs, due to regular examination of the firefighting system.

Pedestal design and anchor arrangement should account for water collection and the surrounding environment to reduce contact with corrosive substances or wash-down water. Therefore, the tops of pedestals should usually be at least 150 mm above the finished paving surface.

Painting the anchor bolt will protect it, but may require more maintenance. ASTM A193/A193M anchors are required to be painted. If the engineer determines that prolonged contact with a corrosive substance is unavoidable, a metallurgist consultant should recommend alternative anchor materials or protective options; the precautions will be discussed in Chapter 9.

8.5.5 Anchor Design Considerations

If the base plate is designed with oversized bolt holes and there is a shear load in excess of what can be transmitted through friction, it is recommended that

a. either shear lugs be used or
b. mechanism to transfer the load from the base plate to the bolt without slippage is incorporated (such as welding washers in place)

If no tensile force is effectively applied to the anchors, they need not be designed for tension. Where the tensile force is adequately transferred to properly designed rebar, there is no requirement to check for the concrete breakout strength of the anchor or anchors in tension.

If the anchors are welded to the attachment to distribute the force to all fasteners, the strength may be based on the strength of the row of anchors furthest from the edge.

8.5.6 Pre-tensioning

Pre-tensioning of anchor bolts can be used to effectively fix the base plate to the foundation. The primary application for pre-tensioning is for large vibrating equipment, tall process columns, and cantilever stacks. Pre-tensioning anchor bolts will virtually eliminate fatigue stresses in the anchor bolts and is also effective in mitigating shear forces on the bolts. Proper anchor bolt detailing, careful installation procedures, and field quality control are essential to ensure that the anchor bolts are successfully tensioned.

The advantages of pre-tensioned anchor bolts include the prevention of stress reversals on anchors susceptible to fatigue weakening, increased dampening for pulsating or vibrating equipment, reduced drift for process towers under wind or seismic loads, and enhanced frictional shear resistance for process towers and other equipment.

The disadvantage of this type of anchor bolt is that it can be a costly process to install accurately, and there is no recognized code authority that provides guidance on the design and installation of pre-tensioned anchors. There is little research in this area, but it depends on the requirements of the manufacturer.

TABLE 8.8
Pre-tension Load and Torque Recommendation[a]

Nominal Bolt Diameter (mm)	Torque (N·m)	Pre-tension Load (kg)
M16	60	3,311
M24	100	7,447
M30	160	18,247
M52	245	37,136

[a] Based on 30 ksi bolt internal stress.

Pre-tensioning should always be carried out when recommended by the equipment manufacturer. The manufacturer's instructions should be followed carefully.

When not otherwise specified, anchors for turbines and reciprocating compressors should be torqued to the values shown in Table 8.8. When pre-tensioning is required, high-strength anchors should be used. Pre-tensioned anchor bolts should also be considered for tall process columns or cantilever stacks, which can develop significant bolt tension under wind or seismic loads.

8.6 DESIGN CALCULATION BASED ON AISC

8.6.1 Tensile and Shear Strength of Bolts and Threaded Parts

The design tensile or shear strength, ϕR_n, with ϕ equal to 0.75 for a snug-tightened or pre-tensioned high-strength bolt or threaded part, shall be determined according to the limit states of tension rupture and shear rupture as follows:

$$R_n = F_n A_b$$

Considering that A_b is the nominal unthreaded body area of the bolt or threaded part, Fn is the nominal tensile stress, F_{nt}, or shear stress, F_{nv}, are presented in this table.

Description of Fasteners	Nominal Tensile Stress, F_{nt}, ksi (MPa)[a][b]	Nominal Shear Stress in Bearing-Type Connections, F_{nv}, ksi (MPa)[c]	
		Threads Not Excluded from Shear Planes—(N)[e]	Threads Excluded from Shear Planes—(X)
A307 bolts	45 (310)	27 (190)[d]	27 (190)[d]
Group 120 (e.g., A325)	90 (620)	54 (370)	68 (470)
Group 144 (e.g., F3148)	108 (750)	65 (450)	81 (560)
Group 150 (e.g., A490)	113 (780)	68 (470)	84 (580)
Group 200 (e.g., F3043)	150 (1 000)	90 (620)[f]	113 (780)[f]
Threaded parts meeting the requirements of Section A3.4	$0.75F_u$	$0.450F_u$	$0.563F_u$

8.6.2 Combined Tension and Shear in Bearing-Type Connections

The available tensile strength of a bolt subjected to combined tension and shear shall be determined according to the limit states of tension and shear rupture as follows:

$$R_n = (1.3\ F_{nt} - (F_{nt}/fF_{nv})\ f_{rv})\ A_b$$

f_{rv} is the required shear stress.

8.6.3 Bearing and Tear-out Strength for Bolt Holes

In calculating the strength of bearing and tear-out, the reduction factor (ϕ) is 0.75.

8.6.3.1 Snug-Tightened or Pre-Tensioned High-Strength Bolt Connection

1. The nominal strength of a connected element at a bolt in a connection with standard, oversized, and short-slotted holes, independent of the direction of loading, or a long-slotted hole with the slot parallel to the direction of the bearing force, shall be the lesser of the following:
   A. If deformation at the bolt hole at service load is a design consideration
      - Bearing strength:

$$R_n = 2.4\ d\ t\ F_u$$

      - Tear-out strength:

$$R_n = 1.2\ l_c\ t\ F_u$$

   B. If deformation at the bolt hole at service load is not a design consideration
      - Bearing strength:

$$R_n = 3.0\ d\ t\ F_u$$

      - Tear-out strength:

$$R_n = 1.5\ l_c\ t\ F_u$$

2. The nominal strength of a connected element at a bolt in a connection with long slotted holes, where the slot is perpendicular to the direction of force, is the lesser of the following:
   - Bearing strength:

$$R_n = 2.0\ d\ t\ F_u$$

   - Tear-out strength:

$$R_n = 1.0\ l_c\ t\ F_u$$

F_u is the specified minimum tensile strength of the connected material, d is the nominal diameter of fastener, l_c is the clear distance, in the direction of the force, between the edge of the hole and the edge of the adjacent hole or the edge of the material, and t is the thickness of connected material.

8.6.4 High-Strength Bolts in Slip-Critical Connections

Slip-critical connections must be designed to prevent slip and meet the limit states required for bearing-type connections. When slip-critical bolts pass through fillers, all surfaces prone to slip must be properly prepared to achieve the specified slip resistance for design purposes. The slip resistance of a single available bolt for the slip limit state shall be determined as follows:

$$R_n = \mu\, D_u\, h_f\, T_b\, n_s$$

The strength reduction factor, ϕ, is equal to 1.0 for standard size and short-slotted holes perpendicular to the direction of the load, 0.85 for oversized and short-slotted holes parallel to the direction of the load, and 0.75 for long-slotted holes.

$$D_u = 1.13$$

T_b is the minimum fastener pre-tension, as presented in Table 8.9.

TABLE 8.9
Minimum Blot Pre-Tension as per AISC

Bolt Size	Group 120[a]	Group 144[b] and Group 150	Group 200[c]
In		**Kips**	
½	12	15	
5/8	19	24	
¾	28	35	
7/8	39	49	
1	51	64	90
1–1/8	64	8	113
1–1/4	81	102	143
1–3/8	97	121	
1–1/2	118	148	
Mm		**kN**	
M12	49	71	
M16	91	114	
M20	142	179	
M22	176	221	
M24	205	257	
M27	267	335	
M30	326	408	
M36	475	595	

[a] Equal to 0.70 times the minimum tensile strength of bolts, as specified in ASTM F3125/F3125M for Grade A325 rounded off.

[b] Equal to 0.70 times the minimum tensile strength of bolts as specified in ASTM F3125/F3125M for Grade A490 rounded off. Group 144 (F3148) assemblies have the same specified minimum pre-tension as Group 150.

[c] Equal to 0.70 times the minimum tensile strength of bolts as specified in ASTM F3043 and F3111 for Grade 2 rounded off.

H_f is the fillers factor, equal to 1.0 for one filler between connected parts and 0.85 for two and more fillers. n_s is the number of slip planes required to permit the connection to slip. The average slip coefficient, μ, has two values based on two classes, A and B. In the case of Class A, where the surfaces are unpainted clean mill scale steel or surfaces with Class A coatings on blast-cleaned steel or hot-dipped galvanized steel (whether as-galvanized or hand roughened), μ is equal to 0.3. For Class B, which includes unpainted blast-cleaned steel surfaces or surfaces with Class B coatings on blast-cleaned steel, μ is equal to 0.5.

Considering that an applied axial tension force results in shear, this tension force will reduce the net clamping force. Therefore, the above equation must be adjusted by multiplying it by the factor k_{sc}, which is calculated as follows:

$$K_{sc} = 1 - \left(T_u / D_u \ T_b \ n_b\right)$$

where T_u is the applied tension force and n_b is the number of bolts carrying the tension force.

8.7 BASE CONNECTION AS PER AISC

When checking the bearing strength of the concrete, there are two options: full concrete support area or partial concrete support area. The general equation is as follows:

$$P_p = 0.85 f_c' A_1 \left(A_2 / A_1 \right)^{0.5}$$

where A_1 is the area of steel bearing concentrically on the concrete support, A_2 is the maximum area of the portion of the supporting surface that is geometrically similar to and concentric with the loaded area, and f_c' is the concrete specified compressive strength.

The bearing strength reduction factor, ϕ, is 0.65. The anchor bolt design will follow ACI 318.

8.8 CHECK WEB AND FLANGE IN CASE OF CONCENTRATED FORCES

It is required to check the flange and web, which are subjected to concentrated loads, to determine if there is a need to add doubler plates or stiffeners.

8.8.1 Check Flange

- For the flange, the strength calculation will be performed as follows, considering that the strength reduction factor is

$$R_n = 6.25 F_{yf} \left(t_f\right)^2$$

8.8.2 Check Web Yielding

- In the case of web yielding strength, it will be calculated using the following equation, considering that the strength reduction factor is 1.0:
  a. When the concentrated force to be resisted is applied at a distance from the member end greater than the full nominal depth of the member, d

$$R_n = F_{yw}t_w\left(5k + l_b\right)$$

  b. When the concentrated force to be resisted is applied at a distance from the member end that is less than or equal to the full nominal depth of the member, d

$$R_n = F_{yw}t_w\left(2.5k + l_b\right)$$

where F_{yw} is the specified minimum yield stress of the web material, k is the distance from the outer face of the flange to the web toe of the fillet, l_b is the length of bearing, and t_w is the thickness of the web.

8.8.3 Check Web Local Crippling

The available strength for the limit state of web local crippling shall be determined as follows, considering that ϕ is equal to 0.75.

The nominal strength will be calculated in two cases: in case one, if the concentrated compressive force to be resisted is applied at a distance from the member end that is greater than or equal to $d/2$.

$$R_n = 0.80t_w^2\left[1+3\left(\frac{l_b}{d}\right)\left(\frac{t_w}{t_f}\right)^{1.5}\right]\sqrt{\frac{EF_{yw}t_f}{t_w}}Q_f$$

In case two, if the concentrated compressive force to be resisted is applied at a distance from the member end that is less than $d/2$.

- For $l_b/d \leq 0.2$,

$$R_n = 0.40t_w^2\left[1+3\left(\frac{l_b}{d}\right)\left(\frac{t_w}{t_f}\right)^{1.5}\right]\sqrt{\frac{EF_{yw}t_f}{t_w}}Q_f$$

- For $l_b/d > 0.2$

$$R_n = 0.40t_w^2\left[1+\left(\frac{4l_b}{d}-0.2\right)\left(\frac{t_w}{t_f}\right)^{1.5}\right]\sqrt{\frac{EF_{yw}t_f}{t_w}}Q_f$$

where Q_f is the chord–stress interaction parameter, equal to 1.0 for wide-flange sections, channels, box sections, and for HSS (connecting surface) in tension, and d is the full nominal depth of the member, in. (mm).

8.8.4 Web Compression Buckling

If a pair of compressive single-concentrated forces or the compressive components in a pair of double-concentrated forces are applied at both flanges of a member at the same location, the available strength for the limit state of web compression buckling shall be determined as follows, considering that ϕ is equal to 0.90:

$$R_n = \left(\frac{24t_w^3\sqrt{EF_{yw}}}{h}\right)Q_f$$

8.8.5 Web Panel-Zone Shear

In the case of double-concentrated forces applied to one or both flanges of a member at the same location, the available strength of the web panel-zone for the limit state of shear yielding shall be determined as follows, considering that the strength reduction factor is 0.90.

1. If the effect of inelastic panel-zone deformation on frame stability is not accounted for in the analysis:
   a. $P_r \leq 0.4P_y$
      $R_n = 0.6\,F_y\,d_c\,t_w$
   b. $P_r > 0.4P_y$
      $R_n = 0.6\,F_y\,d_c\,t_w\,(1.4 - (P_r/P_y))$
2. If the effect of inelastic panel-zone deformation on frame stability is accounted for in the analysis.
   a. $P_r \leq 0.75P_y$
      $R_n = 0.6\,F_y\,d_c\,t_w\,(1 + 3b_{cf}\,(t_{cf})^2/d_b\,d_c\,t_w)$
   b. $P_r > 0.75P_y$
      $R_n = 0.6\,F_y\,d_c\,t_w\,(1 + 3b_{cf}\,(t_{cf})^2/d_b\,d_c\,t_w)(1.9 - (1.2P_r/P_y))$

Example 8.5: Bracing Connection Design as per AISC

The normal factored action force, $P_x = 80$ KN, and the bracing consists of two equal angle sections of 100 ×100×10 mm.

Solution:

This calculation was performed using an Excel sheet. You can contact me or download it from the website www.elreedyma.com

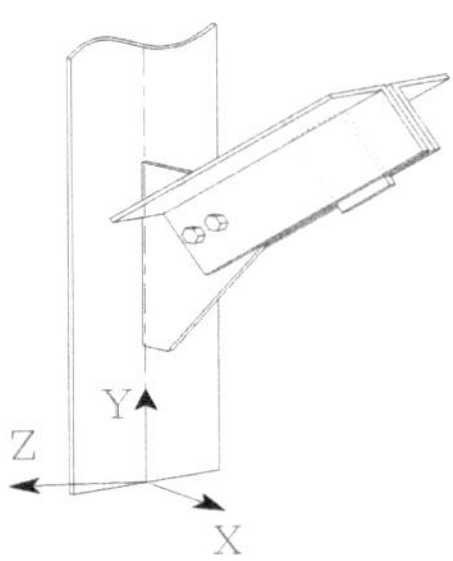

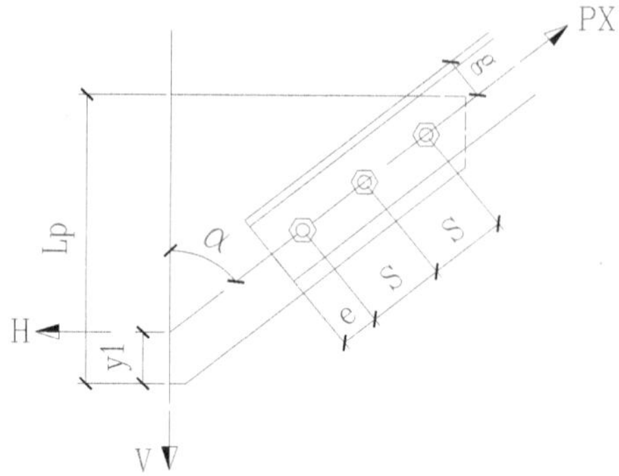

Section properties:

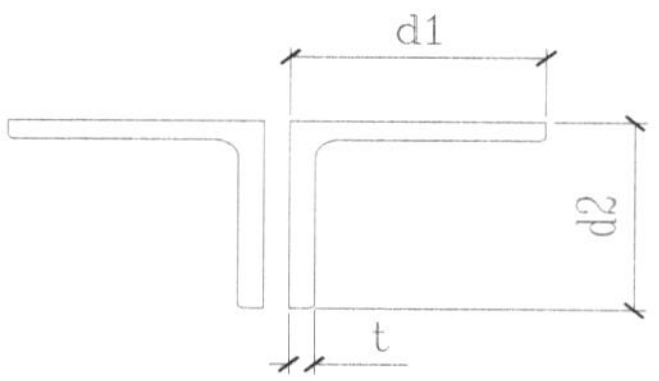

$d_1 = 100$ mm
$d_2 = 100$ mm
$t = 10$ mm
$Ag = 1{,}920$ mm^2

The steel structural member and the connection plates are of Grade S275.

$F_u = 430$ N/mm^2 ultimate strength
$F_y = 275$ N/mm^2 yield strength

- Connection bolts

Choose bolt grade ASTM A325 Grade 5.6

$d = 20$ mm	bolt diameter
$d_h = 22$ mm	hole diameter
A_b gross $= 314$ mm^2	bolt gross area
A_b net $= 245$ mm^2	bolt net area
$f_{nv} = 278$ N/mm^2	shear strength
$f_{nt} = 465$ N/mm^2	tensile strength
$f_p = 95.04$ kN	max. bearing force
$f_{tt} = 135$ kN	max. tear-out strength

- Geometry

$e = 35$ mm	edge distance
$S = 70$ mm	spacing between bolts
$g = 55$ mm	gauge of bolts
$N_b = 2$	number of bolts
$t_p = 12$ mm	gusset plate thickness

$L_p = 200\,mm$	gusset plate length
$Y_1 = 0$	
$L_1 = 120\,mm$	buckling length
$K = 0.5$	effective length factor
$\alpha = 30°$	
$S_w = 8\,mm$	weld size for gusset plate

- Check stress for bolts

Bolt shear strength

$Fnv = f_{nv} \times A_b$ net
$F_{nv} = 278 \times 245/1{,}000 = 68.11$ kN
F_{rv}(act) actual shear force per bolt $= P_x/N_b$
F_{rv}(act) = 40 KN $< F_{nv}$ *Ok safe*

- Check for brace
  1. Stress on effective are
     A_{net}, net area for brace section= $2(A_g - txd_h)$
     $A_{net} = 3{,}400\,mm^2$
     $F_{nt} = f_{nt}\,A_{net} = 465 \times 3{,}400/1{,}000 = 1{,}581$ kN
     $> P_x$ *ok Safe*

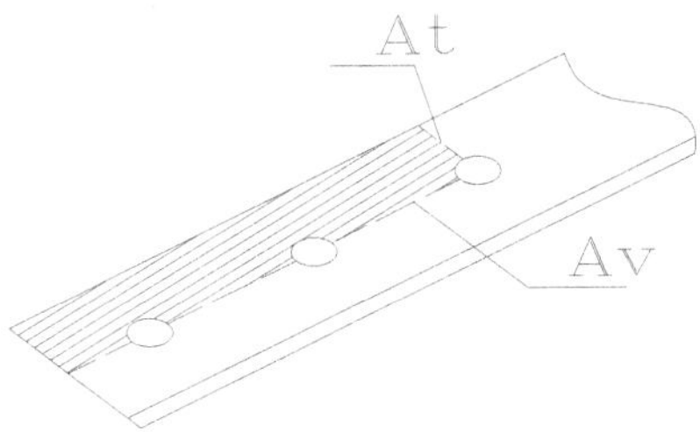

  2. Block Shear Capacity
     Shear area $A_{nv} = 2 \times t \times [(e + (N_b - 1) \times s - (N_b - 0.5) \times d_h]$
     $A_{nv} = 1{,}440\,mm^2$
     Tension area $A_{nt} = 2 \times t \times [d_1 - g - 0.5 \times d_h]$
     $A_{nt} = 680\,mm^2$
     $f_{rv} = 40 \times 1{,}000/1{,}440 = 27.8$ N/mm²
     $R_n = (1.3\ F_{nt} - (F_{nt}/\phi\ F_{nv})\ f_{rv})\ A_{bnet}$
     $R_n = 0.75 \times 245(1.3 \times 465 - ((465/(0.75 \times 278) = (27.8)) = 99{,}684/100 =$
     99.7kN >80 kN ok safe

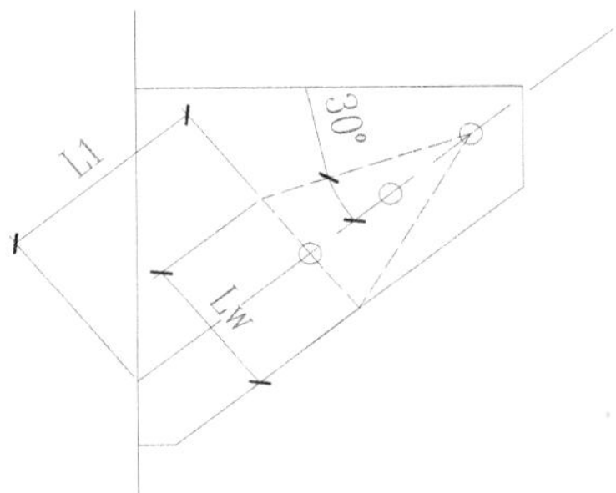

3. Stress on plate section at contact with column
   ecc = Eccentricity is the distance from the point where the axial force is acting to the centre of the plate length, L_p centre
   $H = P_x \sin \alpha$
   $V = P_x \cos \alpha$
   $M = ecc \times H$
4. Stress on plate section at contact with column
   ecc, Distance from the point where the axial force is acting to the plate length L_p centre
   $H = P_x \sin \alpha$
   $V = P_x \cos \alpha$
   $M = ecc \times H$

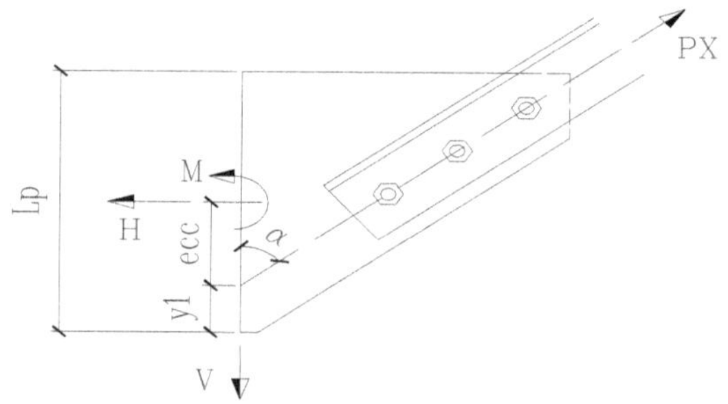

f_t(act), actual tensile stress $= H/(L_p \times t_p) + M/(t_p \times Lp^2/6)$
$f_t(act) = 67\ N/mm^2 > \phi\ F_y$ Ok safe

5. Block shear capacity
   Shear area $A_{nv} = t_p \times [(e + (N_b - 1) \times s - (N_b - 0.5) \times d_h]$
   $A_v = 864\ mm^2$
   Tension area $A_{nt} = t_p \times [d_1 - g - 0.5 \times d_h]$
   $A_{nt} = 408\ mm^2$

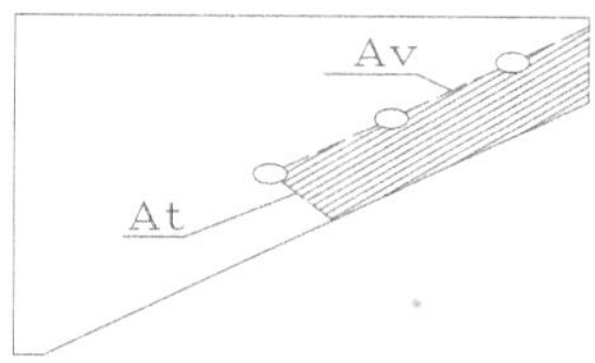

$\phi R_{nt} = 0.9 \times 275 \times 408/1{,}000 = 100.98$ kN $> P_x$ ok safe
$\phi R_v = 1.0 \times 0.6 \times 430 \times 864 = 222.9$kN $> P_x$ Ok safe

- Check for welds
  We will use Electrode E70
  $\phi R_{nw} = 0.8 \times 0.6 \times 483 = 231.84\ N/mm^2$
  weld stress components
  $f_1 = H/(2 \times Lp \times sw) + M/2 \times (sw \times Lp^2/6)$
  $f_1 = 50\ N/mm^2$
  $f_2 = V/(2 \times L_p \times sw)$
  $f_2 = 21.7\ N/mm^2$

$$f_r = \sqrt{f_1^2 + f_2^2}$$
$$f_r = 54.5 \text{ N/mm}^2$$
$$f_{vwd} > f_r \quad \text{OK SAFE}$$

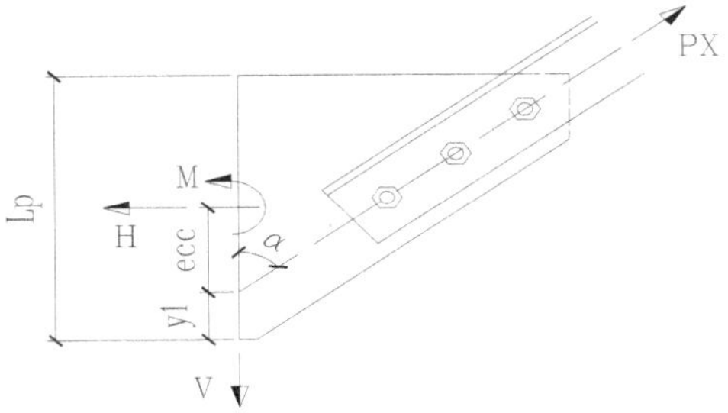

REFERENCES

Kulak, G.L., Fisher, J.W., and Struik, J.H.A. (1987). *Guide to Design Criteria for Bolted and Rivetted Joints*, 2nd edition, John Wiley and Sons, New York.

Lindner, J. (2008). Old and New Solutions for Contact Splices in Columns. *Journal of Constructional Steel Research*, Vol. 64, No. 7–8, pp. 833–844.

Trahair, N.S., Bradford, M.A., Nethercot, D.A., and Gardner, L. (2008). *The Behaviour and Design of Steel Structures to EC3*, 4th edition, Taylor Francis, USA.

Vayas, I., Ermopoulos, J., and Ioannidis, G. (2019). *Design of Steel Structures to Eurocodes*, Springer, Cham.

INDUSTRY CODES AND STANDARDS

AASHTO (n.d.). *AASHTO Standard Specifications for Highway Bridges*, American Association of State Highway and Transportation Officials (AASHTO), USA.

ACI (n.d.a). *Building Code Requirements for Masonry Structures*, ACI 530/ASCE 5, American Concrete Institute (ACI), USA.

ACI (n.d.b). *Building Code Requirements for Structural Concrete and Commentary*, ACI 318/318R, American Concrete Institute (ACI), USA.

ACI (n.d.c). *Environmental Engineering Concrete Structures*, ACI 350R, American Concrete Institute (ACI).

AISC (n.d.a). *AISC Manual of Steel Construction - Allowable Stress Design (ASD)*, American Institute of Steel Construction (AISC), USA.

AISC (n.d.b). *AISC Manual of Steel Construction - Load and Resistance Factor Design (LRFD)*, American Institute of Steel Construction (AISC).

AISC (n.d.c). *Specification for Structural Joints Using ASTM A325 or A490 Bolts*, American Institute of Steel Construction (AISC), USA.

AISI (n.d.a). *Commentary on the Load and Resistance Factor Design Specification for Cold-Formed Steel Structural Members*, AISI SG 913, Part II, American Iron and Steel Institute (AISI).

AISI (n.d.b). *Commentary on the Specification for the Design for Cold-Formed Steel Structural Members*, AISI SG 673, Part II, American Iron and Steel Institute (AISI).

AISI (n.d.c). *Load and Resistance Factor Design Specification for Cold-Formed Steel Structural Members*, AISI SG 913, Part I, American Iron and Steel Institute (AISI).

AISI (n.d.d). *Seismic Provisions for Structural Steel Buildings*, ANSI/AISC 341-02, American Iron and Steel Institute (AISI).

AISI (n.d.e). *Specification for the Design for Cold-Formed Steel Structural Members,* AISI SG 673, Part I, American Iron and Steel Institute (AISI).

American Forest and Paper Association (n.d.a). *National Design Specification for Wood Construction (NDS),* American Forest and Paper Association.

American Forest and Paper Association (n.d.b). *NDS Supplement* - Design Values for Wood Construction, American Forest and Paper Association.

API (n.d.). *Welded Steel Tanks for Oil Storage*, API Standard 650. American Petroleum Institute (API).

ASCE (n.d.a). *ASCE Design of Blast Resistant Buildings in Petrochemical Facilities*, American Society of Civil Engineers (ASCE).

ASCE (n.d.b). *ASCE Guidelines for Seismic Evaluation and Design of Petrochemical Facilities*, American Society of Civil Engineers (ASCE).

ASCE (n.d.c). *ASCE Guidelines for Wind Loads and Anchor Bolt Design for Petrochemical Facilities*, American Society of Civil Engineers (ASCE).

ASCE (n.d.d). *Design Loads on Structures during Construction*, SEI/ ASCE 37-02, American Society of Civil Engineers (ASCE).

ASCE (n.d.e). *Minimum Design Loads for Buildings and Other Structures*, SEI/ASCE 7-02, American Society of Civil Engineers (ASCE).

ASME (n.d.a). *Safety Code for Elevators and Escalators*, ASME A17.1, American Society of Mechanical Engineers (ASME).

ASTM (n.d.b). *Standard Specification for Alloy-Steel and Stainless Steel Bolting Materials for High-Temperature Service*, ASTM A193/A193M, American Society for Testing and Materials (ASTM).

ASTM (n.d.c). *Standard Specification for Anchor Bolts, Steel, 36, 55, and* 105 *ksi Yield Strength*, ASTM F1554, American Society for Testing and Materials (ASTM).

ASTM (n.d.d). *Standard Specification for Carbon Steel Bolts and Studs, 60,000 psi Tensile Strength*, ASTM A307, American Society for Testing and Materials (ASTM).

ASTM (n.d.e). *Standard Specification for Carbon Structural Steel*, ASTM A36/A36M, American Society for Testing and Materials (ASTM).

ASTM (n.d.f). *Standard Specification for Deformed and Plain Billet-Steel Bars for Concrete Reinforcement*, ASTM A615/A615M, American Society for Testing and Materials (ASTM).

ASTM (n.d.g). *Standard Specification for Heat-Treated Steel Structural Bolts, 150 ksi Minimum Tensile Strength*, ASTM 490/490M, American Society for Testing and Materials (ASTM).

ASTM (n.d.h). *Standard Specification for High-Strength Bolts for Structural Steel Joints [Metric]*, ASTM A325M, American Society for Testing and Materials (ASTM).

ASTM (n.d.i). *Standard Specification for Low-Alloy Steel Deformed Bars for Concrete Reinforcement*, ASTM A706/A706M, American Society for Testing and Materials (ASTM).

ASTM (n.d.j). *Standard Specification for Quenched and Tempered Alloy Steel Bolts, Studs, and Other Externally Threaded Fasteners*, ASTM A354, American Society for Testing and Materials (ASTM).

ASTM (n.d.k). *Standard Specification for Steel for Structural Shapes for Use in Building Framing*, ASTM A992/A992M, American Society for Testing and Materials (ASTM).

ASTM (n.d.l). *Standard Specification for Steel Welded Wire Fabric, Plain, for Concrete Reinforcement*, ASTM A185, American Society for Testing and Materials (ASTM).

ASTM (n.d.m). *Standard Specification for Steel Wire, Plain, for Concrete Reinforcement*, ASTM A82, American Society for Testing and Materials (ASTM).

ASTM (n.d.n). *Standard Specification for Structural Bolts, Steel, Heat Treated, 120/105 ksi Minimum Tensile Strength*, ASTM A325, American Society for Testing and Materials (ASTM).

AWS (n.d.). *Structural Welding Code – Steel*, AWS D1.1, American Welding Society (AWS).
CMAA (n.d.a). *Specifications for Top Running and Under Running Single Girder Overhead Electric Traveling Cranes Utilizing Under Running Trolley Hoist*, CMAA No. 74, Crane Manufacturers Association of America (CMAA).
CMAA (n.d.b). *Specifications for Top Running Bridge and Gantry Type Multiple Girder Overhead Electric Traveling Cranes*, CMAA No. 70, Crane Manufacturers Association of America (CMAA).
PCI (n.d.). *Design Handbook - Precast and Prestressed Concrete*, PCI MNL 120, Precast/Prestressed Concrete Institute (PCI).
SJI (n.d.). *SJI Standard Specifications and Load Tables,* Steel Joist Institute (SJI).
OSHA (n.d.a). *Occupational Safety and Health Standards*, OSHA 29 CFR 1910, U.S. Department of Labor, Occupational Safety and Health Administration (OSHA).
OSHA (n.d.b). *Safety and Health Regulations for Construction*, OSHA 29 CFR 1926, U.S. Department of Labor, Occupational Safety and Health Administration (OSHA).

FURTHER READING

CEN (2005). *Eurocode 3: Design of Steel Structures. Part 1.8.* Design of Joints, EN1993-1-8, European Committee for Standardization, CEN, BS, European Union.
CEN (2008). *Execution of Steel Structures and Aluminium Structures – Part 2: Technical Requirements for the Execution of Steel Structures*, EN 1090-2, CEN, BS, European Union.
Lui, E.M. (1999). Structural Steel Design, in *Structural Engineering Handbook,* Chen, W.-F., ed., CRC Press LLC, Boca Raton.
Piluso, V., Faella, C., and Rizzano G. (2001) Ultimate Behavior of Bolted T-Stubs. I: Theoretical Model, *Journal of Structural Engineering*, Vol. 127, No. 6, pp. 686–693.
PIP (2010). *Recommended Practices for Machinery Installation and Installation Design*, PIP REIE 686, Process Industry Practices, university of Texas, Austin.
Witteveen, J., Stark, J.W.B., Bijlaard, F.S.K., and Zoetemeijer, P. (1982). Welded and Bolted Beam-to-Column Connections, *Journal of the Structural Division*, Vol. 108, No. ST2, pp. 433–455.
Zoetemeijer, P. (1974). A Design Method for the Tension Side of Statically Loaded Bolted Beam-to-Column Connections, *Heron*, Vol. 20, No. 1, pp. 1–59.

9 Steel Structure Quality Control

9.1 INTRODUCTION

The quality management system is a crucial issue for the contractor, consultant, and owner to avoid wasting time and money, and to prevent any negative impact on the contractor's reputation.

Before starting the project, the contractor should submit the quality plan for the entire project and the Inspection Test Plan (ITP) for approval by the consultant or the owner's project team.

For the welding process, there are additional documents that should also be issued, such as welding procedure specifications (WPS). The WPS is a document that serves as a guide for the welding process during construction, and it must comply with all applicable code requirements and production standards. The WPS contains details necessary to create the desired weld, including information such as base metal grade, filler metal classification, amperage range, shielding gas composition, and pre-heat and interpass temperatures. The idea is that if a group of welders adhere to all the details in a WPS, they should each be able to produce welds with reasonably similar mechanical properties. The welding procedures will be illustrated in the following sections.

The welding procedure qualification report (WPQR) is considered the initial procedure used to create a series of welds that are then mechanically tested to ensure they meet all applicable requirements. This document outlines the procedures for creating and testing the sample welds, along with the final results. If the test results are acceptable, the WPQR is approved and can then serve as the foundation upon which one or more WPSs are drafted. In short, a PQR serves as evidence that a given WPS can, in fact, be used to produce an acceptable weld.

9.2 INSPECTION AND TESTING

All NDT procedures shall be submitted to the Inspection Department for review by the client or the consultant engineer hired by the client. The inspector should be ASNT Level III for acceptance prior to work commencing.

The quality control inspector has the right to inspect all materials and workmanship and shall have unrestricted access to the fabricator's shop at all times while work is being performed. The buyer may reject improper, inferior, defective, or unsuitable materials and workmanship. All materials and workmanship that are rejected shall be repaired or replaced by the fabricator as directed by the buyer.

The ITP shall be provided for review and approval by the company inspector.

DOI: 10.1201/9781003307303-9

Welding procedures and individual welders shall be qualified in accordance with the requirements of ANSI/AWS D1.1/D1.1M or ANSI/AWS D1.3, as appropriate. All welding procedures and welding performance qualification records shall be made available to the buyer's inspector for review.

Inspection of welding shall be performed in accordance with the Structural Welding Code, ANSI/AWS D1.1/D1.1M. Ultrasonic testing may be substituted for radiography if approved by the client inspector. In most cases, the following criteria should be considered as a guideline:

- All welds shall undergo 100% visual inspection performed per ANSI/AWS D1.1. In addition, any strikes, gouges, and other indications of careless workmanship (such as surface porosity) shall be removed by grinding.
- Pipes used as piling (circumferential welds) and tubular structures shall be randomly radiographed at a rate of 10% (one weld of every ten shall be 100% radiographed). The specified amount of random radiography shall include X-ray samples from each welder's daily production.
- The butt-welded flange sections for all primary load-bearing members shall be 100% radiographed at the weld. Primary load-bearing members shall be defined as the main frames and any members that are part of the lateral load-carrying system.
- Secondary members such as purlins, girts, or rafters that are not part of the main sway frame shall have the butt welds tested as follows: 10% of the butt welds from each day's production (randomly selected) shall undergo radiographic testing.
- All lifting lug connections shall be 100% radiographed. Procedures and operator qualifications shall meet ANSI/AWS D1.1 and be submitted to the Company Inspector for review and approval. For skid-mounted equipment with a total lift weight of less than 3,000 kg, it is acceptable to use fillet-welded lugs with dye penetrant inspection.
- The minimum percentage of coverage for the specified NDT method may be increased (at any level up to 100%) if, in the opinion of the Company Inspector, the welds are of questionable workmanship or if NDT indicates an excessive number of defects. Additional or alternative NDT methods may be used at the discretion of the Company Inspector to assist in determining the type or extent of defects.

Material test reports for each heat of structural steel and each lot of high-strength and ASTM A307 bolts shall be available for review by the quality control inspector. Material test reports shall conform to EN 10204, Type 3.1.

The client representative may require a number of representative samples of bolt assemblies, which the Fabricator shall supply to the buyer for testing. Testing in accordance with ASTM F606 or ASTM F606M will be at the buyer's expense.

If any damage is discovered, or if any parts, components, or documentation are missing or otherwise defective, the occurrence shall be immediately reported to the buyer in writing.

Shop inspection may include, but is not limited to, the following:

TABLE 9.1
Example for Check List for Steel Structure Alignment

"Owner Name" Test Report	Owner Doc. No.:	Date:	QC Number
Structural steel alignment report			Construction Mgr.:
Project title			Contractor/subcontractor
Layout drawing No.:		Rev. No.:	Floor No.:
Schedule test date and time:		Actual date and time:	
Structure:		Area:	

Structural steel alignment report

Item	Location Element	Horizontal Alignment	Vertical Alignment	Misalignment Vert. & Hor.

Contractor/Third Party		Owner
Technician perform test	Date	Representative
QC inspector	Date	Supervisor
QC supervisor	Date	Const. Mgr.:

Verification of the conformance of materials with this Specification and the drawings. The limits of acceptability and repair of surface imperfections for structural steel shall be in accordance with ASTM A6/ASTM A6M.

Inspection of high-strength bolted connections shall be in accordance with the AISC Specification for Structural Joints Using ASTM A325 or ASTM A490 Bolts.

The quality control inspector reserves the right to inspect and reject all galvanized steel in accordance with ASTM A123 and ASTM E376.

All tests and inspections should be conducted in accordance with a checklist included in the quality manual for the project. An example of a checklist for steel structure alignment is shown in Table 9.1, and the inspection for the foundation prior to steel erection is presented in Table 9.2.

9.3 CONNECTION

Connections are a crucial aspect of the steel structure, as they account for 45% to 50% of the project cost, from design and drafting through to the construction phase. The safety of the steel structure mainly depends on quality control during the design and drafting stages. Quality assurance measures must be implemented by any engineering firm delivering the steel structure design. A review of steel structure failures

TABLE 9.2
Example for Check List for Steel Structure Foundation

"Owner Name" Test Report	**Owner Doc. No.:**	**Date:**	**QC Number**
Structural steel alignment report			Construction Mgr.:
Project title			Contractor/subcontractor
Layout drawing No.:		Rev. No.:	Floor No.:
Schedule test date and time:		Actual date and time:	
Structure:		Area:	

Structural steel foundation

Item	Acceptance Criteria	Pass	Fail	Re-Insp Date
1	Before erecting steel, the erector shall check foundations and other connection points to confirm their location, orientation, elevation, and condition (readiness, anchor bolts, curing, clean, etc…)			
2	The top of concrete shall be a minimum of 150 mm above finished grade			
3	The design concrete compressive strength of concrete shall be 27.6 MPa (4,000 psi. at 28 days). Concrete shall be cured before release for structural installation.			
4	The minimum thickness of grout shall be 25 mm.			

Contractor/Third Party		Owner
Technician perform test	Date	Representative:
QC inspector:	Date	Supervisor
QC supervisor:	date	Const. Mgr.:

indicates that over 50% are attributed to issues with connection design, construction, or poor maintenance during the operational phase.

In steel structures, there are two main types of connections: bolted connections and welded connections.

9.3.1 Bolted Connection

Bolted connections are used in most steel structure projects, with the exception of offshore structures where they are not preferred (Figure 9.1).

Bolting has the following advantages:

- Bolting is generally a faster operation than welding

FIGURE 9.1 Fixing HSB by torque wrench.

TABLE 9.3
Properties and Strength of High Strength Bolt (Grade 10.9)

Bolt Diameter mm	Bolt Area (mm^2)	Stress Area (cm^2)	Penetration Force (ton)	Required Torque (kg.m)
M12	113	84	5.29	12
M16	201	157	9.89	31
M20	314	245	15.43	62
M22	380	303	19.08	84
M24	452	353	22.23	107
M27	573	459	28.91	157
M30	706	561	35.34	213
M36	1018	817	51.47	372

For HSB grade 8.8, the above values shall be reduced by 30%.

- Bolting does not have the temperature and weather condition requirements that are associated with welding
- Unexpected weather changes may delay welding operations

The properties of high-strength bolts are presented in Table 9.3.

9.3.2 Welding

Welding is commonly used in steel structures, noting that the welding design will be performed by the structural engineer, and its quality will be ensured during the quality control phase of engineering. Additionally, the drawings should be reviewed by the construction team for constructability.

In general, there are two types of welding from a design point of view: the fillet weld and the butt weld. The fillet weld should be designed by defining the welding thickness and length; on the other hand, the butt weld, in most cases, will not require specific design since the strength of the weld is equivalent to the strength of the original steel in the structural member. Therefore, if the steel member can carry the forces, the weld can carry those forces too. The butt weld will vary according to the groove prepared for the welding. For the fillet weld, two steel plates are simply joined, so there is no need for a groove, as will be discussed in the following sections.

The types of welding processes such as shielded metal arc welding (SMAW), gas metal arc welding (GMAW), also known as Metal Inert Gas MIG welding, flux cored arc welding, gas tungsten arc welding, which is also known as Tungsten Inert Gas TIG, and submerged arc welding (SAW), as well as plasma arc welding, are not typically used in normal steel structures but are commonly used in different application.

For steel structure projects, SMAW is most often used; however, for welding similar types of joints in mass production within a workshop, GMAW may be employed.

The primary code for welding in steel structures is AWS D1.1, which is structural welding code – teel, set by the American Welding Society. This code covers the design of welded connections, prequalification of welders, and the requirements for fabrication and inspection.

9.3.2.1 SMAW

The primary method of welding for the steel structure is SMAW , also known as stick welding. The process, depicted in Figure 7.5, involves creating an arc between the electrode and the steel workpiece as a result of the flow of electricity. This arc produces heat, which melts the base metal of the electrode. The electrode coating provides shielding during the welding process. After welding, a solidified weld metal is left covered by a layer of slag, formed from the converted flux. If the welding process is poor, this flux may remain on the weld and must be identified through non-destructive testing.

The SMAW electrode is designated by the symbol EXXXX E, where the first two or three digits refer to the minimum tensile strength of the weld metal in thousands of pounds per square inch. For instance, E70 indicates 70,000 psi (482 MPa). The next number indicates the positions in which the electrode can be used. These numbers are 1, 2, and 4; number 3 is no longer in use. Their meanings are as follows:

- "1" means that the electrode can be used in any position
- "2" indicates that the molten metal is fluid enough for the electrode to be used in the flat position for all welding types and in the horizontal position for fillet welds
- "4" means that the electrode is suitable for downhill welding positions.

The last number describes the composition of the coating on the electrode, corresponding to the type of electrical current: alternative current, direct current with positive electrode (DCEP), or direct current with negative electrode. Notably, electrodes ending with 5, 6, or 8 are classified as low-hydrogen types.

The equipment used in SMAW welding is shown in Figure 9.2.

When applying the SMAW technique for welding, quality control should consider that the following discontinuities may exist and must be kept in mind during inspection (Figure 9.3). It is the main responsibility of the welding inspector to avoid or

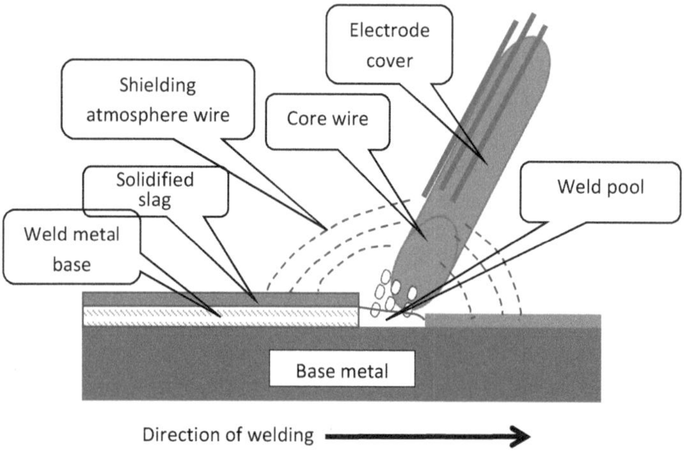

FIGURE 9.2 SMAW welding process.

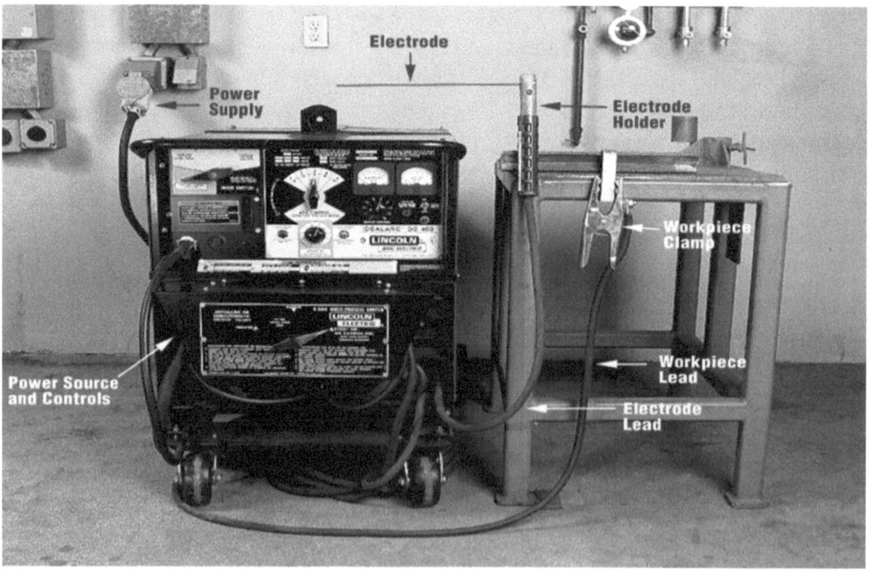

FIGURE 9.3 SMAW welding tools.

reject these discontinuities if they violate code allowances. The traditional discontinuities are as follows:

- Porosity
- Slag inclusions
- Spatter
- Incomplete joint penetration
- Incomplete fusion

9.3.2.2 GMAW

GMAW is primarily used in workshops as it needs special equipment and has some limitations. However, other methods are not typically used in steel structures.

The limitations of GMAW are as follows:

- Unsuitable for windy conditions
- Little tolerance for contamination
- Usually limited to shop welding
- Equipment is more complex
- Consumables – liners and contact tips

The discontinuities usually present in the GMAW welding process are as follows:

- Porosity
- Incomplete fusion
- Incomplete joint penetration

9.4 WELDING TYPES

In general, there are two types of welding used in steel structures: the fillet weld and the butt weld.

9.4.1 Fillet Weld

For the fillet weld, there are only two types: convex and concave fillet. It is important to note that the design of the weld, defined by the structural engineer, dictates the leg length. Therefore, maintaining the leg length as specified in the drawings is crucial for achieving the desired results.

Figures 9.4 and 9.5 present the fillet weld in convex and concave types, respectively, and the following terminology is illustrated in the drawings:

- Leg, size equal
- Convexity or concavity
- Actual throat
- Effective throat
- Theoretical throat

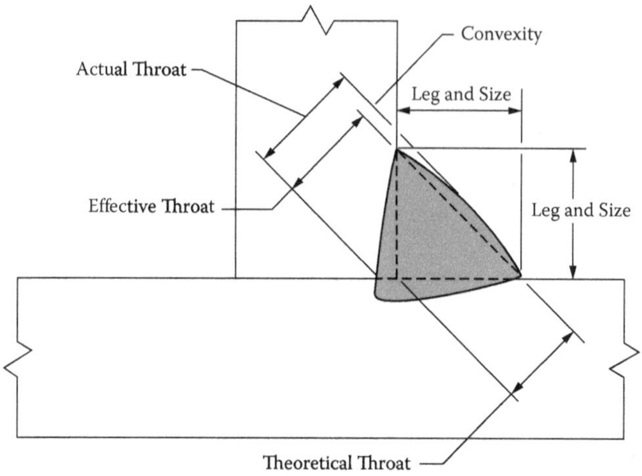

FIGURE 9.4 Fillet weld convex shape.

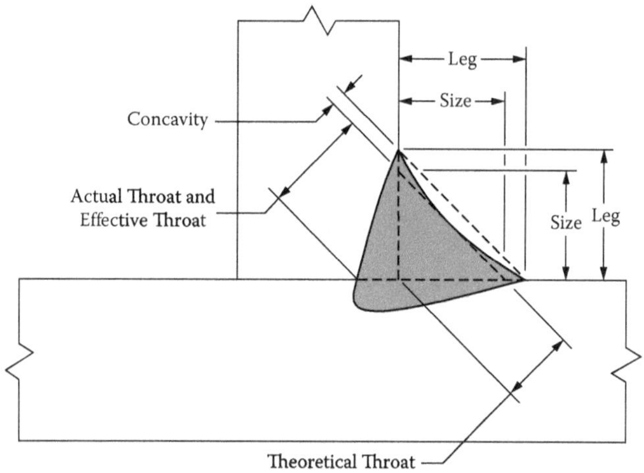

FIGURE 9.5 Fillet weld concave shape.

9.4.2 Butt Weld

9.4.2.1 Parts of Butt Weld

The butt weld consists of the following parts, with the terminology presented in Figure 9.6.

- Toe
- Face
- Face reinforcement
- Root reinforcement
- Root surface

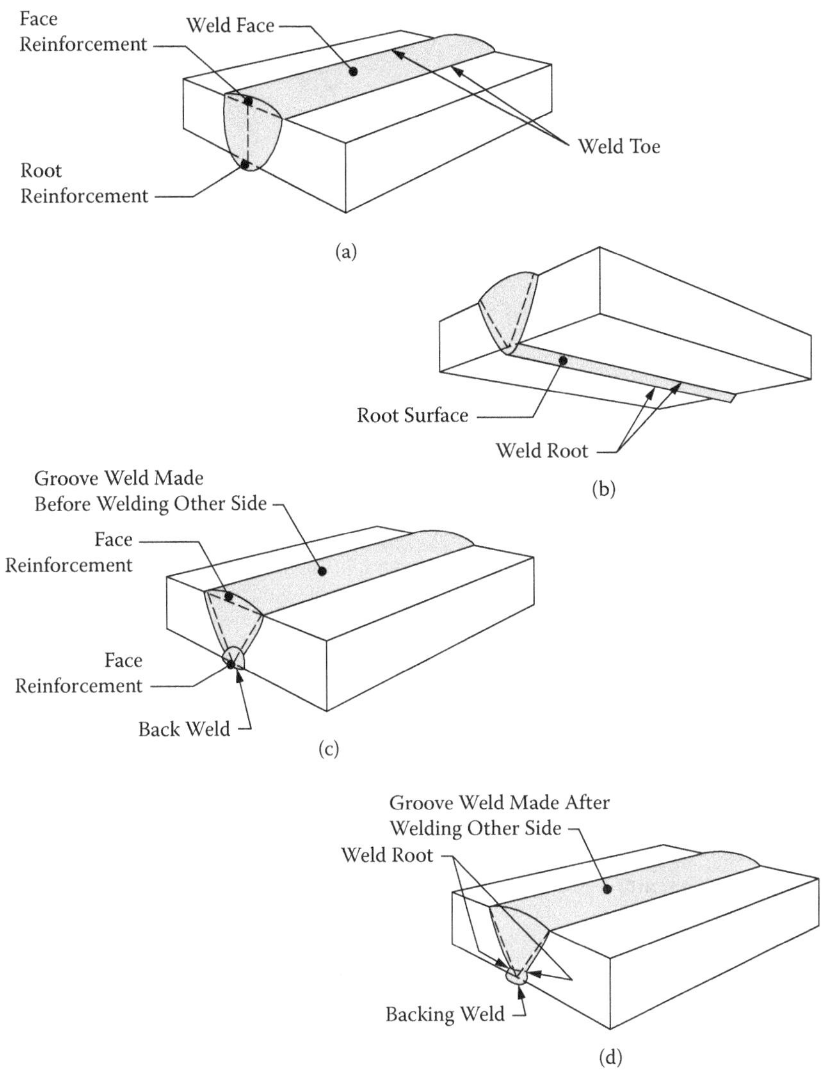

FIGURE 9.6 Welding terminology.

- Weld root
- Bead
- Back weld
- Backing weld
- Spacer
- Heat-affected zone (HAZ)
- Weld root
- Leg, face, toe
- Spacer
- Weld metal

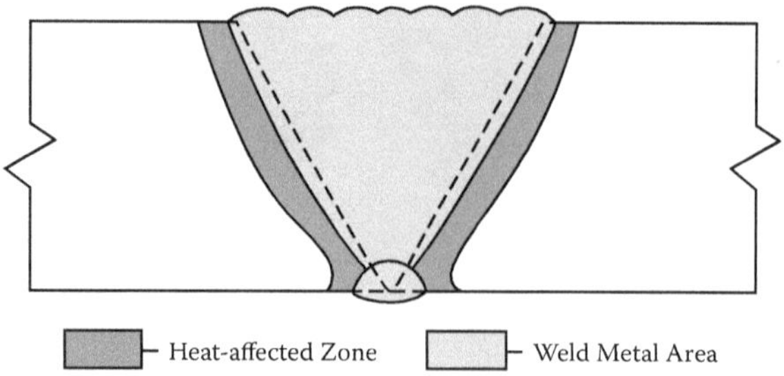

FIGURE 9.7 Heat-affected zone.

The HAZ is a critical and potentially dangerous part regarding the integrity of any welded steel member. The welding metal is depicted in the following Figure 9.7. The microscopic structure of the steel will change in this area due to the effect of heat. Consequently, corrosion will often initiate in this part.

9.4.2.2 Welding Groove Configurations

The following outlines the main configurations for welding plates. It is noted that different types of grooves each have distinct symbol notations, as shown below:

1. Square

   According to AISC, in this type of welding, if the base metal thickness is 0.25 in a maximum of 0.25 inches, the root opening is half the base metal thickness. The shape of the square groove is shown in Figure 9.8.
2. Single V-Groove

   This weld can be used for any thickness of base metal. The root opening depends on the groove angle: for angles of 45°, 30°, 20°, the corresponding root openings are 0.25, 3/8, and 0.5 in (6, 9.5, 13 mm), respectively (Figure 9.9).

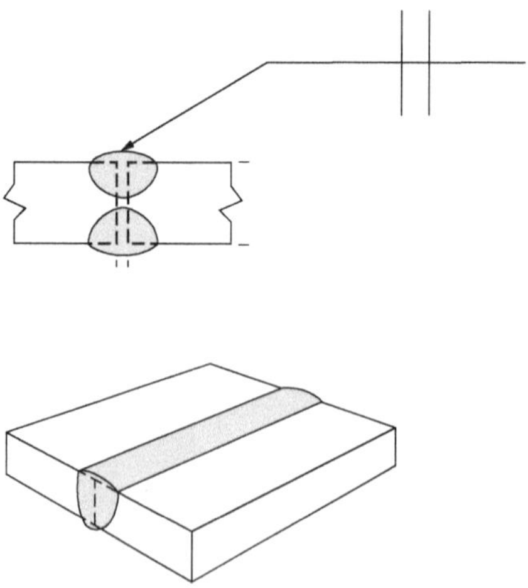

FIGURE 9.8 Square groove.

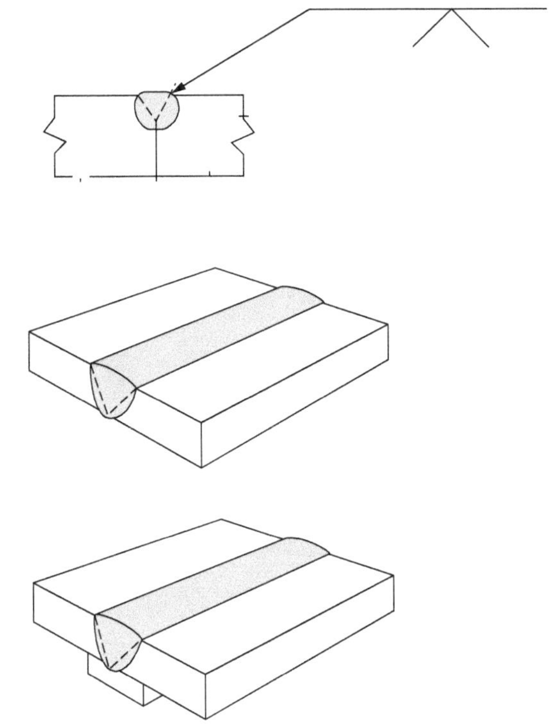

FIGURE 9.9 Single V-groove.

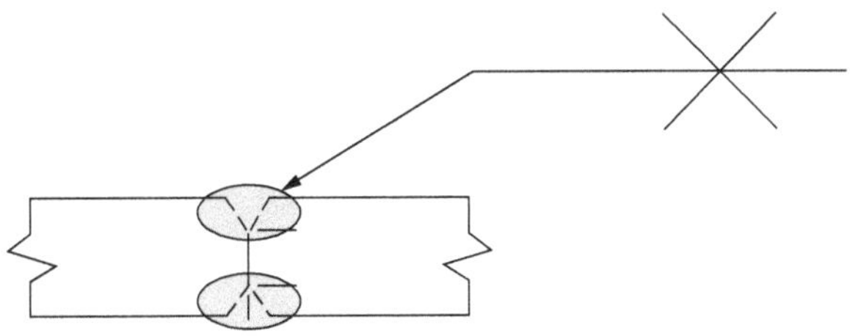

FIGURE 9.10 Double V-groove.

3. Double V-Grove

   The double groove has the same root opening with respect to the groove angle as discussed above (Figure 9.10).

4. Single Bevel Groove

   This type is applicable for all base metal thicknesses, with the root opening measuring 6 m and 9.5 mm (0.25 and 0.375 in) corresponding to groove angles of 45° and 30°, respectively (Figure 9.11).

5. Double Bevel Groove

   The double bevel groove and its welding symbol is shown in Figure 9.12.

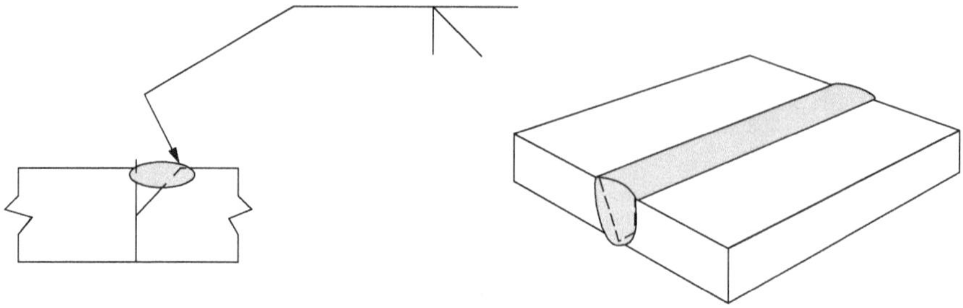

FIGURE 9.11 Single bevel groove.

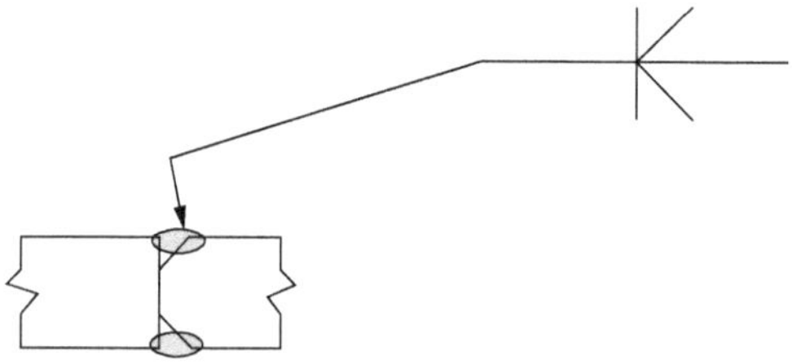

FIGURE 9.12 Double bevel groove.

6. Single U-Groove

   In the case of a U-joint, the root opening ranges from 0 to 3 mm (0 to 0.125 in), with groove angles typically being either 45° or 20°. This type of welding can be applied for all thicknesses of base metal, as shown in Figure 9.13.

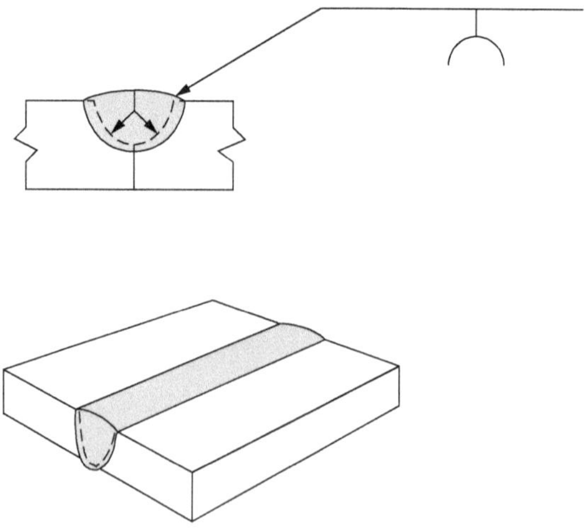

FIGURE 9.13 Single U-groove.

7. Double U-Groove

   In most cases, the groove angle is 45°, and this type can be applied for all metal thicknesses. With a root opening of 6 mm (0.25 in), the U groove is depicted in Figure 9.14.
8. Single J-Groove

   Both single and double J-grooves, shown in Figure 9.15, are suitable for any base metal thickness, with a root opening ranging from 0 to 0.125 in and a groove angle of 45°.
9. Double J-Groove

   The double J-groove is presented in Figure 9.16, along with its welding symbol.
10. Flare Bevel

    For the flare bevel groove, the root opening is zero; its symbol is shown in Figure 9.17.

In general, all the groove shapes are presented in Figure 9.18.

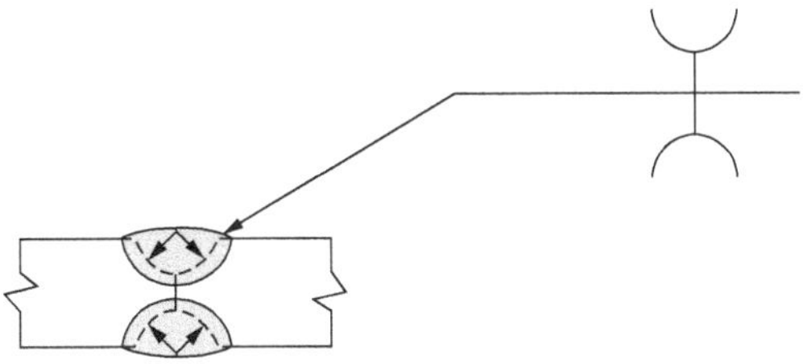

FIGURE 9.14 Double U-groove.

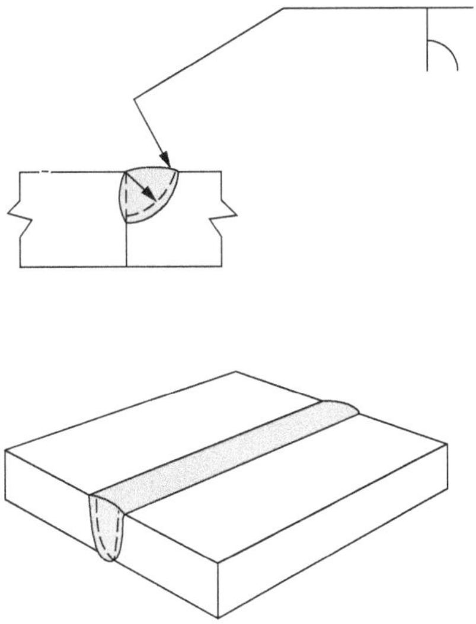

FIGURE 9.15 Single J-groove.

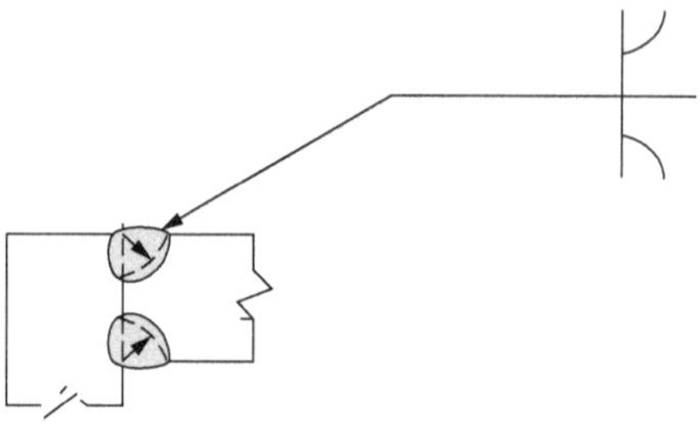

FIGURE 9.16 Double J-groove.

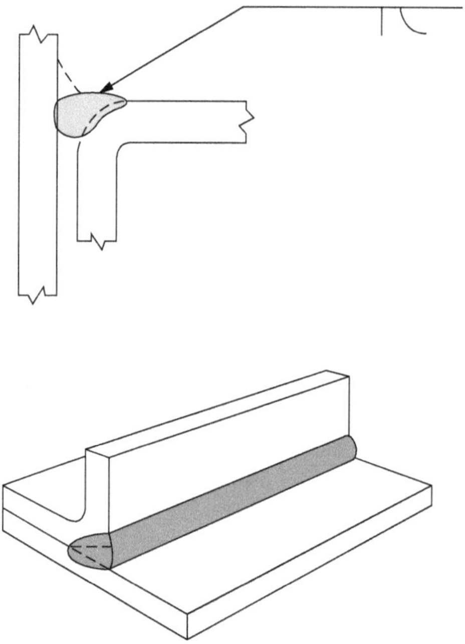

FIGURE 9.17 Flare bevel.

9.4.2.3 Connection between Steel Plates

There are different positions for welding steel plates, and each position has a specific type of welding groove suited to it.

A. Back to Back

Welding between two steel plates aligned approximately in the same plane is shown in Figure 9.19.

In this position, a fillet weld is typically applied, and the following are the main types of grooves that can be used in the welding:

- Bevel groove
- Flare bevel groove

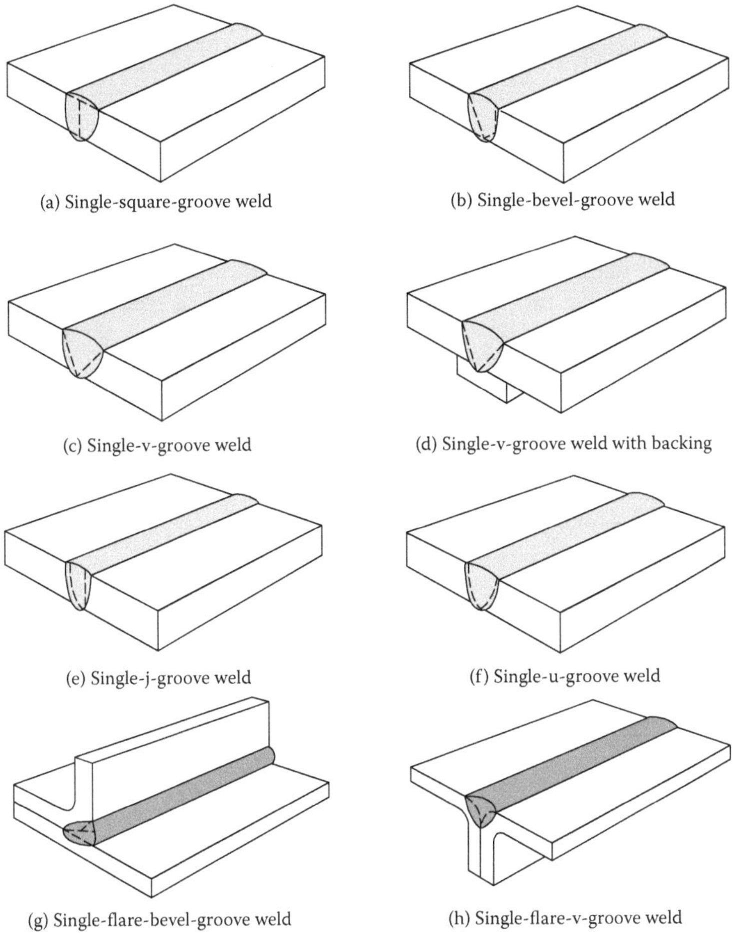

FIGURE 9.18 Different groove shapes.

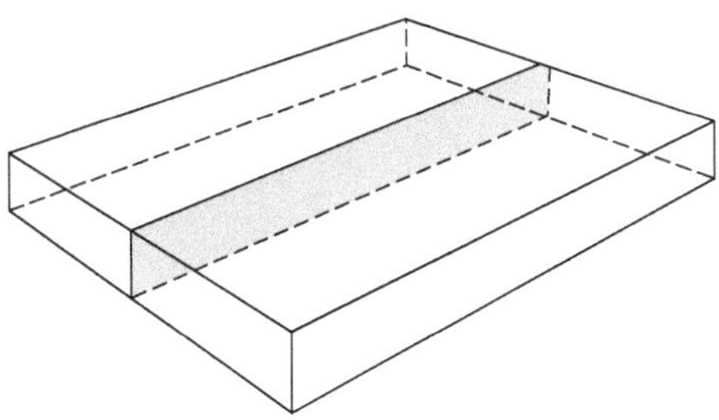

FIGURE 9.19 Face to face plates.

- Flare V-groove
- J-groove
- Square groove
- U-groove
- V-groove
- Edge flange

B. Corner Joint

Welding between steel plates located approximately at right angles to each other in the form of an L is shown in Figure 9.20.

The following are the main types of grooves that can be used in the welding:

- Fillet
- Bevel groove
- Flare bevel groove
- Flare V-groove
- J-groove
- Square groove
  - U-groove
  - V-groove

C. T-Joint

A joint between two members located at approximately right angles to each other in the form of a T (Figure 9.21).

The following are the main types of grooves that can be used in the welding:

- Fillet
- Bevel groove
- Flare bevel groove
- J-groove
- Square groove

D. Lap Joint

The two steel plates will overlap each other as shown in Figure 9.22.

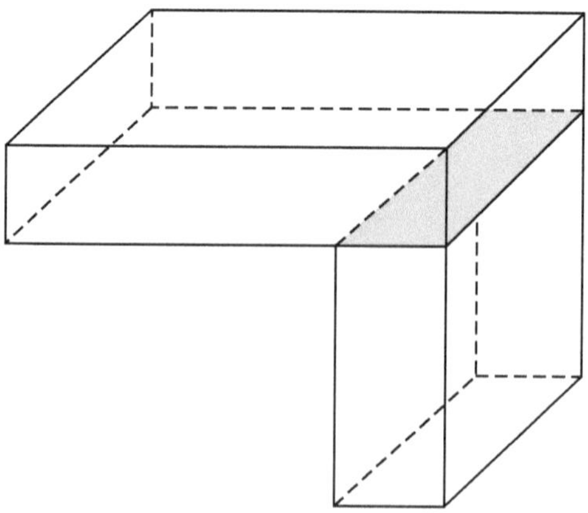

FIGURE 9.20 Corner plates.

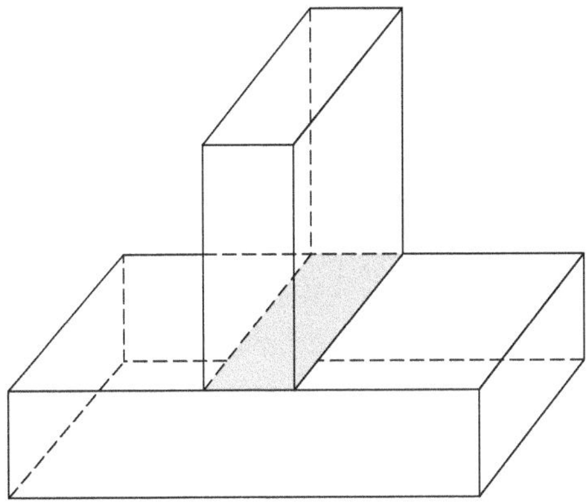

FIGURE 9.21 T-connection.

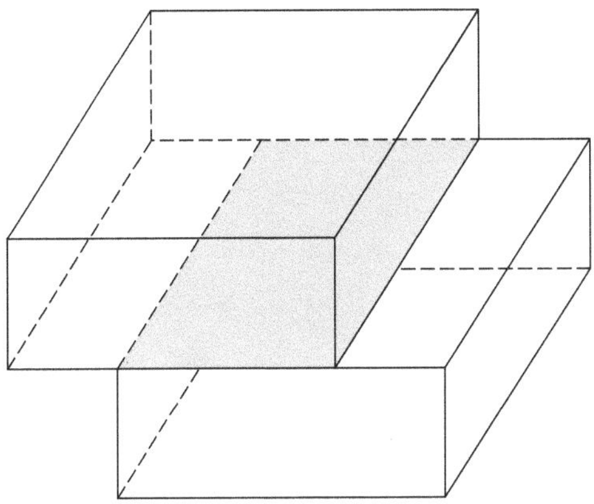

FIGURE 9.22 Lab joint.

The following are the main types of grooves that can be used in the welding:

- Bevel groove
- Flare bevel groove
- Flare V-groove
- J-groove
- Square groove
- U-groove
- V-groove
- Edge
- Seam

9.4.2.4 Welding Groove

The welding groove terminology should be well understood by the team responsible for quality control on site, as shown in Figure 9.23. It is important to note that the details regarding bevel angle, depth of bevel, root opening, and other specifications

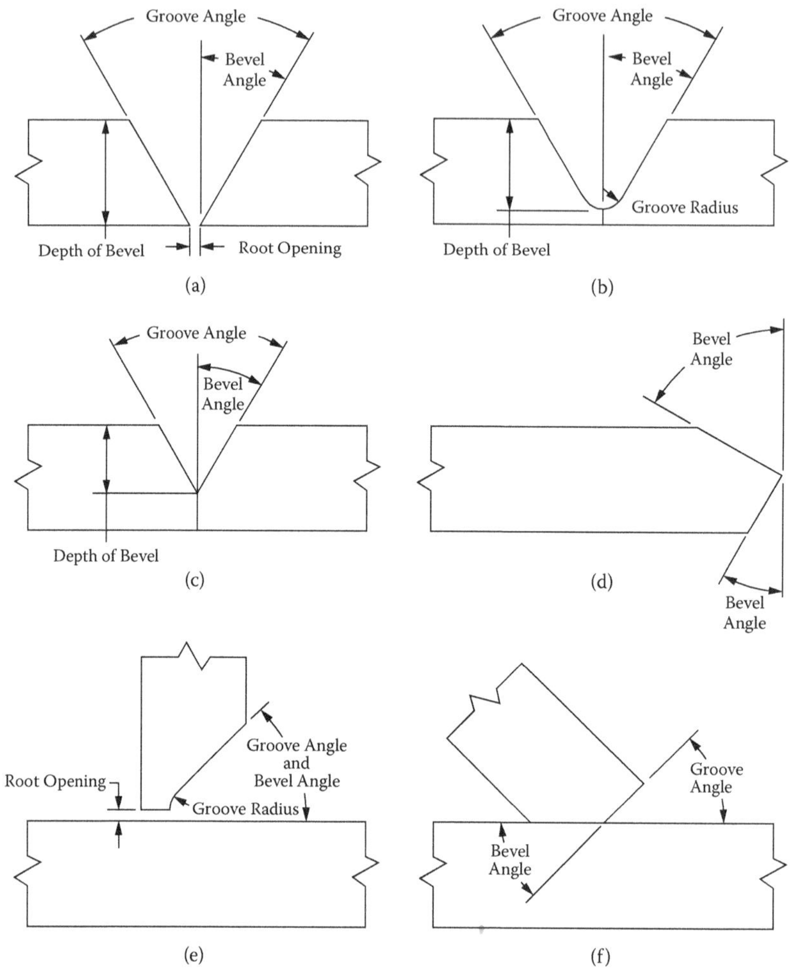

FIGURE 9.23 Welding groove dimension requirement.

should be included in the workshop drawings. These work shop drawings should be reviewed and approved by the client or the consulting engineer representing the client.

9.4.3 Problems in Welding

There are many problems that may occur during the welding process, which should be addressed through an excellent quality control and quality assurance system.

Incomplete joint penetration is one of the main problems in the welding process, as shown in Figure 9.24a and b.

Incomplete penetration in a fillet weld is presented in Figure 9.25.

Incomplete fusion is shown in Figure 9.26 for groove welds and Figure 9.27 for fillet welds.

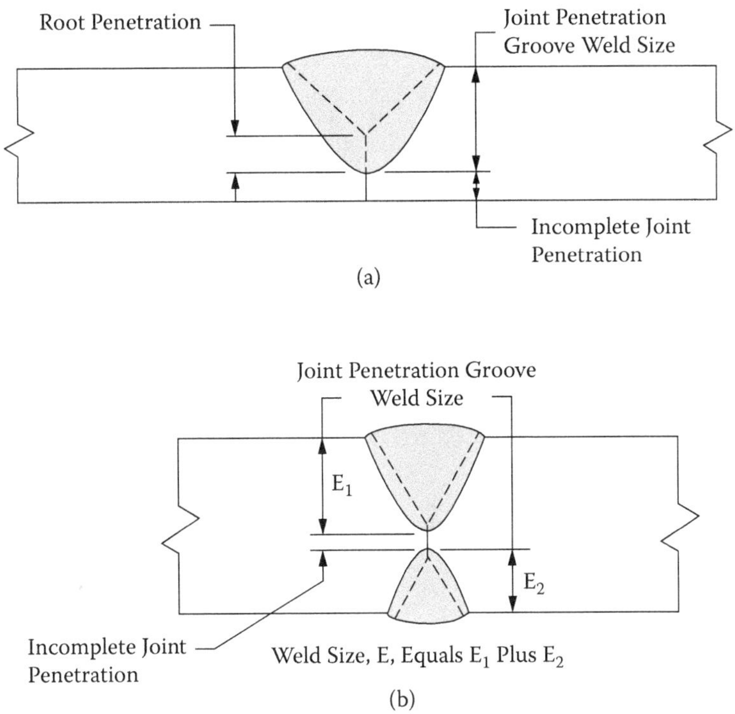

FIGURE 9.24 Butt weld incomplete penetration.

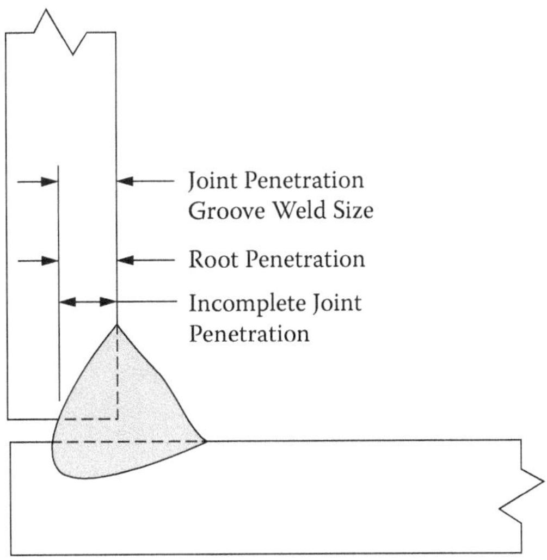

FIGURE 9.25 Incomplete penetration filet weld.

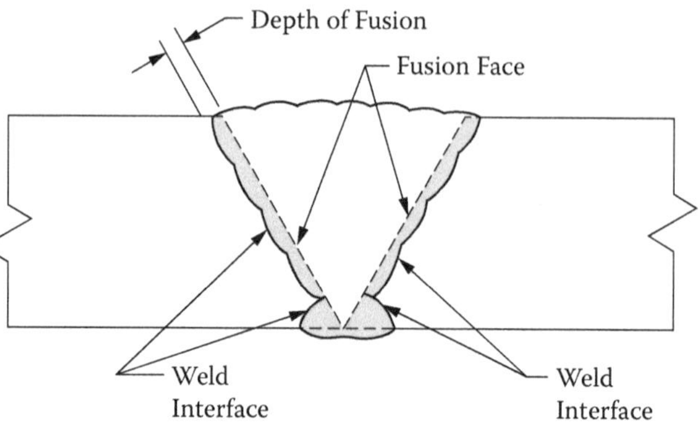

FIGURE 9.26 Fusion faces in butt weld.

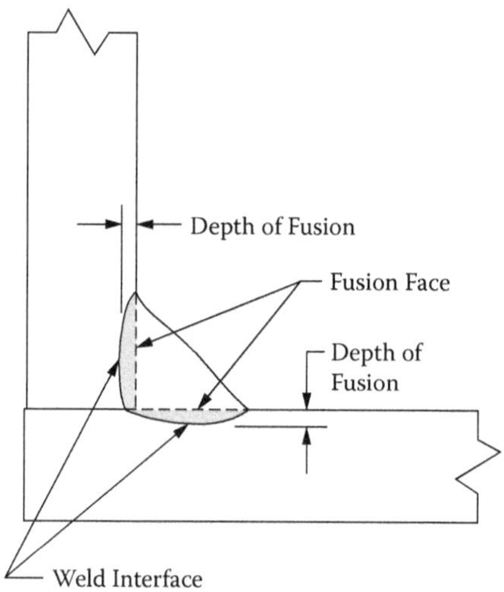

FIGURE 9.27 Fusion faces in fillet weld.

The other well-known problem in welding discontinuities is incomplete fusion. The location of the fusion face is shown in Figures 9.26 and 9.27. The problems caused by incomplete fusion are shown in Figures 9.28 and 9.29.

9.4.4 Weld and Welding Symbol

A weld symbol indicates the type of weld and may be part of a larger context. This welding symbol provides the complete information needed to convey the necessary welding details to all parties, including the welder. It contains several elements, as outlined in Tables 9.4 and 9.5.

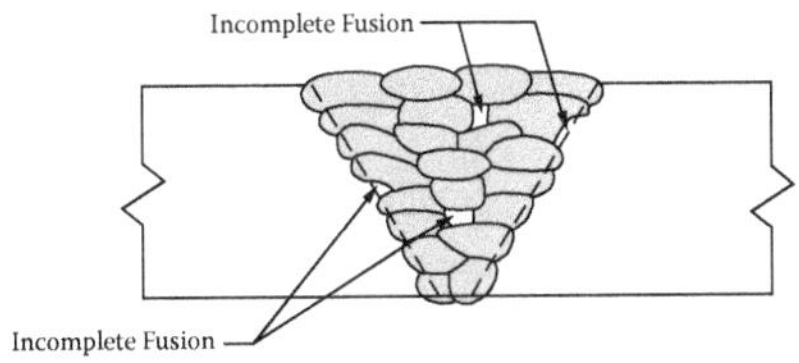

FIGURE 9.28 In complete fusion in butt weld.

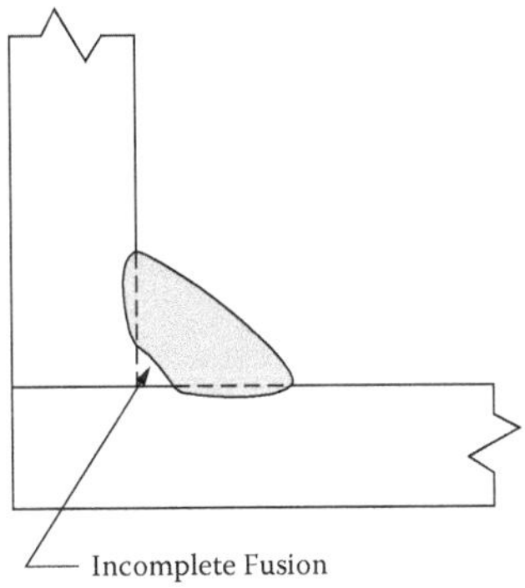

FIGURE 9.29 In complete fusion in fillet weld.

TABLE 9.4
Welding Symbol

Groove							
Square	Scarf	V	Bevel	U	J	Flare-V	Flare-Bevel

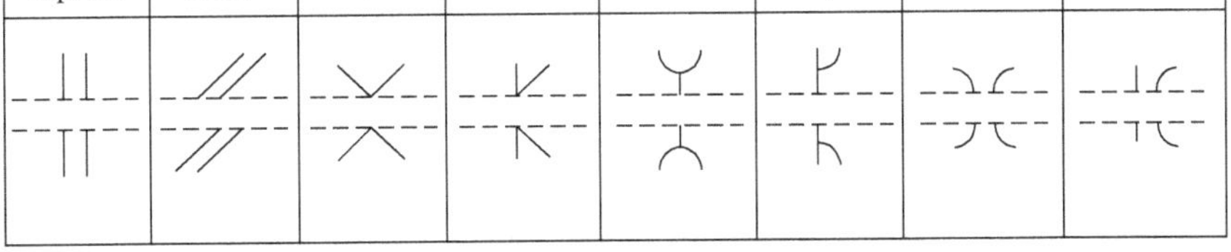

Fillet	Plug or Slot	Stud	Spot or Projection	Seam	Back or Backing	Surfacing	Edge

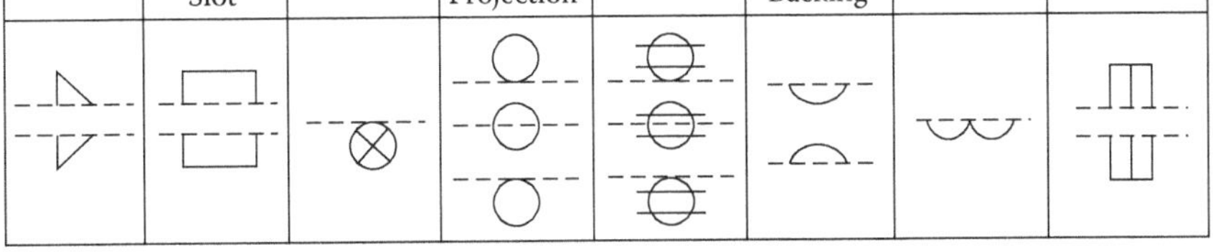

TABLE 9.5
Welding Arrow Symbol

Weld All Around	Field Weld	Melt Through	Consumable Insert (square)	Backing or Spacer (Rectangle)	Contour		
					Flush or Flat	Convex	Concave

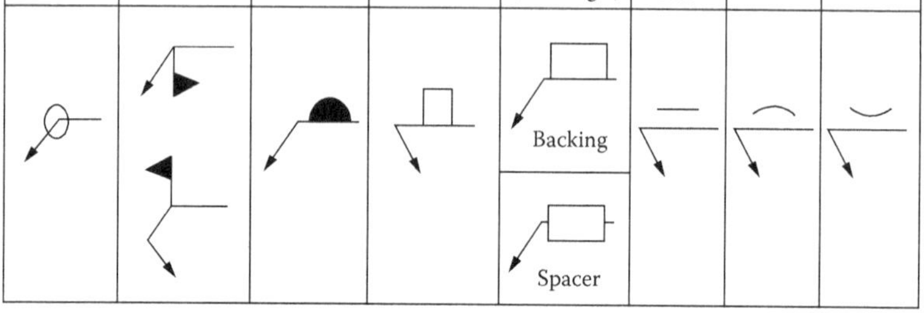

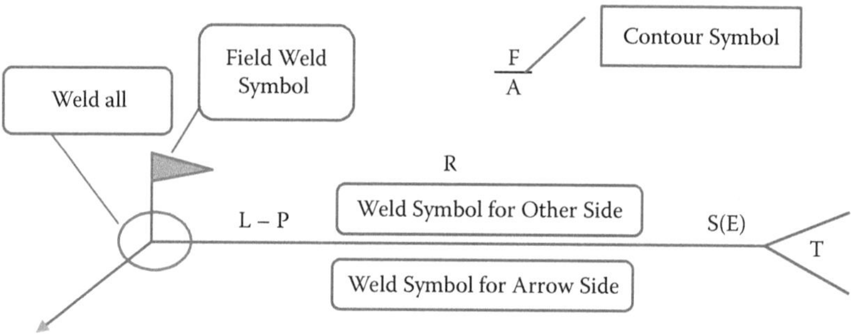

FIGURE 9.30 Welding arrow configuration.

Notably, the weld symbol rule, as mentioned in AWS D1.1, states that symbols with a perpendicular leg should have the perpendicular leg drawn on the left side of the symbol (for fillet, bevel groove, J-groove, and flare bevel groove). The symbols for stud and surfacing welds should always be drawn below the reference line (Figure 9.30).

The direction of the welding symbol with respect to the steel plate is shown in Figure 9.31:

9.4.4.1 Elements of Welding Symbols

The structural engineer in the design phase and the construction engineer on site should be familiar with welding symbols and understand them clearly.

The elements of welding symbols are as follows:

1. Required Elements
   - Reference Line – *always horizontal*
   - Arrow
2. Optional Elements
   - Multiple reference lines
   - Tail

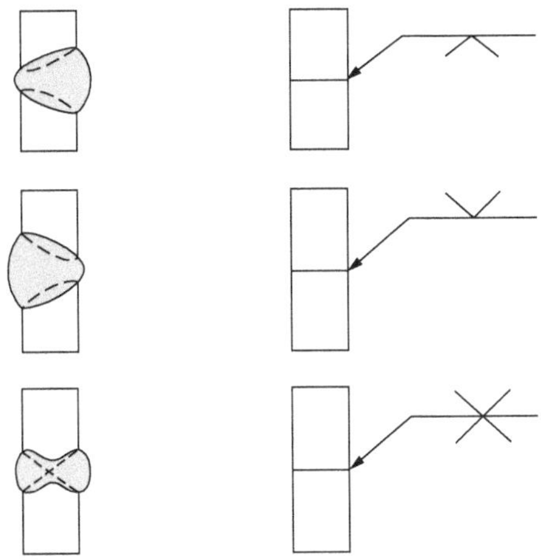

FIGURE 9.31 Welding symbol direction.

- Weld symbol
- Dimensions
- Supplementary symbols
- Finish
- Specification process

3. Arrow Notation
    - Information applicable to the arrow side of the joint is placed below the reference line
    - Information applicable to the other side of a joint is placed above the reference line

Figure 9.32 presents the arrow side with the symbol.

The weld symbol straddles the reference line when there is no side significance

- Resistance Spot Weld
- Resistance Seam Weld
- Flash Weld

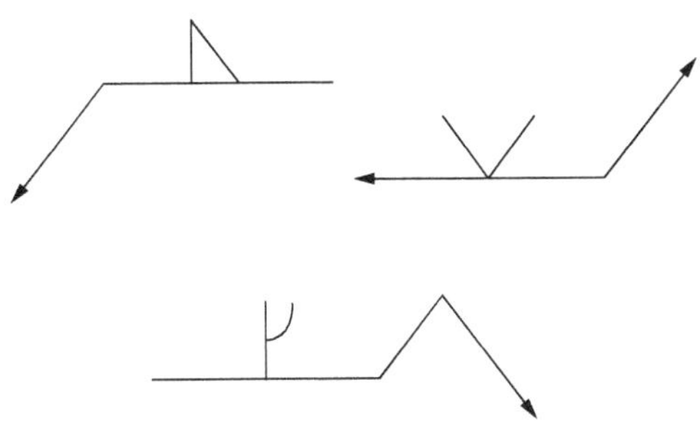

FIGURE 9.32 Arrow side weld symbols.

A break in the arrow line signifies that the member to which the arrow points is the member receiving the edge preparation.

The actual welding drawings with the symbols are shown in Figures 9.33 and 9.34.

For joints requiring more than one weld type, a weld symbol shall be used to specify each weld as presented in Figure 9.35.

Two or more arrows may be used for a single reference line to denote locations where identical welds are specified.

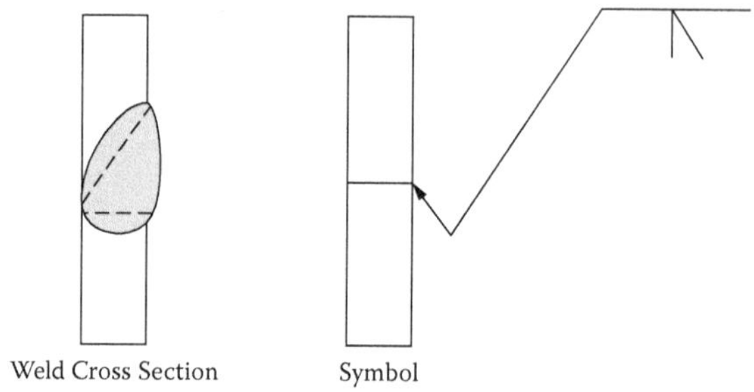

FIGURE 9.33 Arrow line.

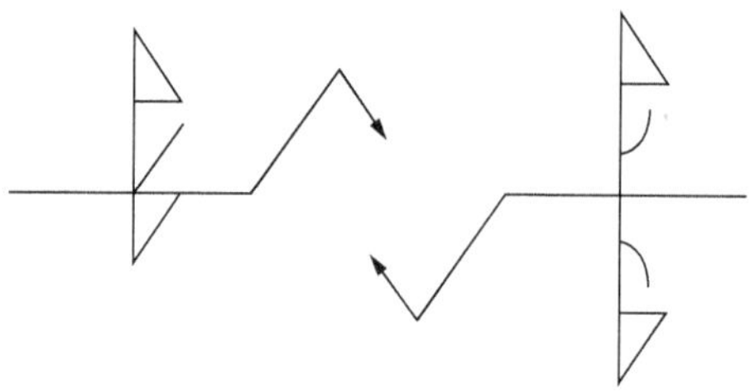

FIGURE 9.34 Two and more arrow symbols.

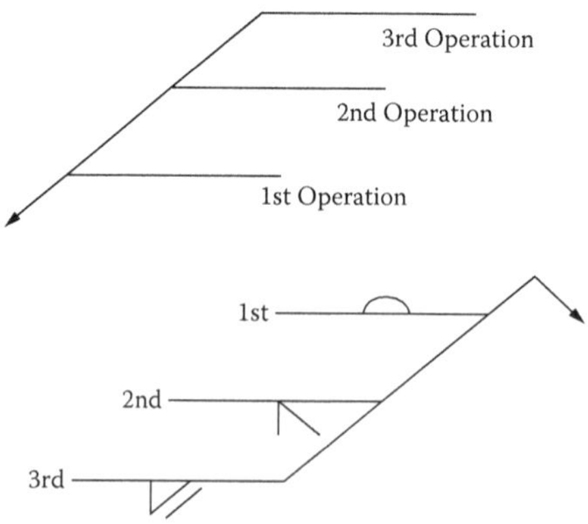

FIGURE 9.35 Operation sequence.

Two or more reference lines may be used to indicate a sequence of operations. The first operation is always on the reference line nearest the arrow as shown in Figure 9.35.

The field weld symbol is a solid flag, as shown in Figure 9.36, and is always located at the junction of the arrow and reference line. The weld all-around symbol is a circle and is located at the junction of the arrow and reference line.

This symbol is very important, as it defines the working procedure and the test requirements during construction, specifying which parts will be welded in the workshop and which will be welded on site.

Figure 9.37 illustrates that the tail is used to specify welding process information or other such details needed to convey the necessary welding specifications.

Weld contours and finishing is presented in Figure 9.38 and is usually included in workshop drawings, which should present the following:

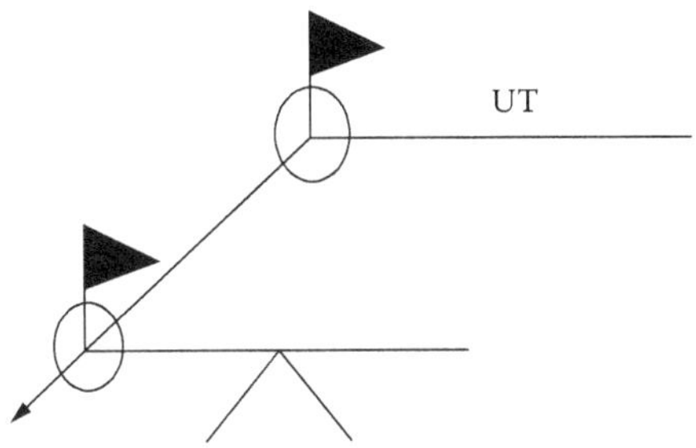

FIGURE 9.36 NDT symbol.

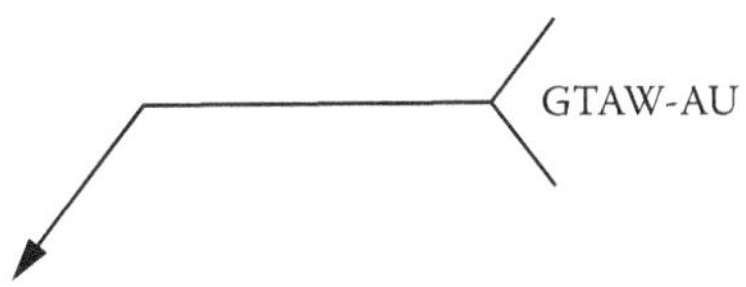

FIGURE 9.37 Welding arrow tail.

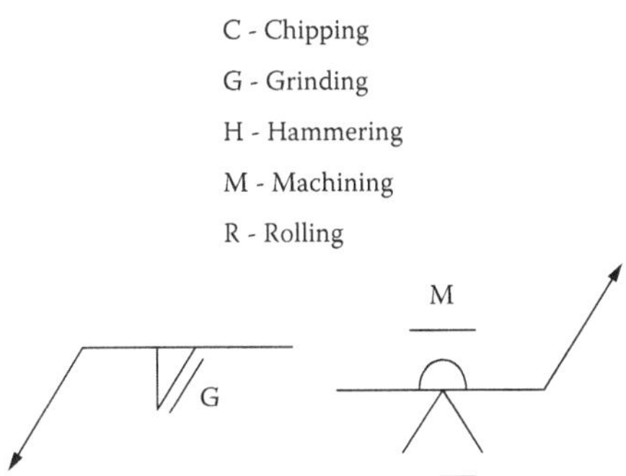

FIGURE 9.38 Welding finishing symbol.

- Weld surfaces can be contoured to several shapes: convex, concave, flush, or flat.
- Finishing methods are abbreviated as shown.
- "U" denotes that the contour method is "Unspecified."
- A contour symbol without a finishing symbol means as-welded.
- Tolerances, if required, are to be placed in the tail.
- Welding symbols are usually drawn without dimension units such as inches or millimetres.
- However, welding symbols used for publications or those requiring high precision should be dimensioned and have the dimensional tolerances noted within the tail (Figures 9.39 and 9.40).

The weld all-around symbol is a very useful symbol, which is shown in Figure 9.41.

The square groove but weld with melt through is presented in Figure 9.42.

The Groove Weld Dimensions are shown in Figure 9.43, with the depth of bevel (S) and size of weld (E) placed to the left of the symbol.

An example of a 6mm (1/4 inch) bevel with a 10mm (3/8 inch) weld is shown in Figure 9.44.

The example for a symbol with the dimensions clearly present is shown in Figure 9.45.

The root opening is placed within the weld symbol or just outside, and only on one side of the reference line, in inches or mm, per shop practice. The symbol for root opening is shown in Figure 9.46.

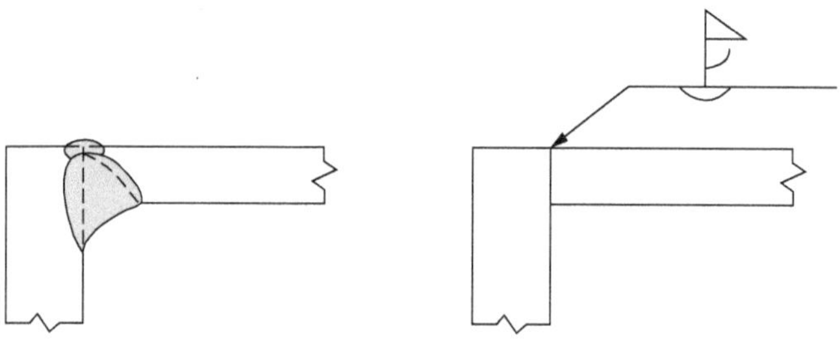

FIGURE 9.39 Finishing welding symbol.

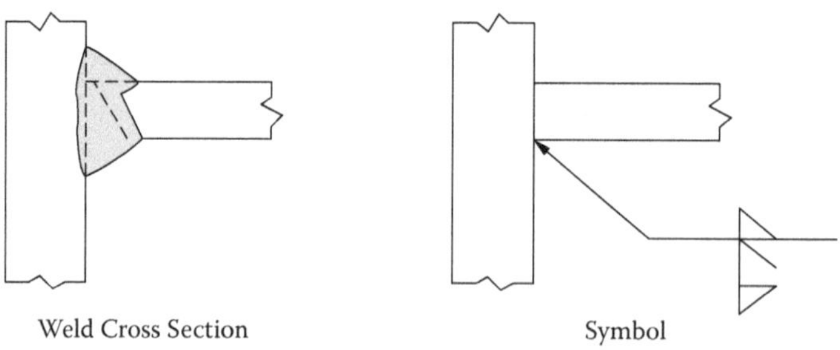

FIGURE 9.40 Single bevel groove, double fillet.

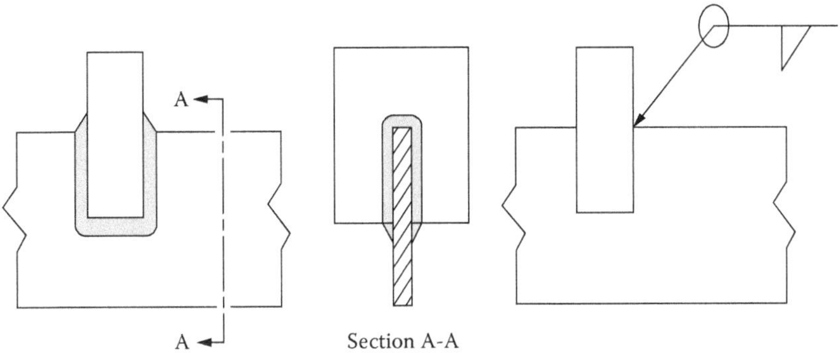

FIGURE 9.41 Weld all around.

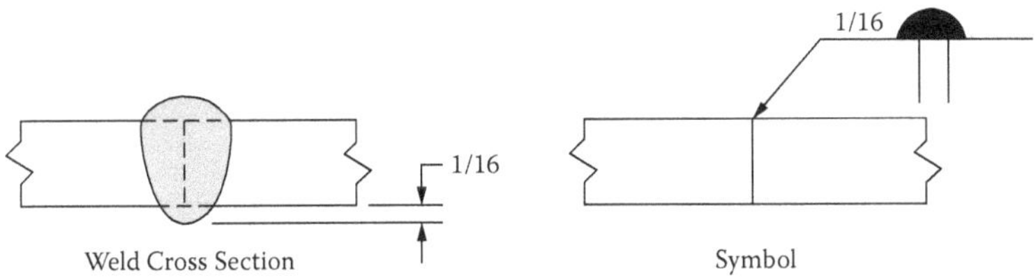

FIGURE 9.42 Square groove butt with melt-through.

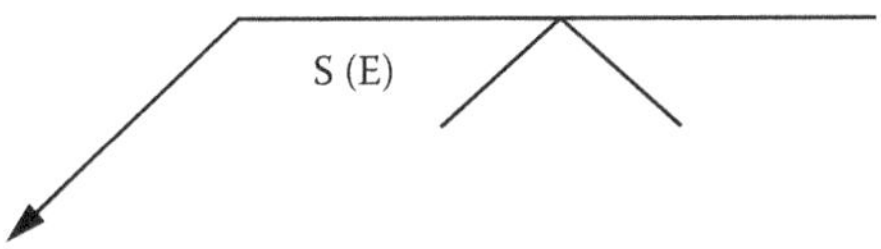

FIGURE 9.43 Dimension on the arrow.

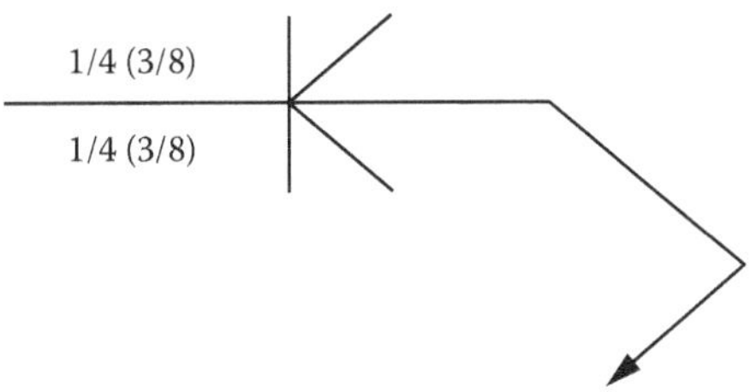

FIGURE 9.44 Dimension example.

The groove angle is placed just outside the weld symbol, as shown in Figure 9.47.

The length of the weld shall be placed to the right of the weld symbol. No dimension is required if the weld is continuous for the full length of the joint.

Pitch is the distance between the centres of adjacent weld segments. The pitch length is shown to the right of the weld length dimension, following a hyphen, as presented in Figure 9.48.

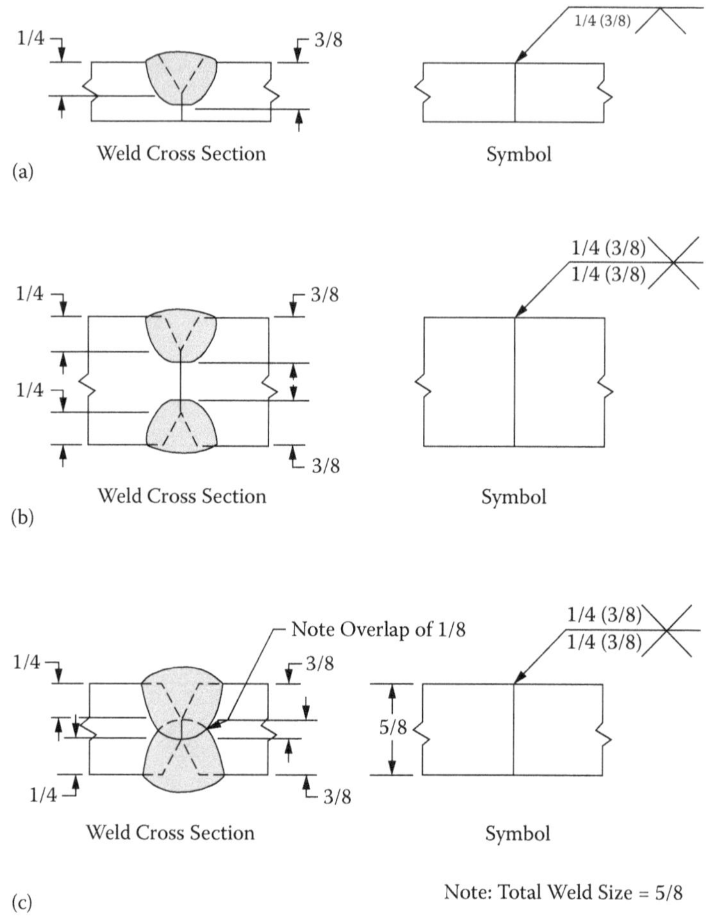

FIGURE 9.45 Dimension for different configuration.

Chain intermittent weld dimensions are to be placed on both sides of the reference line and opposite each other.

Staggered intermittent weld dimensions are to be placed on both sides of the reference line and offset from each other, as shown in Figure 9.49.

To present Back and Backing Welds, the back and backing weld symbols are identical. The sequence of welding determines which designation applies: a back weld is made after a groove weld, while a backing weld is made before the groove weld.

The back weld symbol is placed on the reference line opposite the groove weld using a single reference line with a note in the tail. Alternatively, multiple reference lines can be used, as shown in Figure 9.50.

The backing weld symbol is also placed on the reference line opposite the groove weld, using either a single reference line with a note in the tail or multiple reference lines.

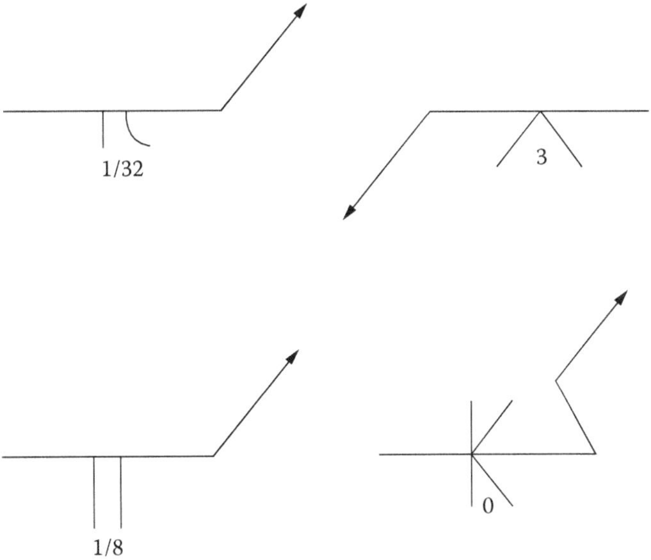

FIGURE 9.46 Root opening dimension.

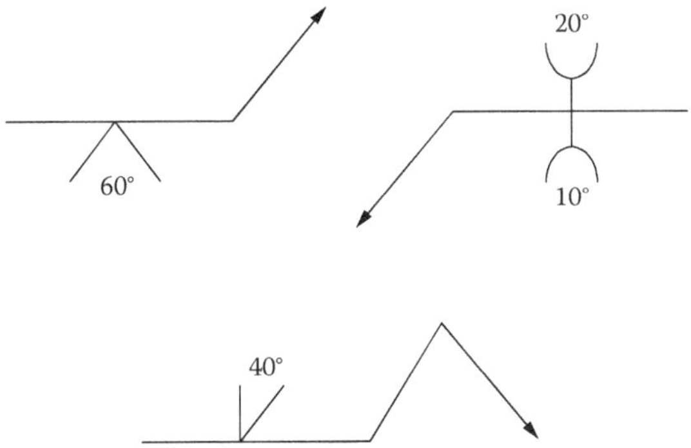

FIGURE 9.47 Groove angle.

The backing symbol is placed on the reference line opposite the weld symbol. The "R" indicates the removal of the backing after welding is completed.

The back welding symbols are presented in Figures 9.51 and 9.52.

9.4.4.2 Fillet Weld Dimensions

The fillet weld size is shown to the left of the symbol, as shown in Figure 9.53. For double fillets, each must be dimensioned.

The fillet weld size for unequal leg fillets is shown to the left of the symbol, with clarifying information provided in the tail.

NDE Symbols are presented in many textbooks to be included in workshop drawings, but in practice, they do not exist, as non-destructive tests will be defined explicitly in the quality control process plan. All conventions used are similar, if not identical,

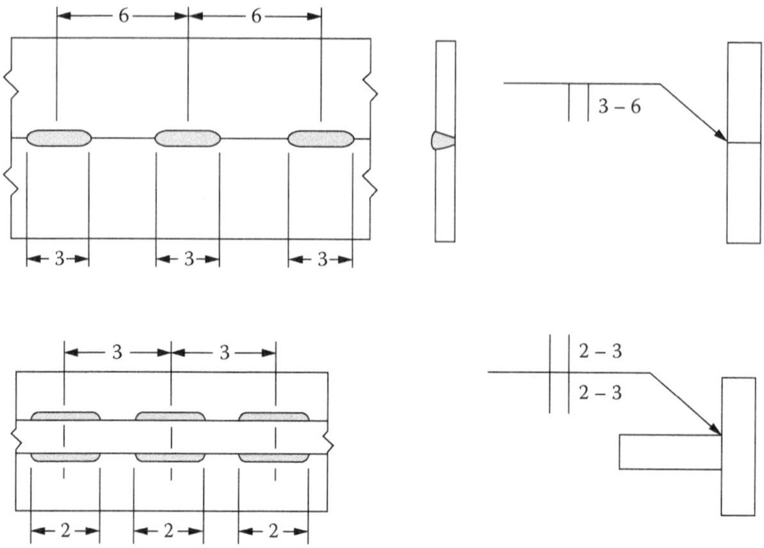

FIGURE 9.48 Symbol for pitch distance.

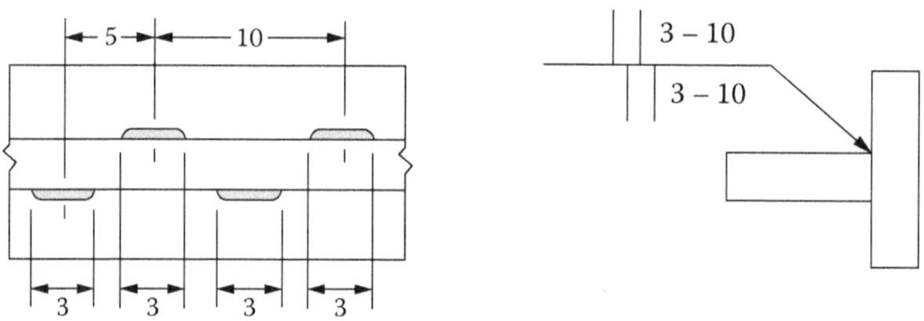

FIGURE 9.49 Staggered pitch symbol.

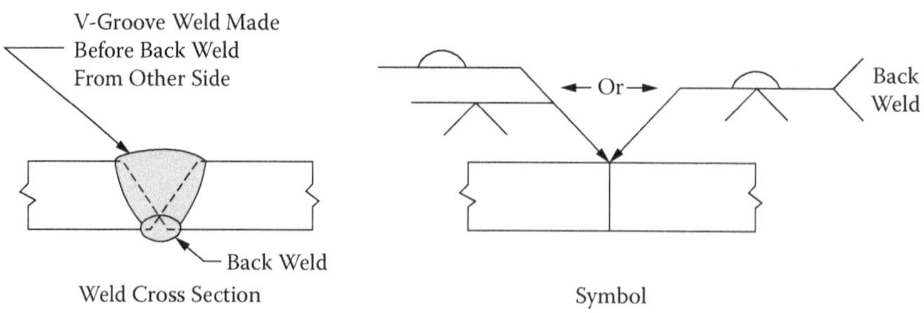

FIGURE 9.50 Back weld symbol.

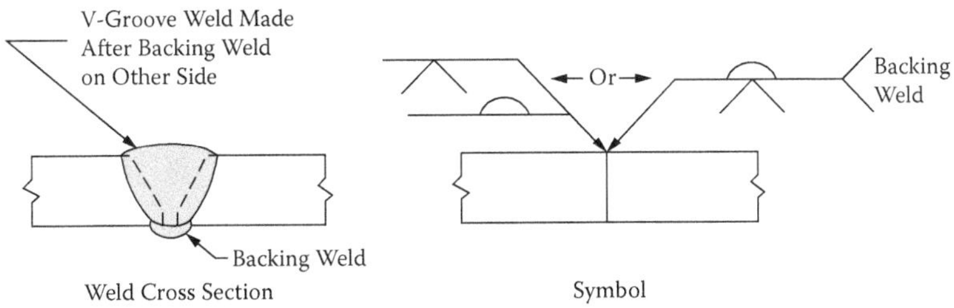

FIGURE 9.51 Backing weld symbol.

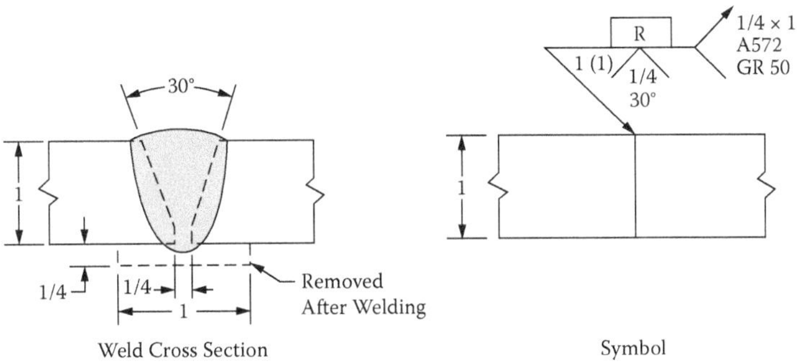

FIGURE 9.52 Backing symbol.

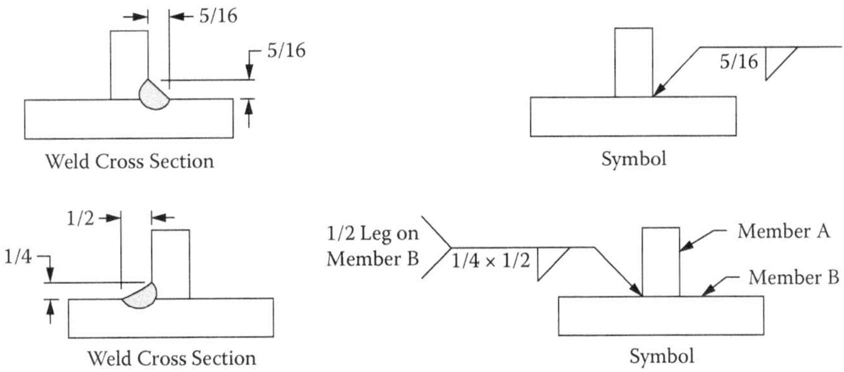

FIGURE 9.53 Examples for filet weld dimensions.

to welding symbols. The inspector must be familiar with NDE abbreviations for test methods, noting that the supplementary symbols have slightly different meanings. Furthermore, the number of tests and the extent of testing must be specified.

NDE abbreviations include

- AET
- ET

- LT
- MT, a magnetic test
- NRT
- PT, a penetration test
- RT, a radiographic test
- UT indicate ultrasonic test
- VT, a visual test

Welding and NDE symbols can be combined on the same reference line or on separate multiple reference lines.

Combining welding and NDE symbols on multiple reference lines often clarifies the exact sequence of operations required.

9.5 STUD WELD

This type of weld is used specifically in the case of composite sections where the stud is placed on the steel deck before pouring the concrete with light reinforcement.

Stud welding is considered an arc welding process because the heat for welding is generated by an arc between the stud and the metal. The process is performed using a mechanical gun connected to a power supply through a control panel. The arc quickly melts the end of the stud and a spot on the workpiece beneath it.

The stud bolts are as shown in Figures 9.54 and 9.55.

Figure 9.56 illustrates the stud welding symbol drawings. The stud weld symbol is always placed on the arrow side. The size (diameter of the stud) is to the left of the symbol, while the pitch is to the right. The number in parentheses appears just outside the symbol.

FIGURE 9.54 Stud bolts.

FIGURE 9.55 Stud bolt in roof slab.

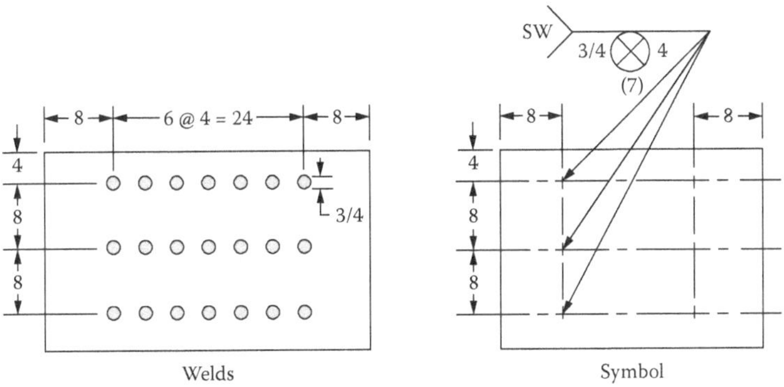

FIGURE 9.56 Stud weld symbol.

9.6 QUALITY CONTROL ON SITE

9.6.1 Quality of Human Resources

As welding is a main element of structural safety and integrity throughout its lifespan, the welding inspector responsible for quality on site should be competent and certified to perform this role by a reputable organization, such as the CWI from the American Welding Society or CSWIP from the UK. According to the American Welding Society, welding inspectors should possess the following knowledge and skills:

1. Drawings and specifications
2. Welding terms and process
3. Knowledge of testing methods
4. Professional attitude
5. Training in engineering and metallurgy
6. Welding experience

7. Reasonable knowledge of safety courses
8. Good physical condition and vision

Communication skills are also essential, as daily interaction with fabricators and other personnel is required. The inspector must be the focal point for communication and should have the ability to express ideas and thoughts clearly.

One of the inspector's main tasks is to select qualified welders by reviewing their history and conducting tests in accordance with AWS D1.1, Section 4. Welders must undergo tests for groove and fillet welds in four positions: flat, horizontal, vertical, and overhead. This is a crucial step to ensure the quality of the welding

9.6.2 Handling, Shipping, and Delivery

Delivery of steel to the site for erection shall be in the order specified in the erection plan.

The delivery sequence for fabricated steel, unless otherwise noted in the contract documents or arranged by the client representative, shall be as follows: anchor bolts, loose base plates, steel embedded in concrete, erection bolts, first-tier columns and framing for all levels (including stairs and handrails), second-tier columns and their framing, etc.

All bolts, washers, and nuts shall be packaged and delivered in rigid, weatherproof containers. Cardboard containers are not acceptable.

Railcars and/or trucks shall be loaded and cribbed to facilitate easy unloading by others. The load shall be arranged in such a manner that continuous drainage occurs.

It is important to ensure that the Fabricator takes care to protect all steel and its coatings from any damage caused by handling, storage, or shipping prior to receipt by the buyer.

The Fabricator shall ensure that adequate protection is provided for the threads on sag rods, anchor bolts, and any other threaded components to prevent damage during shipping and handling. Care shall be taken to ensure that holes and surfaces prepared for connection, as well as the threads of bolts, anchors, mortises, etc., remain clean and undamaged. The bottom of base plates shall be clean and free from any rust or corrosion.

The Fabricator is responsible for delivering all materials and documentation to the job site in good condition. All materials and documentation will be inspected immediately upon receipt by the buyer to verify that all items included in the Bill of Materials have been supplied, that all documentation has been received, and to check for any damage.

All materials designated for the care, custody, and control of the Erector shall be received, unloaded, stored, and otherwise handled in a manner that prevents distortion, deterioration, or damage.

9.6.3 Erection and Shop Drawings

In general, shop drawings and erection drawings shall be prepared in accordance with the AISC documents listed in this specification. When reviewing and checking the shop drawings, the following criteria should be assessed:

- Erection drawings shall reference the corresponding design drawings. Every steel piece on the shop drawings shall reference the appropriate erection drawing.
- Erection and shop drawings shall be grouped in sets and identified separately for each structure or yard area.
- Erection drawings shall clearly show the mark number and position for each member.
- All fabricated steel sections shall be match-marked for field assembly with designating numbers or letters corresponding to the field erection drawings. Match-marking of steel shall be done with suitable paint, waterproof ink, or pressed metal tags.
- In addition to the fabricator's identification marks, each item or bundle of walkways and platforms shall be marked with a unique tag number to clearly indicate its associated equipment. Each item or bundle of walkways and platforms shall also be shown on the erection drawings.
- The shop drawings shall state the welding procedure to be used.
- Shop drawings shall clearly show the specification and grade of steel to be used.
- Surface preparation and shop-applied coatings, including areas to be masked, shall be noted on the shop drawings.
- The fabricator shall provide a bolt list and a list of other fasteners showing the number, grade, size, and length of field bolts for each connection. These lists may be shown on either the shop drawings or on separate sheets.
- In the event that drawing revisions are necessary, the fabricator shall clearly flag all changes on the shop drawings, noting the latest revisions.

FURTHER READING

American Institute of Steel Construction (AISC)-ASD (1989). *Specification for Structural Steel Buildings—Allowable Stress Design and Plastic Design*, 9th ed. American Institute of Steel Construction, Chicago, IL.

American Welding Society (1972). *Structural Welding Code*, AWS D1.1, 1st ed. American Welding Society, Miami, FL.

ASTM (2010). *Standard Specification for Structural Bolts, Steel, Heat Treated, 120/105 ksi Minimum Tensile Strength*, ASTM A325, ASTM International, West Conshohocken, PA.

ASTM (2011). *Standard Test Methods for Determining the Mechanical Properties of Externally and Internally Threaded Fasteners, Washers, Direct Tension Indicators, and Rivets*, ASTM F606, ASTM International, West Conshohocken, PA.

ASTM (2012a). *Standard Specification for General Requirements for Rolled Structural Steel Bars, Plates, Shapes, and Sheet Piling*, ASTM A6/A6M, ASTM International, West Conshohocken, PA.

ASTM (2012b). *Standard Specification for Structural Bolts, Alloy Steel, Heat Treated, 150 ksi Minimum Tensile Strength*, ASTM A490, ASTM International, West Conshohocken, PA.

ASTM (2012c). *Standard Specification for Zinc (Hot-Dip Galvanized) Coatings on Iron and Steel Products*, ASTM A123/A123 M, ASTM International, West Conshohocken, PA.

ASTM (2011). *Standard Practice for Measuring Coating Thickness by Magnetic-Field or Eddy-Current (Electromagnetic) Testing Methods*, ASTM E376-11, ASTM International, West Conshohocken, PA.

ASTM (2012). *Standard Specification for Carbon Steel Bolts, Studs, and Threaded Rod 60 000 PSI Tensile Strength*, ASTM A307-12, ASTM International, West Conshohocken, PA.

10 Non-Destructive Testing for Steel Structure

10.1 INTRODUCTION

Non-destructive testing is an essential technique in the steel industry, as applying these methods can guarantee that the work meets the required standards,. The main activity that needs special focus is welding; therefore, it is necessary to define the appropriate techniques that can be used according to the standards to ensure that the product conforms to the specifications. The individuals who perform these tests should be certified, and according to the ASNT association, there are three levels of certification.

10.2 VISUAL TEST

Some precautions are stated when conducting visual inspections by ASME V. Direct visual examination may usually be made when access is sufficient to place the eye within 24 inches (600 mm) of the surface to be examined, and at an angle not less than 30° to that surface. Mirrors may be used to improve the angle of vision, and aids such as magnifying lenses may be used to assist examinations.

Illumination (natural or supplemental white light) for the specific part, component, vessel, or section being examined is required. The minimum light intensity at the examination surface shall be 100 foot candles (1,000 lux).

The light source, technique used, and light level verification is required to be demonstrated one time, documented, and maintained on file.

Visual inspection is usually performed by eye, relying on the experience of the inspector. Therefore, it depends largely on the inspector's capability and extensive experience in welding. In addition, some tools can be used on site to check whether the welding meets the standards and project specifications.

As shown in Figure 10.1, the fillet weld set is a rectangular piece of metal with the shape of a fillet weld at each corner. The inspector will move this piece over the steel to verify if the welding thickness matches the gauge value and thus meets the required thickness.

The other tool, illustrated in Figure 10.2, is used to measure the fillet weld for both concavity and convexity of the fillet weld shape. Additionally, this tool is also used to measure the butt weld reinforcement.

Caliper vernier is an essential tool, as shown in Figure 10.3, for measuring the thickness of the base metal and weld with good accuracy.

The Vernier caliper is an extremely precise measuring instrument, with a reading error of 1/20 mm = 0.05 mm.

DOI: 10.1201/9781003307303-10

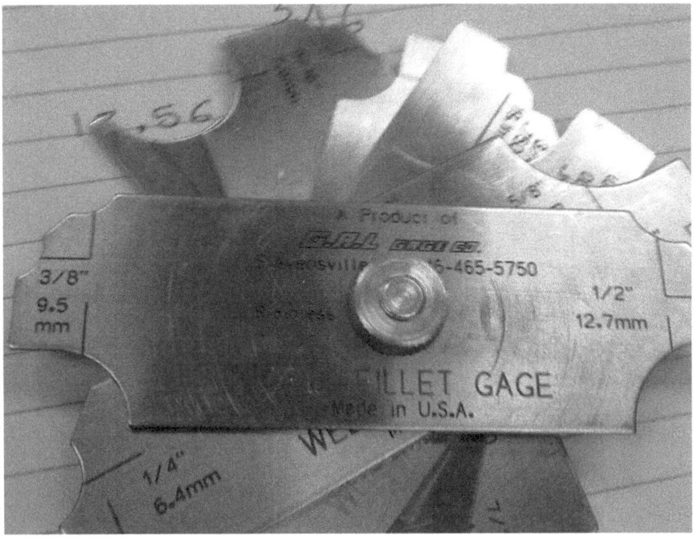

FIGURE 10.1 Fillet weld thickness test.

FIGURE 10.2 Fillet weld thickness.

If there are pits in the base metal due to corrosion or between the base metal and the weld due to a poor welding process, such as undercutting, the pit measurement tools shown in Figure 10.4 will be used.

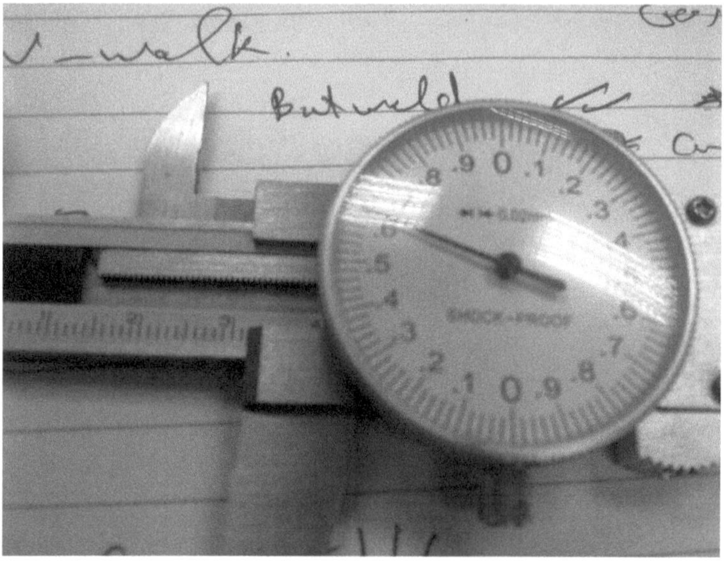

FIGURE 10.3 Caliper vernier.

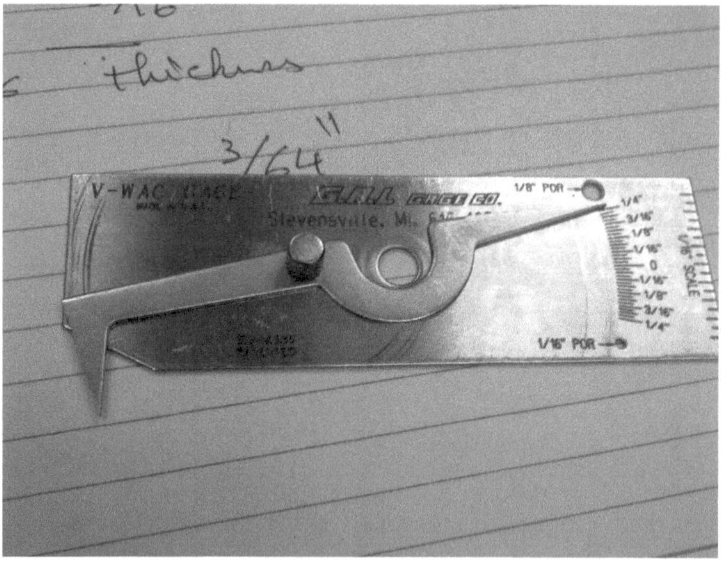

FIGURE 10.4 Steel pit measurement tool.

The tool in Figure 10.5 is an example of a tool that can measure the angle of preparation, excess weld metal, depth of undercut, depth of pitting, weld throat size, and more.

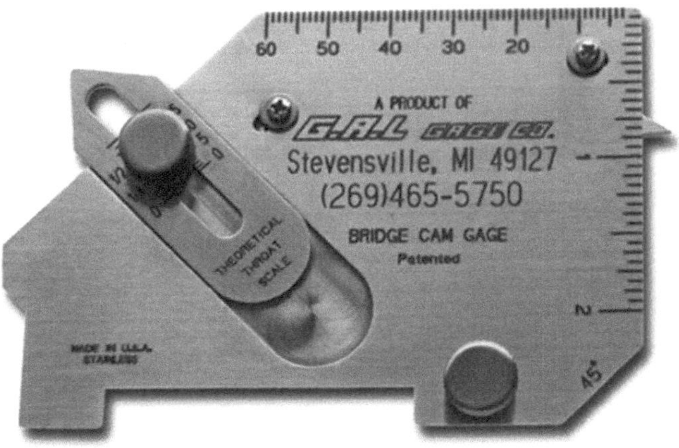

FIGURE 10.5 Multi-functional tool for welding.

FIGURE 10.6 X-ray machine.

10.3 RADIOGRAPHIC TEST

10.3.1 Principle of Radiographic Test

The most important technique in NDT is the radiographic test, which is more accurate and well-documented. However, it poses significant health risks if the user does not follow safety precautions and restrictions. The basis of this test is the use of short-wavelength electromagnetic radiation with high-energy photons to penetrate materials.

Either an X-ray machine, as shown in Figure 10.6, or a radioactive source such as Ir-192, Co-60, or, in rare cases, Cs-137 can be used as sources of photons, as shown in Figure 10.7. In most steel structures, gamma radiation is used because it can be portable, whereas X-ray equipment should be fixed in place.

The main principle of this test is to apply gamma or X-ray radiation to the steel component, which will absorb the radiation depending on its thickness. The intensity

FIGURE 10.7 Radiographic test by gamma ray.

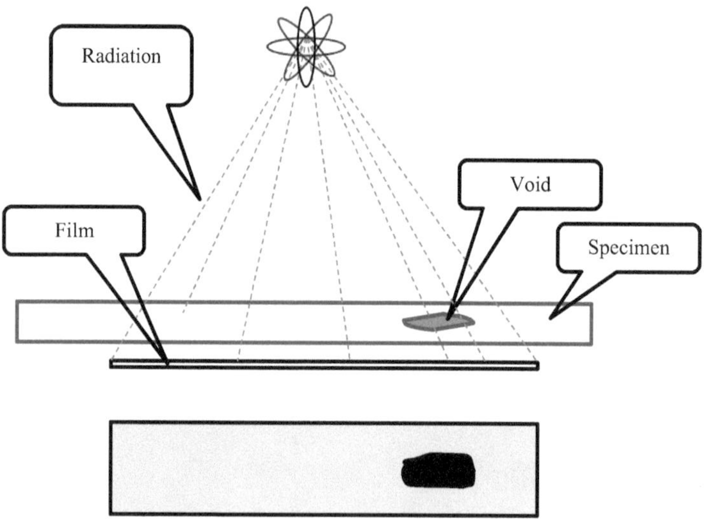

FIGURE 10.8 Reflect discontinuity on the film.

of the radiation is then reflected onto a film placed on the other side of the test specimen, as shown in Figure 10.8. Therefore, for this test to be effective, the specimen must have accessibility from both sides. The latent image produced on the film becomes a shadow picture of the specimen once the film is processed.

Penetrating radiations are those confined to that part of the electromagnetic spectrum with a wavelength of less than approximately 10 nanometres.

10.3.2 Isotope Decay Rate (Half-life)

The curie is the basic unit for radioactive material. The specific activity of any radioactive source is defined as the activity in curies per gram. The curie was originally defined as the amount of any radioactive material that disintegrates at the same rate as 1 g of pure radium. It has since been defined more precisely as a quantity of radioactive material in which $3.7 \times 1{,}010$ atoms disintegrate per second. Specific activity is defined as the concentration of radioactivity, or the relationship between the mass of radioactive material and its activity, and is expressed as the number of curies per unit mass or volume. For example, each gram of Cobalt-60 contains approximately 50 curies, while Iridium-192 contains 350 curies per gram of material. The shorter the half-life, the smaller the amount of material required to produce a given activity, or curies. The higher specific activity of Iridium results in physically smaller sources. Radioactive sources decay over time. The half-life of an isotope is defined as the time it takes for half of the atoms to decay or disintegrate.

The half-lives of some common radioisotopes are given in Table 10.1:

As shown in the table, each radionuclide decays at its own unique rate, which cannot be altered by any chemical or physical process. A useful measure of this rate is the half-life of the radionuclide, defined as the time required for the activity of any particular radionuclide to decrease to one half of its initial value.

For selecting a suitable isotope, refer to Table 10.2.

TABLE 10.1
Half-Lives of Radio Isotopes

Radioactive	Time (years)
Radium 226 (RA-226)	1,620 years
Cesium 137 (Cs-137)	30 years
Cobalt 60 (Co-60)	5.3 years
Thulium 170 (TM-170)	130 days
Iridium 192 (IR-192)	75 days

TABLE 10.2
Properties for Different Isotope Materials

Factors	Cobalt	Radium	Cesium	Iridium	Thulium
Half-life	5.3 years	1,600 years	30 years	75 days	130 days
Radiation level RHF/CUIRE	14.5	9.0	4.2	5.9	0.03
Hald value layer (lead)	0.5 in	0.5 in	0.3 in	0.2 in	0.05 in
Source size	Small	Radium large Radon small	Relatively large	Very small	Small
Specific activity	Medium	Very low	Low	Very high	Highest
Cost	Low	High to buy low to rent	Medium	Relatively low	High

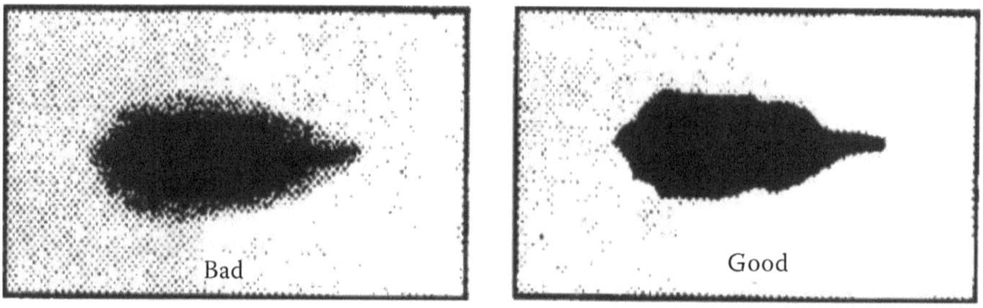

FIGURE 10.9 Presents good and bad definition.

10.3.3 Radiographic Sensitivity

The measure of the accuracy of a radiograph is called the "Sensitivity" of a radiograph. Sensitivity is a function of the contrast and definition of the radiograph; it involves the comparison between film densities for different areas of the radiograph. Definition refers to the line of demarcation between areas of different densities, as shown in Figure 10.9.

Before discussing these concepts, it is important to define certain parameters. Contrast is the comparison between film densities for different areas of the radiograph, while definition is the line of demarcation between areas of different densities. For example, if the image on the film is clear and sharp, it indicates that the radiograph has good definition.

10.3.4 Geometric Unsharpness

Geometric unsharpness occurs if the basic principles of shadow formation are not followed. One cause of geometric unsharpness relates to the size of the radiographic source; when the source is not a point but rather a small area, the image cast is not perfectly sharp, and the fuzzy edges of the image are referred to as the penumbra.

The penumbra cannot be completely eliminated because a point source cannot be obtained.

Geometric unsharpness refers to the loss of definition that results from the geometric factors of the radiographic equipment and setup. It occurs because the radiation does not originate from a single point but rather over an area. Consider the images below, which show two sources of different sizes, the paths of the radiation from each edge of the source to each edge of the feature of the sample, the locations where this radiation will expose the film, and the density profile across the film. In the first image, the radiation originates from a very small source. Since all of the radiation originates from basically the same point, very little geometric unsharpness is produced in the image. In the second image, the source size is larger, and the different paths that the rays of radiation can take from their point of origin in the source cause the edges of the notch to be less defined.

There are three factors affecting the unsharpness: source size, source-to-object distance, and object-to-detector distance. The source size will be specified by the manufacturer's specifications for the gamma ray source. It is important to note that as

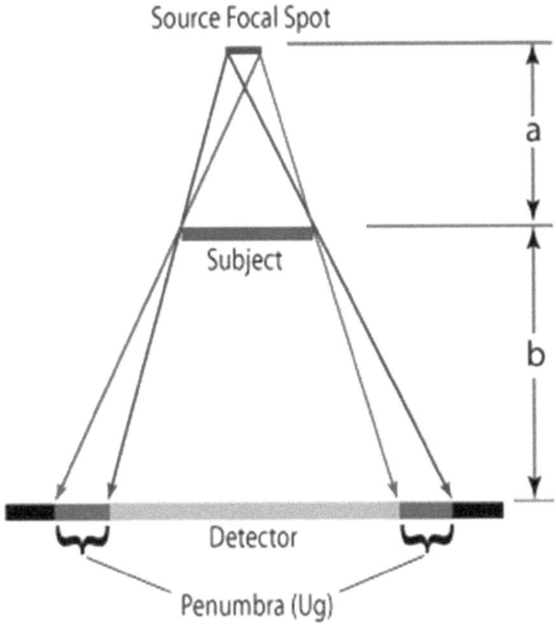

FIGURE 10.10 Causes of penumbra.

the source size decreases, the geometric unsharpness also decreases. For a given size source, the unsharpness can also be decreased by increasing the source-to-object distance, but this comes with a reduction in radiation intensity.

In most cases, the distance from the object to the detector is kept as small as possible to help minimise unsharpness. However, there are situations, such as when using geometric enlargement, where the object is separated from the detector. This separation can reduce definition. The area of varying density at the edge of a feature that results from geometric factors is called the penumbra, as shown in Figure 10.10. From this figure, one can define the relationship between unsharpness, the focal spot size of the source, and the distances between the source and the specimen, and between the specimen and the film. It is important to note that when the distance between the film and the specimen is reduced, the penumbra will also decrease.

Codes and standards used in industrial radiography require that geometric unsharpness be limited. Generally, the allowable amount is 1/100 of the material thickness, up to a maximum of 0.040 inch. These values refer to the degree of penumbral shadow in a radiographic image.

10.3.5 Scatter Radiation

Another problem that should be considered in radiographic testing is scatter radiation. The scattered photons cause a loss of contrast and definition.

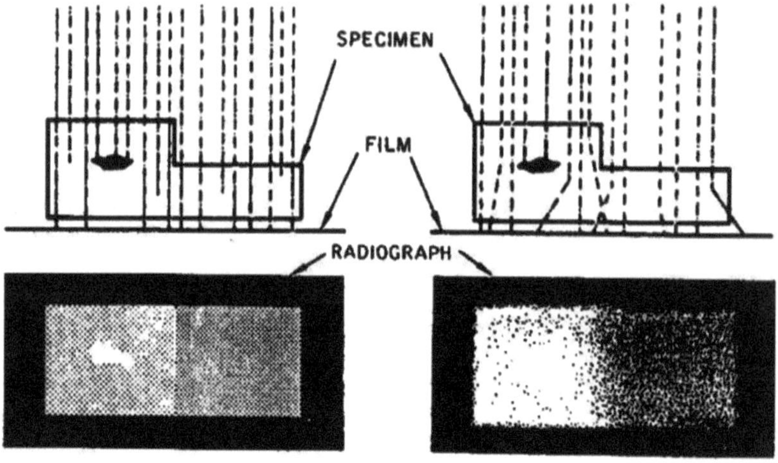

FIGURE 10.11 Internal scatter.

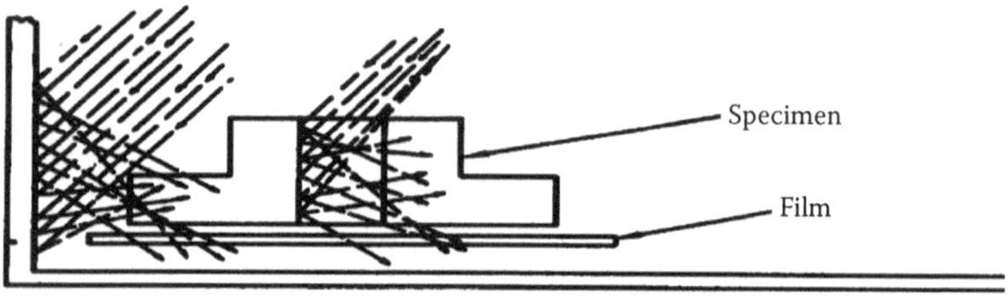

FIGURE 10.12 Side scatter.

Scattered radiation is usually described with reference to its origin:

1. Internal Scatter:
   The internal scatter is as shown in Figure 10.11. As shown in case (b), the definition is very poor.
   This is caused by free electrons generated by the radiographic ray as it passes through the film.
   - The scattering of free electrons through the film causes the film to be exposed wherever the electrons travel.
   - The scattering causes some degree of fuzzy edges on the image that cannot be avoided.
2. Side Scatter:
   Side scatter is shown in Figure 10.12, where the scattering occurs due to walls or any other objects near the path of the radiation ray.
3. Back Scatter:
   Back scatter is often identified by placing the letter B on the back side of the film cassette.

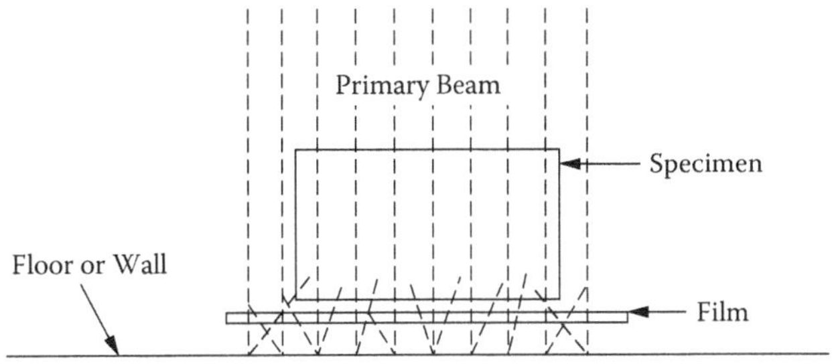

FIGURE 10.13 Back scatter.

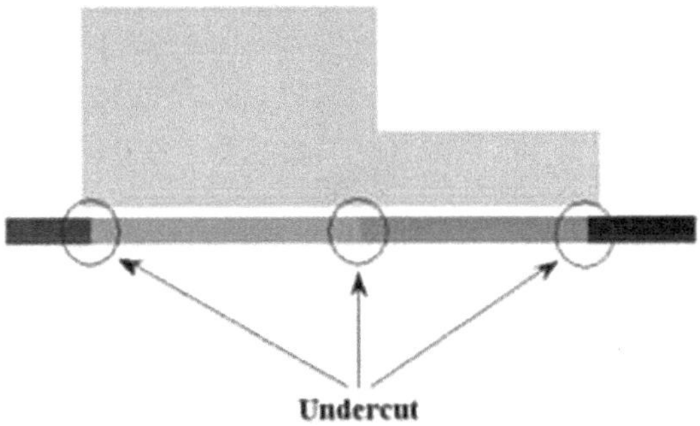

FIGURE 10.14 Locations of undercut.

This occurs due to any object, such as walls, floors, tables, or similar objects, located behind the film, as shown in Figure 10.13.

4. Undercut:

   The undercut phenomenon, as illustrated in Figure 10.14, should be controlled when producing a radiograph. Parts with holes, hollow areas, or abrupt changes in thickness are likely to suffer from undercut if controls are not implemented. Undercut appears as a darkening of the radiograph in the area of the thickness transition, resulting in a loss of resolution or blurring at the transition area.

   It is worth mentioning that the undercut occurs due to scattering within the film. At the edges of a part or in areas where the part transitions from thick to thin, the intensity of the radiation reaching the film is much greater than in the thicker areas of the part.

   The high level of radiation intensity reaching the film results in a significant amount of scattering within the film. It should also be noted that the faster the film speed, the more undercut is likely to occur. Scattering from within the walls of the part also contributes to undercut, but research has shown that scattering within the film is the primary cause. Masks are used to control undercut; sheets of lead are cut to fill holes or surround the part, and metallic shot and liquid absorbers are often used as masks.

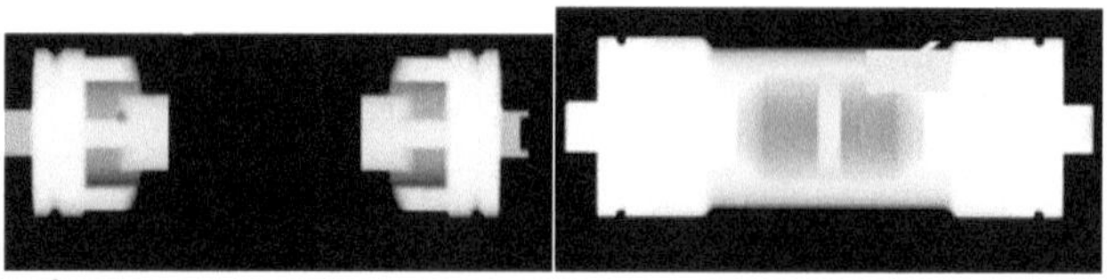

FIGURE 10.15 Effect of high- and low-energy radiation on the image.

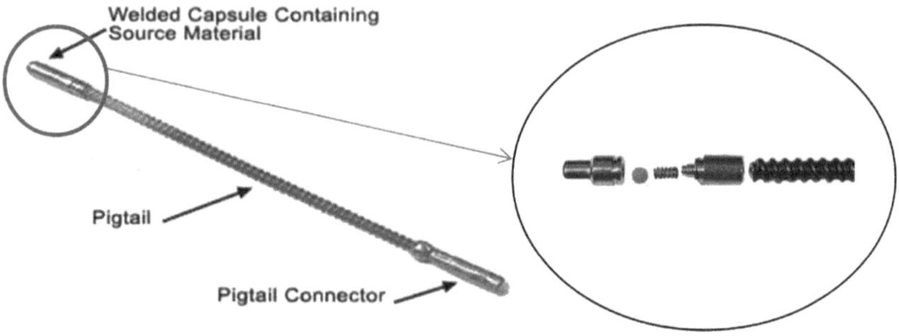

FIGURE 10.16 Location of the isotope source.

The relationship between high and low-energy radiation on the thinner sections and its representation in the radiographic image is presented in Figure 10.15.

10.3.6 Radio Isotope (Gamma) Sources

For the gamma-ray source, the physical size of isotope materials varies between manufacturers; however, generally, an isotope material is a pellet that measures 1.5 mm × 1.5 mm. Depending on the level of activity desired, one or more pellets are loaded into a stainless steel capsule and sealed by welding. The capsule is attached to a short flexible cable called a pigtail, as in Figure 10.16.

All precautions should be taken to ensure that the source is away from those performing the test. Therefore, the source is located in shielding materials, and the exposure device for iridium-192 and cobalt-60 sources will contain 45 pounds and 500 pounds of shielding materials, respectively. The location of the inspector relative to the vessel is shown in Figure 10.17.

Cobalt cameras are often fixed to a trailer and transported to and from inspection sites. When the source is not being used to make an exposure, it is locked inside the exposure device.

The main principle for operating with gamma rays is the safety of personnel. Therefore, a crank-out mechanism and a guide tube are attached to opposite ends of the exposure device, as shown in Figure 10.18. The guide tube often has a collimator at the end to shield the radiation, except in the direction necessary to make the exposure. The end of the guide tube is secured in the location where the radiation source needs to be to produce the radiograph. The crank-out cable is stretched as far

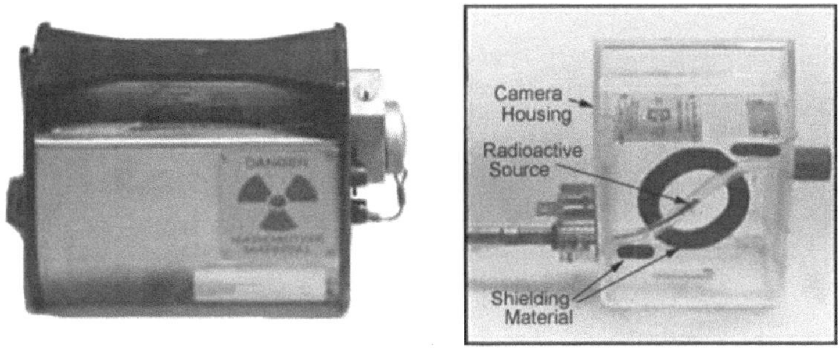

FIGURE 10.17 UT equipment source storage.

FIGURE 10.18 UT operation.

as possible to maximise the distance between the exposure device and the inspector. The isotope is shown in the stored position and at the end position in Figures 10.19 and 10.20, respectively.

10.3.7 Radiographic Film

The radiographic film consists of five layers: the core is a film sheet, with lead screens on both the front and back, and cardboard on both sides. The lead screens are required to filter out low-energy radiation and to increase the photographic action on the film.

The film contains radiation-sensitive silver halide crystals, such as silver bromide or silver chloride, and a flexible, transparent, blue-tinted base. The emulsion differs from those used in other types of photographic films to account for the distinct characteristics of gamma rays and X-rays, although X-ray films are sensitive to light. Typically, the emulsion is coated on both sides of the base in layers about 0.0005 inch thick. Applying emulsion to both sides of the base doubles the amount of radiation-sensitive silver halide, thereby increasing the film speed.

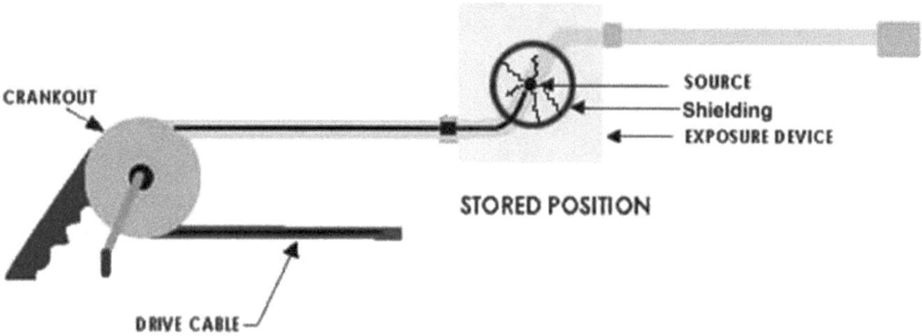

FIGURE 10.19 Isotope in the stored position.

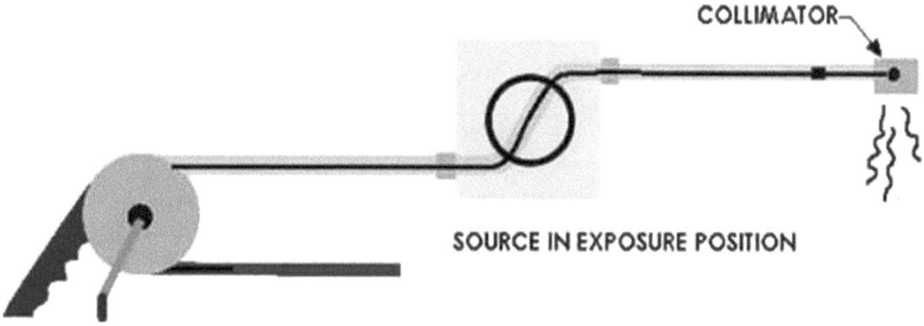

FIGURE 10.20 Isotope in the end position.

Radiographic film is available in different packaging options. The most basic form is as individual sheets in a box. In preparation for use, each sheet must be loaded into a cassette or film holder in the darkroom to protect it from exposure to light.

A rip strip makes it easy to remove the film in the darkroom for processing. This form of packaging has the advantage of eliminating the process of loading the film holders in the darkroom. The shape of the film is shown shown in Figure 10.21. The film should be protected from fingerprints and dirt until it is removed from the envelope for processing.

Packaged film is also available in rolls, which allows the radiographer to cut the film to any length. The ends of the packaging are sealed with electrical tape in the darkroom. In applications such as the radiography of circumferential welds and the examination of long joints on an aircraft fuselage, long lengths of film offer significant economic advantages. The film is wrapped around the outside of a structure, and the radiation source is positioned on axis inside, allowing for examination of a large area with a single exposure.

Envelope-packaged film can be purchased with the film sandwiched between two lead oxide screens. The screens function to reduce scatter radiation at energy levels below 150 keV and act as intensification screens above 150 keV.

Radiographic film should be handled carefully to avoid physical strains, such as pressure, creasing, buckling, friction, or other sources of film deterioration.

FIGURE 10.21 Shape of film.

There are numbers and words in lead, as shown in Figure 10.22, to be included on the film to indicate the date and film number for easy identification.

To define the sensitivity of the film, ASTM and ASME recommend that penetrameters have an identification number representing their thickness. A penetrameter contains three holes drilled into the penny, with diameters of 1, 2, and 4 times the thickness of the penetrameter.

10.3.7.1 New Films

There is a new type of film on the market that has the same size as the conventional one but functions like a memory stick. After taking a photo, it can be converted directly to a computer and reused many times, with a lifespan of up to 1000 uses. The computer, along with the input device, facilitates easy handling and transferring of welding photos by email. The procedures for dealing with the film, along with all calculations, precautions, and constraints, are the same as for conventional film.

For conventional film, scanners are also available on the market to convert conventional photos to a computer for easier use.

FIGURE 10.22 Fixing the identification number.

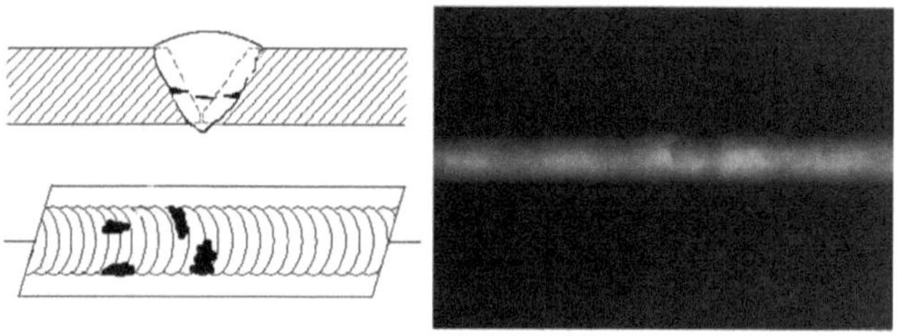

FIGURE 10.23 Cold lap.

10.3.8 General Welding Discontinuities

To identify discontinuities by examining the film, inspectors require training. The following are types of discontinuities and the corresponding film images for reference.

In the case of cold lap, it is a condition where the weld filler metal does not properly fuse with the base metal or the material of the previous weld pass. This occurs because the arc does not melt the base metal sufficiently, causing the slightly molten puddle to flow into the base material without bonding. Figure 10.23 presents the cold lap with a sketch and radiographic film.

The porosity problem is due to gas entrapment in the solidifying metal. Porosity can take many forms on a radiograph but often appears as dark, round, or irregular spots or specks, sometimes appearing individually, as shown in Figure 10.24. Sometimes, porosity may be elongated and appear to have a tail; this is the result of gas attempting to escape while the metal is still in a liquid state, and it is called wormhole porosity. All porosity is a void in the material and will exhibit a higher radiographic density than the surrounding area.

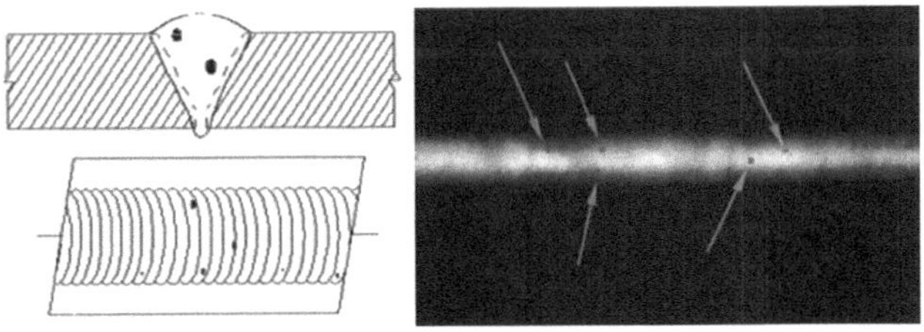

FIGURE 10.24 Porosity.

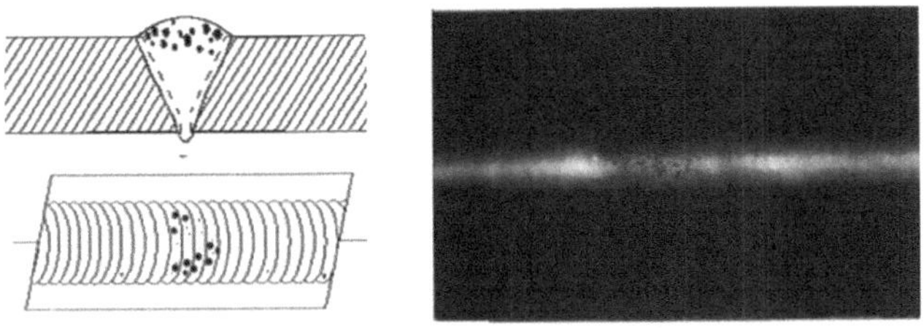

FIGURE 10.25 Cluster porosity.

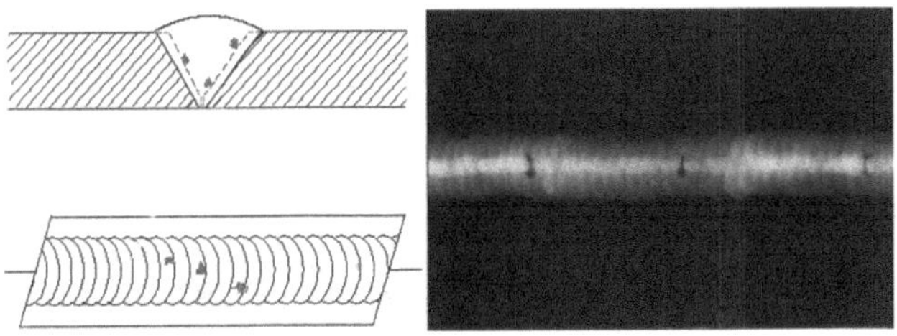

FIGURE 10.26 Slag inclusions.

Figure 10.25 presents cluster porosity, which occurs when flux-coated electrodes are contaminated with moisture. The moisture turns into gas when heated and becomes trapped in the weld during the welding process. The shape of cluster porosity appears similar to regular porosity on the radiograph, but the indications will be grouped closely together.

Slag inclusions are non-metallic solid materials entrapped in the weld metal or between the weld and base metal. In a radiograph, dark, jagged, asymmetrical shapes within the weld or along the weld joint areas are indicative of slag inclusions, as shown in Figure 10.26.

In case of incomplete penetration,it occurs when the weld metal fails to penetrate the joint. It is one of the most objectionable weld discontinuities. Lack of penetration creates a natural stress riser from which a crack may propagate. The appearance on a radiograph is a dark area with well-defined, straight edges that follows the land or root face down the centre of the weldment, as shown in Figure 10.27.

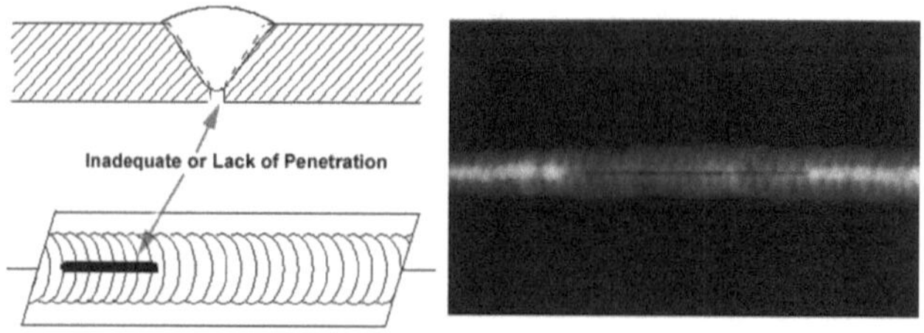

FIGURE 10.27 Incomplete penetration.

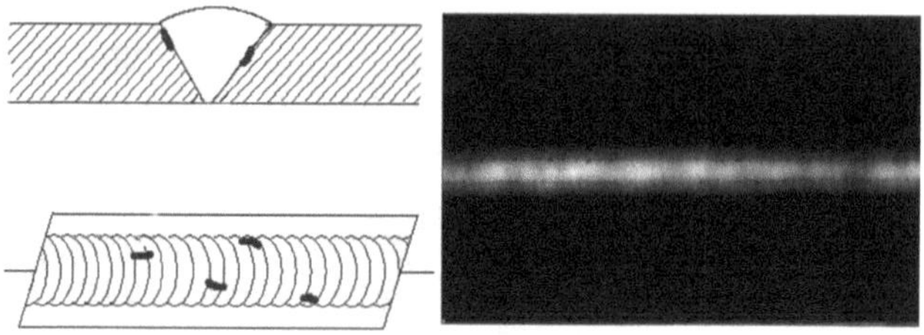

FIGURE 10.28 Incomplete fusion.

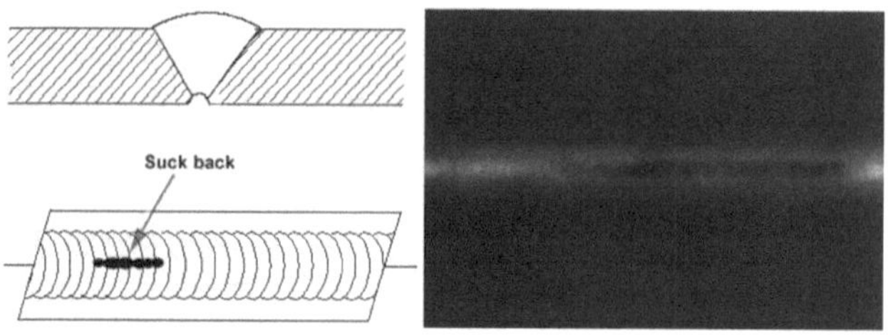

FIGURE 10.29 Internal concavity.

Incomplete fusion is shown in Figure 10.28, where the weld filler metal does not properly fuse with the base metal. Appearance on radiograph: it usually appears as a dark line or lines oriented in the direction of the weld seam along the weld preparation or joining area.

In the case of internal concavity or suck back, it occurs when the weld metal has contracted as it cools and has been drawn up into the root of the weld. On a radiograph, it looks similar to a lack of penetration, but the line has irregular edges and is often quite wide in the centre of the weld image (Figure 10.29).

In the case of undercut, near the crown or the root, it is shown in Figure 10.30 for cases A and B, respectively.

In the case of offset or mismatch between the two plates that are welded, this is shown in Figure 10.31. The radiographic image shows a noticeable difference in density between the two pieces. The difference in density is caused by the difference in material thickness. The dark, straight line is caused by the failure of the weld metal to fuse with the land area.

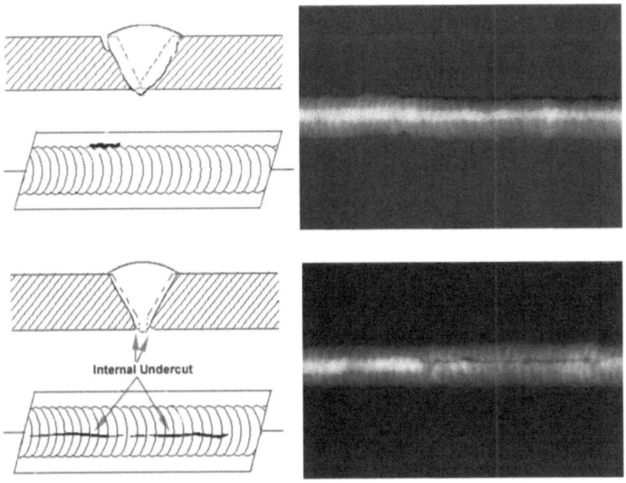

FIGURE 10.30 Undercut for crown (a) and the root (b).

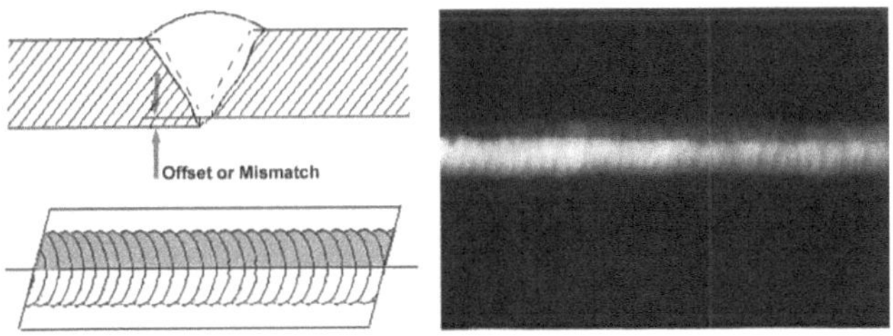

FIGURE 10.31 Offset between two plates.

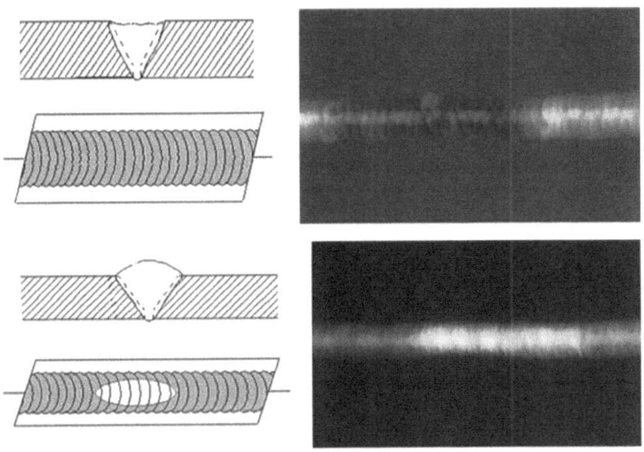

FIGURE 10.32 Less or more welding reinforcement.

In the case of inadequate weld reinforcement or excess weld reinforcement, this is shown in Figure 10.32a and b, where the thickness of the weld metal deposited is less than or greater than that specified in the engineering drawings.

10.4 ULTRASONIC TEST

The main principle of this technique is to send sound waves through the materials and receive the sound waves, which allows for the detection of cracks or other flaws in the steel and also measurement of the steel thickness.

In 1929 and 1935, Sokolov studied the use of ultrasonic waves in detecting metal objects. Mulhauser, in 1931, obtained a patent for using ultrasonic waves, employing two transducers to detect flaws in solids. Firestone and Simons developed pulsed ultrasonic testing using a pulse-echo technique in 1940 and 1945 respectively.

Ultrasonic testing (UT) employs very short ultrasonic pulse waves with centre frequencies ranging 0.1–15MHz and occasionally up to 50MHz, penetrating materials to detect internal flaws or to characterize materials. A common example is ultrasonic thickness measurement, used for measuring the webs and flanges thickness of existing steel structures.

10.4.1 Wave Propagation

Ultrasonic testing is based on time-varying vibrations in materials, commonly referred to as acoustics. This test applies ultrasonic sound to a specimen to determine its soundness, thickness, or other physical properties.

There are two types of sound waves: longitudinal waves and shear waves, along with surface waves and, in thin materials, plate waves. The longitudinal and shear waves are the two modes of propagation most widely used in ultrasonic testing.

As shown in Figure 10.33, longitudinal waves propagate in the longitudinal direction. In contrast, for transverse or shear waves, the particles oscillate at a right angle, or transversely, to the direction of propagation. Shear waves require an acoustically solid material for effective propagation and, therefore, are not effectively propagated in materials such as liquids or gases. Shear waves are relatively weak when compared to longitudinal waves; in fact, shear waves are usually generated in materials using some of the energy from longitudinal waves.

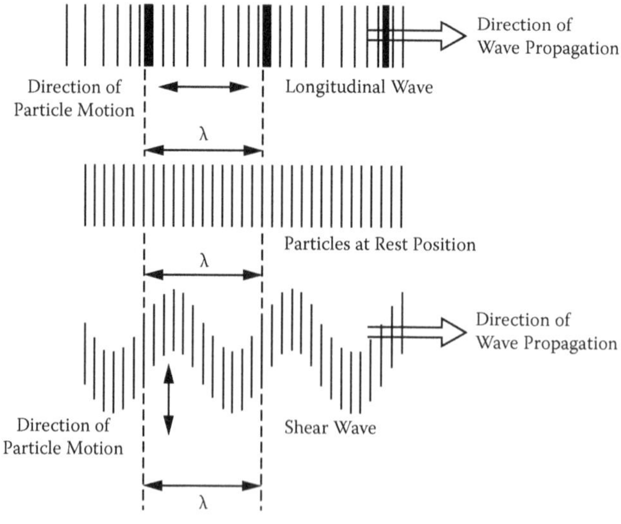

FIGURE 10.33 Types of waves.

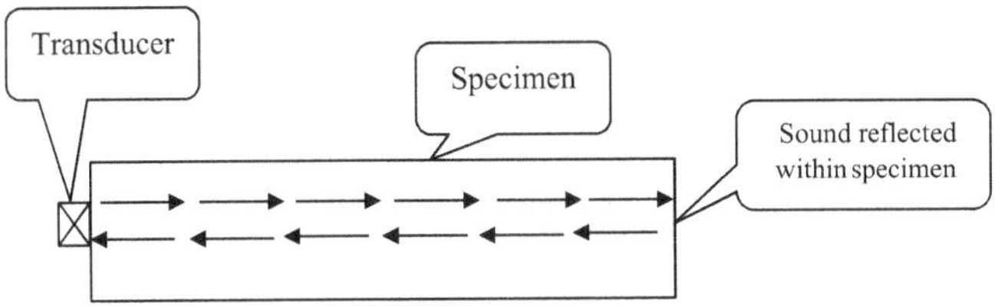

FIGURE 10.34 Sound propagation.

The sending and receiving waves of UT in the case of steel structuresdifferfrom those in concrete, as discussed previously, with the receiving and sending waves occurring from a single transducer, as shown in Figure 10.34.

Among the properties of waves propagating in isotropic solid materials are wavelength, frequency, and velocity. Wavelength is directly proportional to the velocity of the wave and inversely proportional to the frequency of the wave.

The relationship between wavelength, sound velocity, and frequency is as follows:

$$\text{Wave length } (l) = \text{Velocity}(V)/\text{frequency}(f)$$

The wavelength can be derived by altering the frequency, as wave velocity is a fixed value depending on the specimen'smaterial. According to ASNT, the smallest discontinuity that can be detected by UT is about half the wavelength (0.5 λ). Note that the frequency must be maintained between 0.1 and 1 MHz (one million cycles per second) and the wave velocity must be between 0.1 and 0.7 cm/μs.

10.4.2 Attenuation of Sound Waves

When sound travels through a medium, its intensity reduces with distance. In idealised materials, sound pressure (signal amplitude) is only reduced by the spreading of the wave. However, natural materials produce an additional effect that further weakens the sound. This weakening results from scattering and absorption. Scattering is the reflection of sound in directions other than its original direction of propagation, while absorption is the conversion of sound energy into other forms of energy. The combined effect of scattering and absorption is called **attenuation**. Ultrasonic attenuation is the decay rate of the wave as it propagates through the material, as shown in Figure 10.35.

The beam of sound energy spreads out as it moves through the specimen, as shown in Figure 10.36. It is found that for the same size of transducer, high-frequency waves produce narrower sound beams than low-frequency transducers.

As shown in Figure 10.36, the near zone varies irregularly due to sound wave interaction close to the transducer, which prevents reliable detection of discontinuities near the surface. In the far zone, the intensity decreases steadily due to both attenuation and beam spread.

It is worth mentioning that the primary reasons for attenuation are the absorption and scattering of ultrasonic energy.

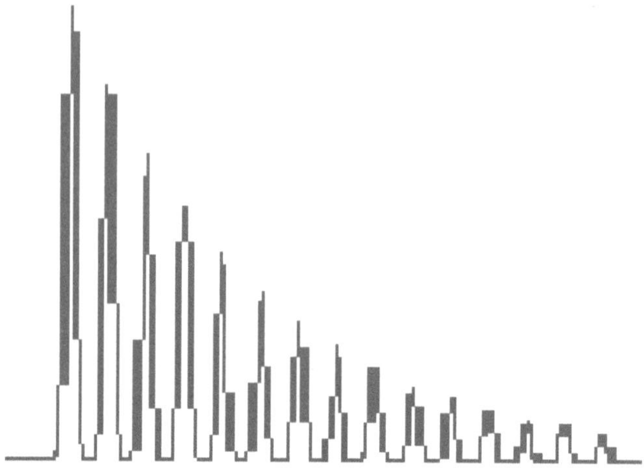

FIGURE 10.35 Attenuation in sound waves.

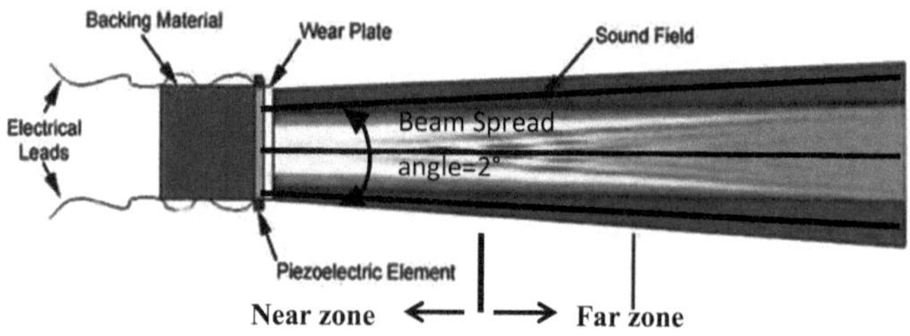

FIGURE 10.36 Beam spread for ultrasonic waves.

The beam spread may be calculated as a function of the transducer diameter (D), frequency (F), and sound velocity (V) in the liquid or solid medium. The following equation can be used to calculate the spread angle.

$$\sin\theta = 1.2\frac{V}{DF}$$

θ = beam divergence angle from the centerline to the point where the signal is at half strength.
V = sound velocity in the material (inch/sec or cm/sec).
D = diameter of the transducer (inch or cm).
F = frequency of the transducer (cycles/second).

10.4.3 Reflection in Sound Waves

The wave velocity is constant through a given medium. There is a parameter called acoustic impedance, which is obtained by multiplying the wave velocity of a certain medium with the density of that medium.

TABLE 10.3
Acoustic Sound

Material	Velocity (cm/sec)	Density (gm/cm^3)	Impedance (gm/cm^2-sec)
Air	0.33×10^5	0.001	0.000033×10^6
Water	1.49×10^5	1.00	0.149×10^6
Aluminum	6.35×10^5	2.71	1.72×10^6
Steel	5.85×10^5	7.8	4.56×10^6

The following table presents the velocity, density, and impedance in different media (Table 10.3).

Acoustic impedance is important in determining acoustic transmission and reflection at the boundary of two materials with different acoustic impedances. It is essential for the design of ultrasonic transducers and for assessing the absorption of sound in a medium.

Ultrasonic waves are reflected at boundaries where there is a difference in acoustic impedances of the materials on either side of the boundary. This difference in impedance is commonly referred to as the impedance mismatch. The greater the impedance mismatch, the larger the percentage of energy that is reflected at the interface or boundary between one medium and another. In our case, we have two media: steel and air.

The reflection and the reasons for attenuation are clearly shown in Figure 10.37. If reflection and transmission at interfaces are followed through the component, only a small percentage of the original energy makes it back to the transducer, even when loss due to attenuation is ignored. For example, consider an immersion inspection of a steel block. The sound energy leaves the transducer, travels through the water, encounters the front surface of the steel, encounters the back surface of the steel, and reflects back through the front surface on its way back to the transducer. At the water–steel interface (front surface), 12% of the energy is transmitted. At the back surface, 88% of the 12% that made it through the front surface is reflected. This amounts to 10.6% of the intensity of the initial incident wave. As the wave exits the part back through the front surface, only 12% of 10.6% or 1.3%.

The reflection factor can be calculated from the following equation:

$$\text{Reflection factor } (R) = \left(\frac{Z_1 - Z_2}{Z_1 + Z_2}\right)$$

where Z is the acoustical impedance. If we apply the equation above for the water–steel interface, the reflection value will be 88%.

10.4.4 Refraction of Sound Wave and Snell's Law

When an ultrasonic wave passes through an interface between two materials at an oblique angle, and the materials have different indices of refraction, both reflected and refracted waves are produced. This also occurs with light, which is why objects seen across an interface appear to be shifted relative to their true positions.

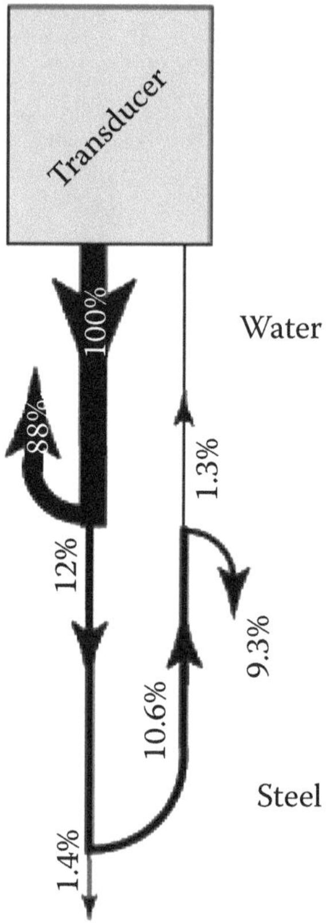

FIGURE 10.37 Reflection.

For example, if you look straight down at an object at the bottom of a glass of water, it appears closer than it actually is. A good way to visualize how light and sound refract is to shine a flashlight into a bowl of slightly cloudy water, noting the refraction angle with respect to the incident angle.

The refraction of sound waves and Snell's law are presented in Figure 10.38, where V_{L1} is the longitudinal wave velocity in material (a) and V_{L2} is the longitudinal wave velocity in material (b).

The velocity of sound in each material is determined by the material properties (elastic modulus and density) of that material. Therefore, when the wave encounters the interface between these two materials, the portion of the wave in the second material is moving faster than the portion of the wave in the first material. This difference in speed causes the wave to bend.

Snell's Law describes the relationship between the angles and the velocities of waves. It equates the ratio of material velocities $\mathbf{V_1}$ and $\mathbf{V_2}$ to the ratio of the sines of the incident (<β>Θ_1</β >) and refracted (<β>Θ_2</β>) angles, as shown in the following equation.

$$\frac{\sin\theta_1}{V_{L_1}} = \frac{\sin\theta_2}{V_{L_2}}$$

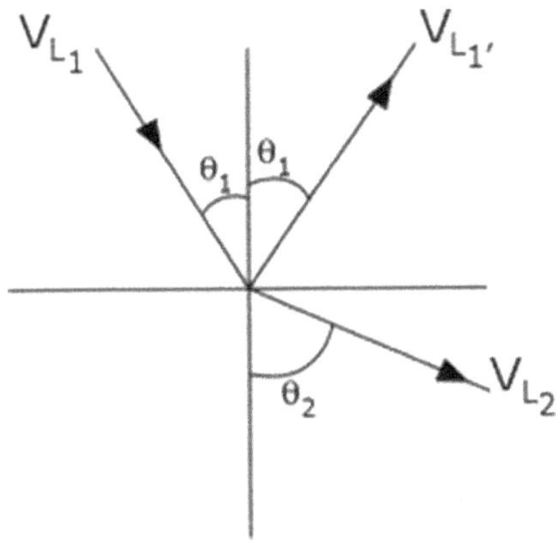

FIGURE 10.38 Reflection angles.

where

It is important to note that the wave is reflected at the same angle as the incident wave because the two waves are travelling in the same material and, therefore, have the same velocities. Although the reflected wave is not central to our explanation of Snell's law, it should be remembered that some of the wave energy is reflected at the interface.

When a longitudinal wave moves from a slower to a faster material, there is an incident angle that makes the angle of refraction for the wave 90°. This is known as the first critical angle. The first critical angle can be determined using Snell's law by setting the angle of the refracted ray to 90°, as shown in Figure 10.38. At the critical angle of incidence, much of the acoustic energy manifests as an inhomogeneous compression wave, which travels along the interface and decays exponentially with depth from the interface. This wave is sometimes referred to as a "creep wave." Due to their inhomogeneous nature and rapid decay, creep waves are not used as extensively as Rayleigh surface waves in NDT. However, they can be more useful than Rayleigh waves in certain contexts because they are less affected by surface irregularities and coarse material microstructures due to their longer wavelengths.

10.4.5 Angle Beams I

Angle Beam Transducers and wedges are commonly employed to introduce a refracted shear wave into the test material, as depicted in Figure 10.39. An angled sound path allows the sound beam to enter from the side, thereby enhancing the detectability of flaws in and around welded areas.

where θ is the angle of refraction and T is the specimen test thickness. The location of the crack can be determined using its surface distance and depth, which can be calculated using the following equations:

$$\text{Surface distance} = \text{sound path} \times \text{Sin } \theta$$

$$\text{Depth} = \text{sound path} \times \text{Cos } \theta$$

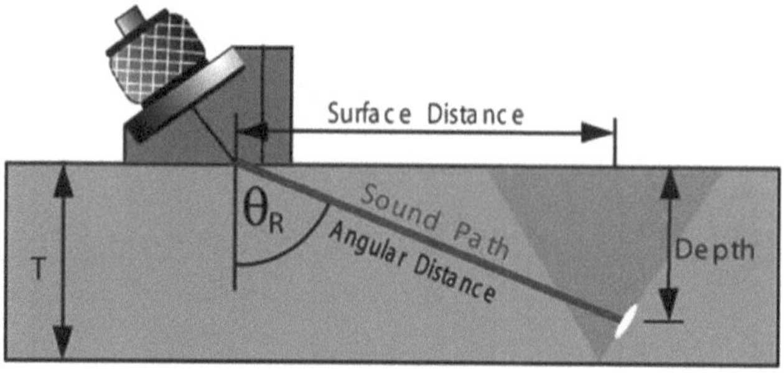

FIGURE 10.39 Angle beam.

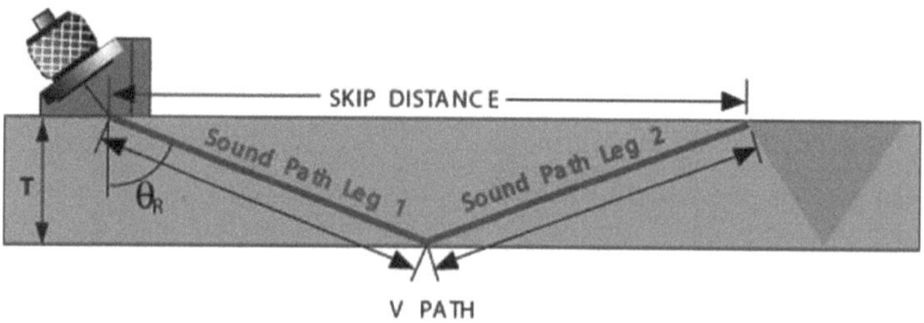

FIGURE 10.40 Inclined probe measurement.

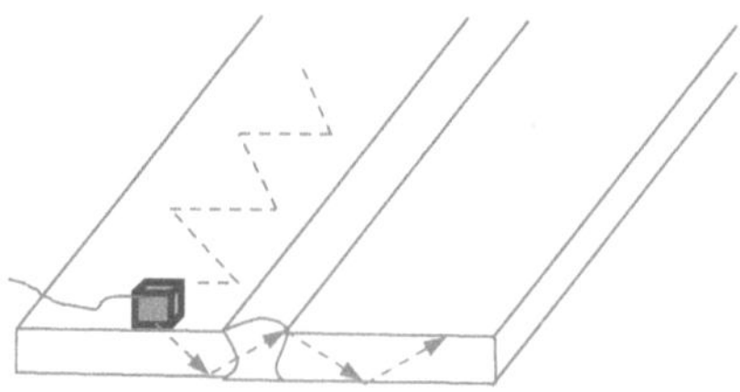

FIGURE 10.41 Angle beam path.

Angle Beam Transducers and wedges are typically employed to introduce a refracted shear wave into the test material. The geometry of the sample, as shown in Figure 10.40, allows the sound beam to be reflected from the back wall, improving the detectability of flaws in and around welded areas.

The leg length, skip distance, and *V*-path can be computed using the following equation:

$$\text{Leg} = T/\text{Cos}\ \theta$$

$$\text{Skip distance} = 2T(\tan\theta)$$

$$V\text{-path} = 2T/\text{Cos}\ \theta$$

To check the welding, the transducer must be moved in the following manner, as shown in Figure 10.41.

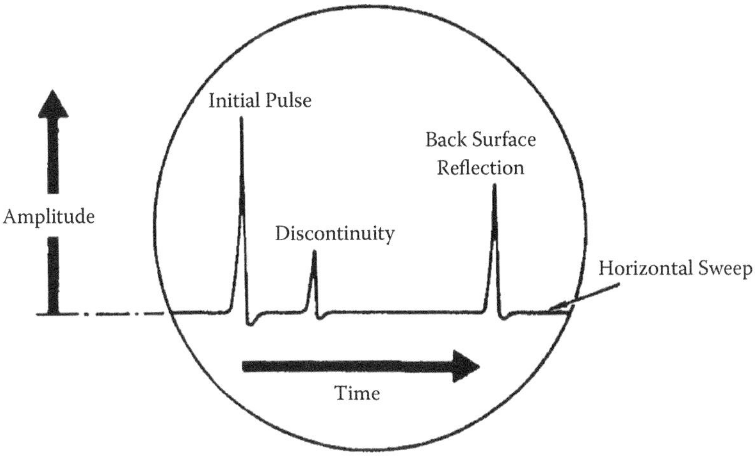

FIGURE 10.42 The shape of UT screen in case of discontinuity.

The skip distance can be calculated using the equation above. From this, you can define the distance needed to check the thickness of the weld.

10.4.6 Wave Interaction or Interference

The inspector should be familiar with and fully capable of understanding all discontinuities on the steel plate as presented on the UT screen. Figure 10.42 illustrates the echoes that will appear on the UT screen.

There are three basic types of visual displays: A-scan, B-scan, and C-scan. The A-scan provides a time versus amplitude display that reveals a discontinuity using a "PIP" on a cathode ray tube. The A-scan is depicted in Figure 10.43, alongside a photo shown in Figure 10.43 the same figure. The B-scan relies on a cross-sectional view of the testing materials, while the C-scan is similar to an X-ray image, providing a planar view, although depth information cannot be obtained. In most cases, the A-scan is used.

The A-scan presentation displays the amount of received ultrasonic energy as a function of time. The relative amount of received energy is plotted along the vertical axis, while the elapsed time is represented on the x-axis. As shown in Figure 10.44, the location of cracks B and C, along with the reduction in thickness at A, is illustrated by the corresponding echoes on the UT screen. It is important to note that the last echo, BW, is due to the back wall echo.

For different crack patterns, the pulses on the UT screen will vary, and inspectors must be well-acquainted with these differences. Examples of various discontinuities with distinct shapes are presented in Figure 10.45.

10.4.7 Transducer Types

The transducer is a crucial component of ultrasonic testing equipment. There are many types of transducers available, as ultrasonic transducers are manufactured for

FIGURE 10.43 UT machine presenting the first echo.

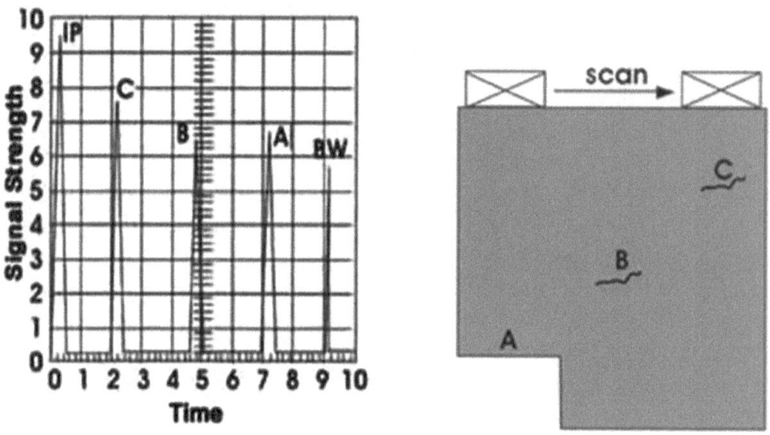

FIGURE 10.44 Reading of the UT screen.

a variety of applications, as shown in Figure 10.46, and can be custom fabricated when necessary.

The main component of the transducer is the piezoceramic material, which converts electrical pulses into mechanical vibrations. The conversion of returned mechanical vibrations back into electrical energy forms the basis for ultrasonic testing. Being a fundamental part of the transducer, it is well protected from damage.

The contact between the transducer and the test specimen is facilitated by the couplant, as shown in Figure 10.47.

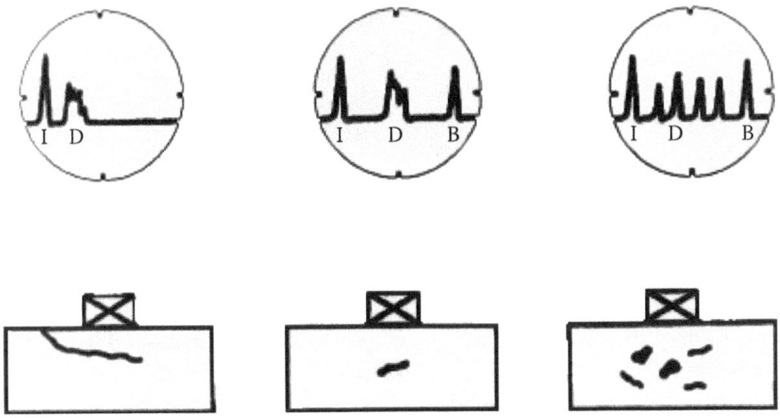

FIGURE 10.45 UT screen for different patterns of discontinuities.

FIGURE 10.46 Types of transducers.

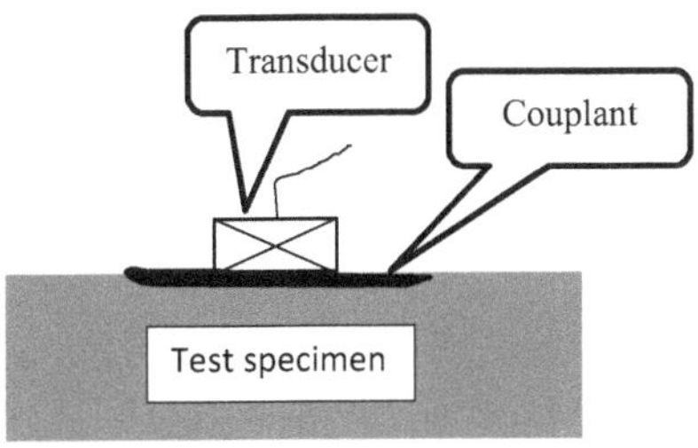

FIGURE 10.47 Location of couplant.

A couplant is typically a liquid or grease (such as oil, glycerin, or water)placed between the transducer and the test specimen. The couplant is generally necessary due to the large acoustic impedance mismatch between air and solids (i.e. the test specimen). It displaces the air, allowing more sound energy to penetrate the test specimen so that a usable ultrasonic signal can be obtained during contact ultrasonic testing.

10.4.8 Calibration Methods

There is a calibration for the UT equipment that should be conducted for both the straight beam transducer and the angle beam transducer.

The steel blocks, as shown in Figure 10.48, are used for calibration; their dimensions, along with any notch or small hole dimensions, are known to calibrate this equipment.

FIGURE 10.48 Calibration tools.

Calibration is required for evaluating and adjusting the precision and accuracy of measurement equipment. In ultrasonic testing, several forms of calibration must be performed. First, the electronics of the equipment must be calibrated to ensure they are functioning as designed.

In ultrasonic testing, there is also a need for reference standards. Reference standards are used to establish a general level of consistency in measurements and to aid in interpreting and quantifying the information contained in the received signal. They ensure that the equipment and setup yield similar results from 1 day to the next and that comparable results are produced by different systems. Reference standards also assist the UT inspector in estimating the size of flaws. In a pulse-echo setup, signal strength depends on both the size of the flaw and the distance between the flaw and the transducer. The inspector can use a reference standard with an artificially induced flaw of known size, located at approximately the same distance from the transducer, to produce a signal. By comparing the signal from the reference standard to that received from the actual flaw, the inspector can estimate the flaw size.

Calibration and reference standards for ultrasonic testing come in various shapes and sizes. The type of standard used depends on the NDE application and the form and shape of the object being evaluated. The material of the reference standard should be the same as the material being inspected, and the artificially induced flaw should closely resemble the actual flaw. This second requirement is a significant limitation of most standard reference samples.

10.5 PENETRATION TEST

The penetration test is a low-cost, easy technique for identifying discontinuities on the surface of steel and in the welding zone. This test relies on a liquid that reveals surface-breaking flaws by bleeding out a coloured or fluorescent dye from the flaw.

After applying the liquid by spray to the steel surface, a period of time known as the "dwell" is observed. Excess surface penetrant is then removed, and a developer is applied. Coloured penetrants require good white light, whereas fluorescent penetrants should be used in darkened conditions with an ultraviolet "black light"; however, they are not used in steel structure construction.

The surface preparation is one of the most critical steps of a liquid penetrant inspection, as the surface must be free of oil, grease, water, or other contaminants that may prevent the penetrant from entering flaws. The sample may also require etching if mechanical operations such as machining, sanding, or grit blasting have been performed. These mechanical operations can smear metal over the flaw opening,preventing the penetrant from entering. Once the surface has been thoroughly cleaned and dried, the penetrant liquid is applied, typically by spraying, although other methods such as brushing or immersing the part in a penetrant bath may be used in some industrial applications.

The penetrant is left on the surface for a sufficient period to allow as much penetrant as possible to be drawn from or to seep into a defect. Penetrant dwell time is the total time that the penetrant is in contact with the part surface. Dwell times are usually recommended by the penetrant producers and can vary depending on the application, the penetrant materials used, the material being inspected, and the type of defect being sought. Minimum dwell times typically range from 5 to 60 minutes. Generally, there is no harm in using a longer penetrant dwell time, provided the penetrant is not allowed to dry. The ideal dwell time is often determined by experimentation and may be very specific to a particular application.

Afterward, a thin layer of developer is applied to the sample to draw the penetrant trapped in flaws back to the surface, where it will be visible. Developers come in a variety of forms and may be applied by dusting (dry powder), dipping, or spraying (wet developers).

The developer is allowed to stand on the part surface for a sufficient period to permit the extraction of the trapped penetrant from any surface flaws. This development time is usually a minimum of 10 minutes, and significantly longer times may be necessary for tight cracks.

Finally, the inspection is performed under appropriate lighting to detect indications of any flaws that may be present. The final step is to clean the part surface to remove the developer from the parts that have been found to be acceptable.

10.5.1 Advantages and Disadvantages of Penetrant Testing

To decide when to use this technique, it is important to clearly understand the advantages and disadvantages of liquid penetrant inspection. The primary advantages and disadvantages, as described in the Dry Penetrate Method in ASTN (1980a,b,c,d), when compared to other NDE methods, are summarized below.

10.5.1.1 Primary Advantages

- The liquid penetrant test has high sensitivity to small surface discontinuities.
- The method has few material limitations; metallic and nonmetallic, magnetic and nonmagnetic, and conductive and nonconductive materials may all be inspected.
- Large areas and large volumes of parts or materials can be inspected rapidly and at low cost, making it suitable for tank inspections.
- Parts with complex geometric shapes can be routinely inspected.
- Indications are produced directly on the surface of the part, providing a visual representation of the flaw.
- It is very portable when using aerosol sprays as penetrant materials.
- Penetrant materials and associated equipment are relatively inexpensive.

10.5.1.2 Primary Disadvantages

- Only surface-breaking defects can be detected.
- The method can be used only on materials with relatively non-porous surfaces, making it recommended for steel structures.
- Surface pre-cleaning is critical, as contaminants can mask defects.
- The inspector must have direct access to the surface being inspected.
- Surface finish and roughness can affect inspection sensitivity.
- Multiple process operations must be performed and controlled.
- Post-cleaning of acceptable materials after inspection is essential.
- Chemical handling and proper disposal are required.

10.5.2 Penetrant Testing Materials

The penetrants used these days are designed to produce the level of sensitivity desired by the inspector. The following characteristics are important for the penetrant:

- It should spread easily over the surface of the material being inspected to provide complete and even coverage.
- It should be drawn into surface breaking defects by capillary action.
- It should remain in the defect but be easily removable from the surface of the part.
- It should remain fluid so that it can be drawn back to the surface of the part during the drying and developing steps.
- It should be highly visible or fluoresce brightly to produce easy-to-see indications.
- It should not be harmful to the material being tested or the inspector.

Figure 10.49 presents the shape of a discontinuity when applying the penetrant.

All penetrant materials do not perform identically and are not designed to do so. Penetrant manufacturers have developed different formulations to address a variety of inspection applications.Some applications require the detection of the smallest possible defects and feature smooth surfaces from which the penetrant is easy to remove. In other cases, the acceptable defect size may be larger, allowing the use of a penetrant formulated to detect larger flaws. However, penetrants designed to detect the smallest defects tend to produce a greater number of irrelevant indications.

Penetrant materials are classified in various industry and government specifications by their physical characteristics and performance. The Aerospace Material Specification (AMS) 2644, Inspection Material, Penetrant, is now the primary specification used in the USA to control penetrant materials.Historically, Military Standard 25135, Inspection Materials, Penetrants, has been the principal document for specifying penetrants, but this is slowly being phased out and replaced by AMS 2644. Other specifications, such as ASTM 1417, Standard Practice for Liquid Penetrant Examinations, may also include information on the classification of penetrant materials, though they generally refer back to MIL-I-25135 or AMS 2644.

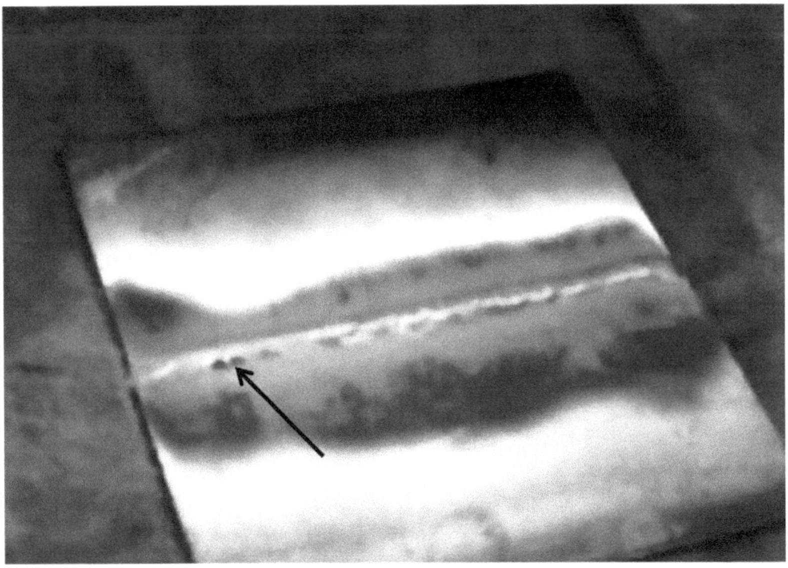

FIGURE 10.49 Shape of discontinuities in liquid penetrant.

Penetrant materials come in two basic types.

- Type 1: Fluorescent Penetrants
- Type 2: Visible Penetrants

Fluorescent penetrants contain a dye or several dyes that fluoresce when exposed to ultraviolet radiation.Visible penetrants contain a red dye that provides high contrast against a white developer background. They are also less vulnerable to contamination from substances, such as cleaning fluids, that can significantly reduce the strength of a fluorescent indication.

Penetrants are further classified by the method used to remove the excess penetrant from the part. The four methods are:

- Water washable
- Post-emulsifiable, lipophilic
- Solvent removable
- Post-emulsifiable, hydrophilic

Water washable penetrants can be removed from the part by rinsing with water alone. These penetrants contain an emulsifying agent (detergent) that allows for their removal from the part surface with water. Water washable penetrants are sometimes referred to as self-emulsifying systems. Post-emulsifiable penetrants come in two varieties: lipophilic and hydrophilic. In post-emulsifiable lipophilic systems, the penetrant is oil-soluble and interacts with the oil-based emulsifier, facilitating removal. Post-emulsifiable hydrophilic systems (Method D) use an emulsifier that is a water-soluble detergent, which lifts the excess penetrant from the surface of the part with a water wash. Solvent removable penetrants require the use of a solvent for their removal from the part.

Penetrants are then classified based on the strength or detectability of the indication that is produced for a number of very small and tight fatigue cracks. The five sensitivity levels are shown below:

- Level ½: ultra-low sensitivity
- Level 1: low sensitivity
- Level 2: medium sensitivity
- Level 3: high sensitivity
- Level 4: ultra-high sensitivity

The major US government and industry specifications currently rely on the US Air Force Materials Laboratory at Wright-Patterson Air Force Base to classify penetrants into one of the five sensitivity levels. This procedure utilises titanium and Inconel specimens with small surface cracks produced in low-cycle fatigue bending to classify penetrant systems. The brightness of the indication produced is measured using a photometer. The sensitivity levels and the test procedure can be found in Military Specification MIL-I-25135 and Aerospace Material Specification 2644, Penetrant Inspection Materials.

An interesting note about the sensitivity levels is that only four levels were originally planned. However, when some penetrants were assessed as having sensitivities significantly lower than most others in the Level 1 category, the ½ level was created. An excellent historical summary of the development of test specimens for evaluating the performance of penetrant materials can be found in the following reference.

10.5.3 Penetrants

The industry and military specifications that govern penetrant materials and their use stipulate certain physical properties that must be met. Some of these requirements address the safe use of the materials, such as toxicity, flash point, and corrosiveness, while others address storage and contamination issues. Additionally, some requirements delineate properties believed to be primarily responsible for the performance or sensitivity of the penetrants. The properties of penetrant materials controlled by AMS 2644 and MIL-I-25135E include flash point, surface wetting capability, viscosity, colour, brightness, ultraviolet stability, thermal stability, water tolerance, and removability.

10.5.3 Developers

The objective of the developer is to draw the trapped penetrant material out of defects and spread it across the surface of the part so it can be seen by an inspector. The fine developer particles both reflect and refract the incident ultraviolet light, allowing more of it to interact with the penetrant, resulting in more efficient fluorescence. The developer also facilitates greater light emission through the same mechanism. This is why indications appear brighter than the penetrant itself under UV light. Additionally, some developers create a white background, enhancing the contrast between the indication and the surrounding area.

AMS 2644 and MIL-I-25135 classify developers into six standard forms. These forms are listed below:

1. Form a: dry powder
2. Form b: water soluble
3. Form c: water suspendable
4. Form d: nonaqueous type 1 fluorescent (solvent-based)
5. Form e: nonaqueous type 2 visible dye (solvent-based)
6. Form f: special applications

The developer classifications are based on the method of application. Developers can be applied as a dry powder or dissolved or suspended in a liquid carrier. Each developer form has its own advantages and disadvantages.

Nonaqueous developers are the most traditional technique used specifically in steel structure projects. Figure 10.50 presents the spray of nonaqueous cleaner, penetrant, and developer. Nonaqueous developers are commonly distributed in aerosol spray cans for portability. The solvent tends to pull the penetrant from indications through solvent action. Since the solvent is highly volatile, forced drying is not required. A nonaqueous developer should be applied to a thoroughly dried part to form a slightly translucent white coating.

When it is not practical to conduct a liquid penetrant examination within the temperature range of 40°F–125°F (5°C–52°C), the examination procedure at the proposed lower or higher temperature range requires qualification of the penetrant materials and procedures.

FIGURE 10.50 Using spray for developer, penetrant, and cleaner.

10.6 MAGNETIC PARTICLE TEST

Magnetic particle inspection (MPI) is a nondestructive testing method used for defect detection. MPI is fast and relatively easy to apply, and part surface preparation is not as critical as it is for some other NDT methods. These characteristics make MPI one of the most widely utilized nondestructive testing methods.

MPI relies on using magnetic fields and small magnetic particles (i.e. iron filings) to detect discontinuities on the surface of materials. The only requirement from an inspectability standpoint is that the component being inspected must be made of a ferromagnetic material, such as iron, nickel, cobalt, or some of their alloys. Ferromagnetic materials are those that can be magnetized to a level that allows the inspection to be effective.

In theory, magnetic particle inspection (MPI) is a relatively simple concept. It can be considered a combination of two nondestructive testing methods: magnetic flux leakage testing and visual testing. Consider the case of a bar magnet. It has a magnetic field in and around the magnet. Any place where a magnetic line of force exits or enters the magnet is called a pole. A pole where a magnetic line of force exits the magnet is called a north pole, while a pole where a line of force enters the magnet is called a south pole. The poles of the magnet are presented in Figure 10.51.

When a magnetic part has a crack, as shown in Figure 10.52, it will form two different magnetic poles; a north and south pole will develop at each edge of the crack. The magnetic field exits at the north pole and re-enters at the south pole. The magnetic field spreads out when it encounters the small air gap created by the crack because air cannot support as much magnetic field per unit volume as the magnet can. When the field spreads out, it appears to leak out of the material, and thus it is called a flux leakage field.

If iron particles are sprinkled on a cracked magnet, they will be attracted to and cluster not only at the poles at the ends of the magnet but also at the edges of the crack. This cluster of particles is much easier to see than the actual crack, and this is the basis for magnetic particle inspection.

The first step in a magnetic particle inspection is to magnetize the component that is to be inspected. If any defects on or near the surface are present, these defects will create a leakage field. After the component has been magnetized, iron particles, either in a dry or wet suspended form, are applied to the surface of the magnetized part. The particles will be attracted to and cluster at the flux leakage fields, thus forming a visible indication that the inspector can detect.

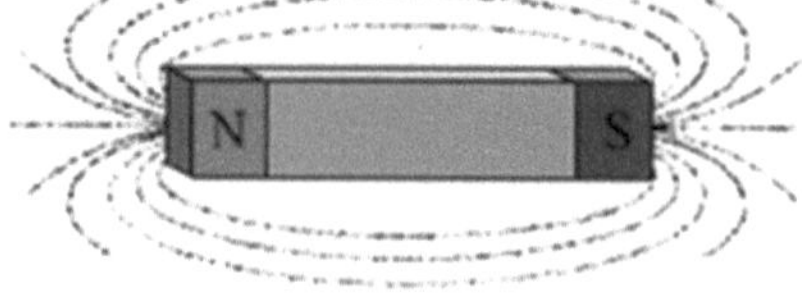

FIGURE 10.51 Magnet poles.

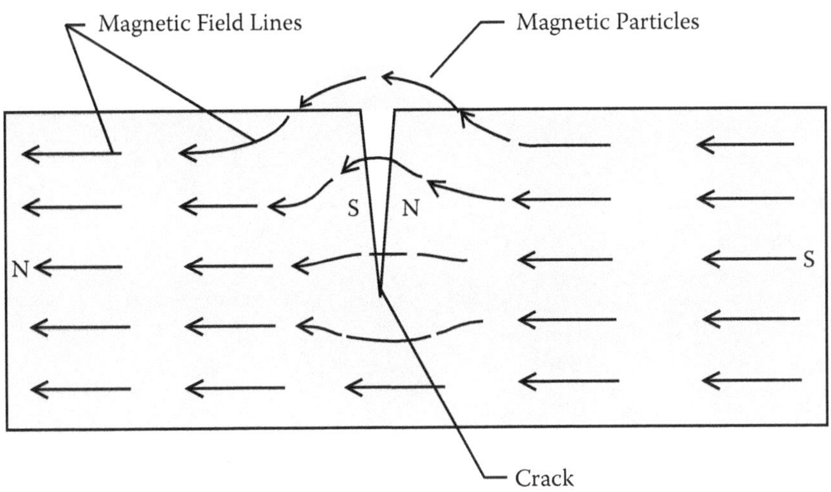

FIGURE 10.52 Magnet poles in metal cracks.

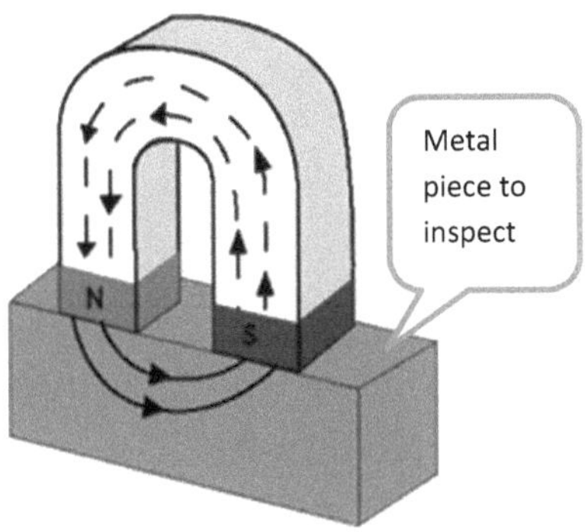

FIGURE 10.53 Direction of propagation of the magnetic field

10.6.1 Magnetic Field Characteristics

Magnets come in different shapes, and the horseshoe (U) magnet, as shown in Figure 10.53, is one of the more common types. The horseshoe magnet has north and south poles just like a bar magnet, but the magnet is curved so that the poles lie in the same plane. The magnetic lines of force flow from pole to pole, similar to those in a bar magnet. However, since the poles are located closer together and there is a more direct path for the lines of flux to travel between the poles, the magnetic field is concentrated between them.

If a bar magnet is placed across the end of a horseshoe magnet, or if a magnet is formed in the shape of a ring, the lines of magnetic force would not even need to enter the air. The value of such a magnet, in which the magnetic field is completely

FIGURE 10.54 Magnetic field in the horizontal plane.

contained within the material, probably has limited use. However, it is important to understand that the magnetic field can flow in loops within a material (see the section on circular magnetism for more information, (Figure 10.54).

Magnetic lines of force have a number of important properties, which include

- They seek the path of least resistance between opposite magnetic poles. In a single bar magnet, as shown to the right, they attempt to form closed loops from pole to pole.
- They never cross one another.
- They all have the same strength.
- Their density decreases (they spread out) when they move from an area of higher permeability to an area of lower permeability.
- Their density decreases with increasing distance from the poles.
- They are considered to have direction as if flowing, though no actual movement occurs.
- They flow from the south pole to the north pole within a material and from the north pole to the south pole in air.

10.6.2 Electromagnetic Fields

The magnetic field forms in a circular pattern around the wire, and its intensity is directly proportional to the amount of current carried by the wire. It was also observed that the strength of the field is strongest next to the wire and diminishes with distance from the conductor until it can no longer be detected. In most conductors, the magnetic field exists only as long as the current is flowing. In ferromagnetic materials, the electric current can cause some or all of the magnetic domains to align, resulting in a residual magnetic field.

It is also found that the direction of the magnetic field depends on the direction of the electrical current in the wire. A three-dimensional representation of the magnetic field is shown below. There is a simple rule known as the **right-hand rule** for remembering the direction of the magnetic field around a conductor, as shown in Figure 10.55.

For the right-hand rule to apply, it must be considered that the current flows from the positive terminal to the negative terminal (Figure 10.56).

The strength of a coil's magnetic field increases not only with increasing current but also with each loop added to the coil. A long, straight coil of wire is called a solenoid and can be used to generate a nearly uniform magnetic field, similar to that of a bar magnet. The concentrated magnetic field inside a coil is very useful for magnetizing ferromagnetic materials during inspection using the magnetic particle testing method. Please note that the field outside the coil is weak and unsuitable for magnetizing ferromagnetic materials.

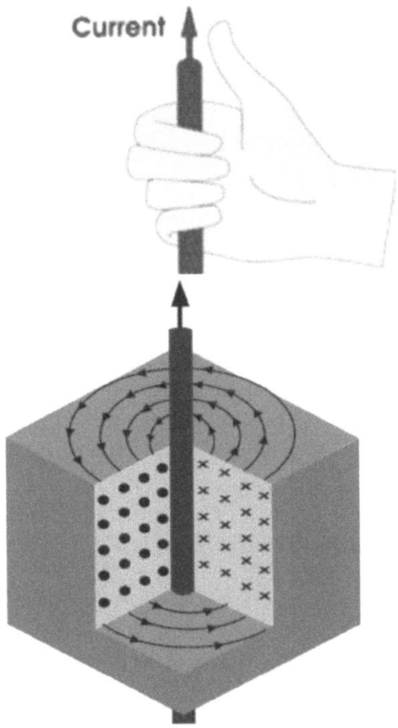

FIGURE 10.55 Right-hand rule.

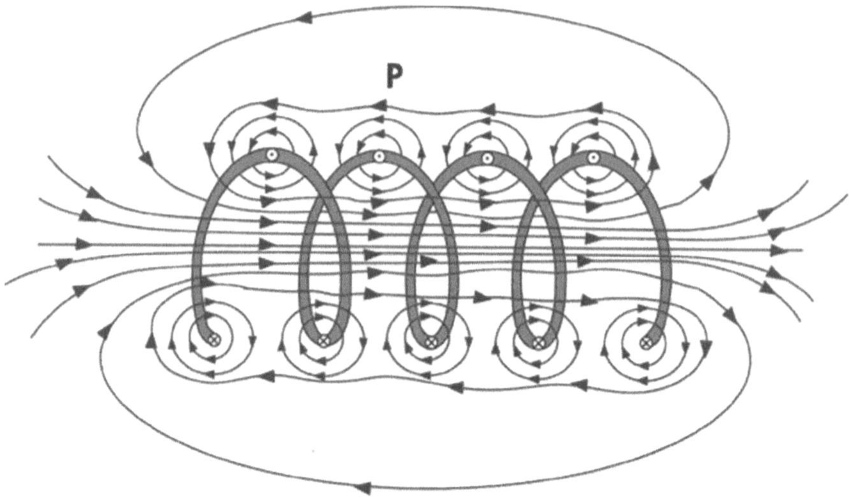

FIGURE 10.56 Direction of magnetic flow.

10.6.3 Magnetic Field Orientation and Flaw Detectability

> To properly inspect a component for cracks or other defects, it is important to understand that the orientation between the magnetic lines of force and the flaw is very important. There are two general types of magnetic fields that can be established within a component.

A longitudinal magnetic field has magnetic lines of force that run parallel to the long axis of the part. Longitudinal magnetization of a component can be achieved using the longitudinal field set up by a coil or solenoid. It can also be accomplished using permanent magnets or electromagnets.

A circular magnetic field has magnetic lines of force that run circumferentially around the perimeter of a part. A circular magnetic field is induced in an object by either passing current through the component or by passing current through a conductor that surrounds the component.

> The type of magnetic field established is determined by the method used to magnetize the specimen. Being able to magnetize the part in two directions is important because the best detection of defects occurs when the lines of magnetic force are established at right angles to the longest dimension of the defect. This orientation creates the largest disruption of the magnetic field within the part and the greatest flux leakage at the surface of the part. As can be seen in the image below, if the magnetic field is parallel to the defect, the field will see little disruption and no flux leakage field will be produced.
>
> An orientation of 45 to 90 degrees between the magnetic field and the defect is necessary to form an indication. Since defects may occur in various and unknown directions, each part is normally magnetized in two directions at right angles to each other. If the component below is considered, it is known that passing current through the part from end to end will establish a circular magnetic field that will be 90 degrees to the direction of the current. Therefore, defects that have a significant dimension in

the direction of the current (longitudinal defects) should be detectable. Alternately, transverse-type defects will not be detectable with circular magnetization.

Watch this short movie showing the effect of field direction on indication visibility. (775 KB mov)

A variety of methods can be used to establish a magnetic field in a component for evaluation using magnetic particle inspection. It is common to classify the magnetizing methods as either direct or indirect.

10.6.4 Tools for Testing

In the case of steel structure sections, it is typical to use prop magnetisation, as shown in Figure 10.57. By applying the hand rule theory, the direction of magnetization is illustrated in the figure, parallel to the welding line.

As a guideline from ASTN, the following table presents the relationship between steel plate thicknesses and the amperage with the prod spacing (Table 10.4).

As mentioned in the following section, half-wave direct current (HWDC) provides better powder mobility than DC; it also consumes less power and correspondingly

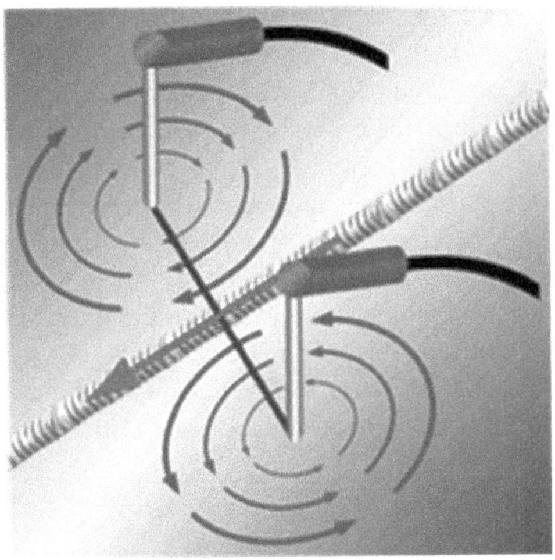

FIGURE 10.57 Magnet probes.

TABLE 10.4
Recommended Current

	Thickness of Plate in Inches	
Prod Spacing, Inches	**Under ¾ Inch**	**¾ Inch and Over**
2 to 4	200 to 300 amperes	300 to 400 amperes
Over 4 to less than 6	300 to 400 amperes	400 to 600 amperes
6 to 8	400to 600 amperes	600 to 800 amperes

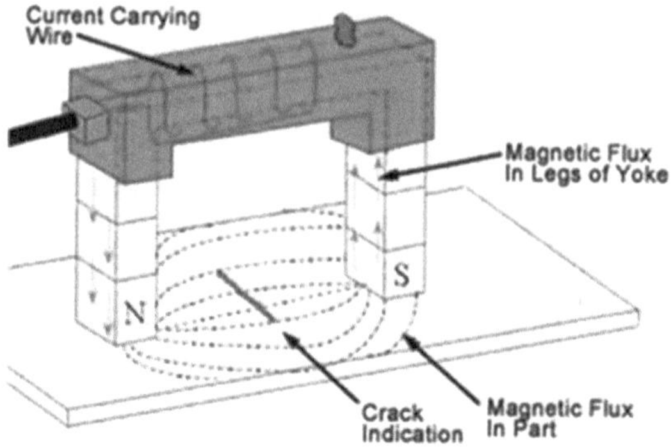

FIGURE 10.58 Yoke tool for MPI.

produces a lower heat effect at the prod contact point, as per ASME V recommendations. To avoid arcing, a remote control switch, which may be built into the prod handles, shall be provided to permit the current to be applied after the prods have been properly positioned.

Another regular method used in the market is the YOKE. This machine is used to magnetize a specimen longitudinally, as shown in Figure 10.58. It consists of a coin-shaped horseshoe magnet made of soft, low retentivity iron, which is magnetized by a small coil wound around its horizontal bar. This machine also follows the right-hand rule to demonstrate the current flow. According to ASME V, this method shall only be applied to detect discontinuities that are open to the surface of the part.

To calibrate the yoke machine, the maximum lifting force of the AC yoke shall be determined at the actual leg separation to be used in the examination, as per ASME V. This may be accomplished by holding the yoke with a 10 lb (4.5 kg) ferromagnetic weight between the legs of the yoke and adding additional weights, calibrated on a postage or other scale, until the ferromagnetic weight is released. The lifting power of the yoke shall be the combined weight of the ferromagnetic material and the added weights before the ferromagnetic weight is released. Other methods, such as a load cell, may also be used.

10.6.5 Magnetizing Current

Alternating current is the most widely used for conducting magnetic particle testing. Because AC continuously reverses direction, the magnetic field has a tendency to agitate the iron particles to make them more mobile.

Current from single-phase 110 V to three-phase 440 V is used when generating an electric field in a component. Current flow is often modified to provide the appropriate field within the part. The type of current used can have an effect on the inspection results, so the types of currents commonly used will be briefly reviewed.

Alternating current (AC) reverses direction at a rate of 50 or 60 cycles per second. In the United States, 60-cycle current is the commercial norm, while 50-cycle current is common in many other countries. Since AC is readily available in most facilities, it is convenient to use it for magnetic particle inspection. However, when

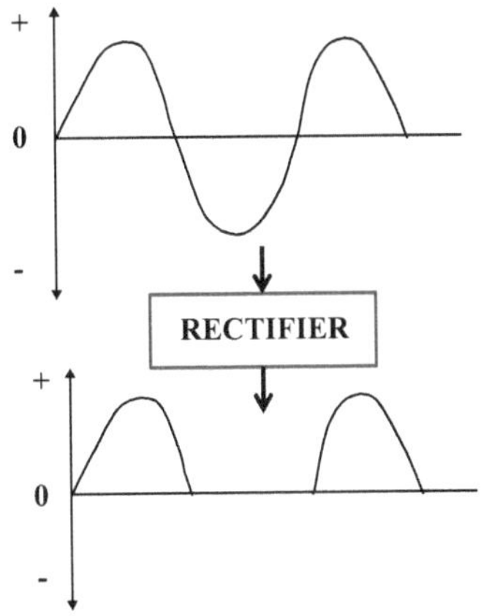

FIGURE 10.59 AC and HWAC.

AC is used to induce a magnetic field in ferromagnetic materials, the magnetic field is limited to a narrow region at the surface of the component. This phenomenon is known as the "skin effect" and occurs because induction is not a spontaneous reaction, and the rapidly reversing current does not allow the domains below the surface time to align. Therefore, it is recommended that AC be used only when the inspection is limited to surface defects.

Single-phase AC can be rectified to produce half-wave alternating current, commonly referred to as HWDC.

When single-phase alternating current passes through a rectifier, the current is allowed to flow in only one direction. The reverse half of each cycle is blocked out, resulting in a one-directional, pulsating current. The current rises from zero to a maximum and then returns to zero, with no current flowing during the time when the reverse cycle is blocked out. The HWAC repeats at the same rate as the unrectified current. Since half of the current is blocked out, the amperage is reduced to half of that of the unaltered AC.

This type of current is often referred to as half-wave DC or pulsating DC. The pulsation of the HWAC facilitates the formation of magnetic particle indications by vibrating the particles and providing them with added mobility. This added mobility is especially important when using dry particles, and the pulsation is reported to significantly improve inspection sensitivity. HWAC is most commonly used to power electromagnetic yokes (Figure 10.59).

10.6.6 The Hysteresis Loop and Magnetic Properties

The hysteresis curve, as shown in Figure 10.60, depends on the electric current that magnetizes the materials. This curve can be drawn by placing a piece of metal in coils through which alternating current is flowing. We can plot the relationship between the magnetizing current on the x-axis and the flux density on the y-axis.

It is worth mentioning that the permeability of a material can be determined by increasing the magnetising force, which has a direct relation to the electric current

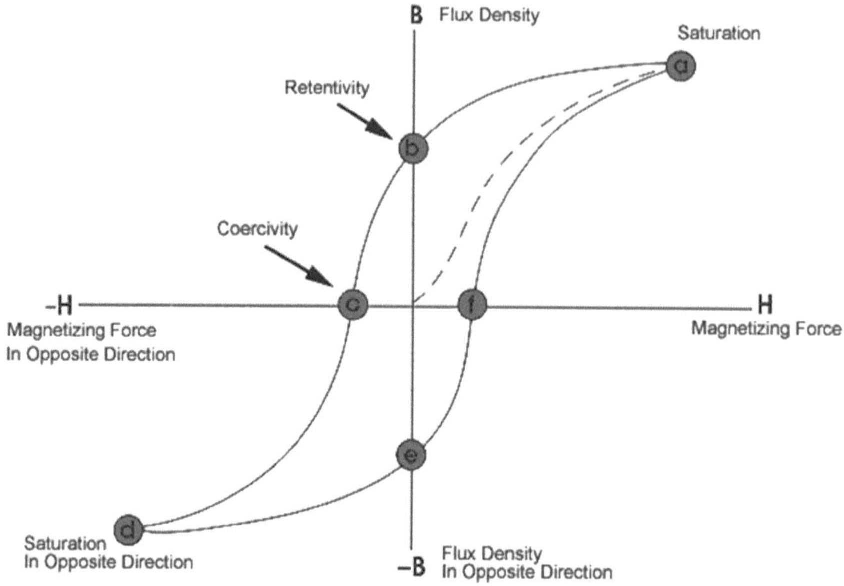

FIGURE 10.60 Hysteresis loop.

density, until the material reaches its saturation point. From this, it can be inferred that every material has a point of maximum saturation or maximum flux density.

The dashed line from the zero origin to the positive maximum saturation at point A, as shown in Figure 10.60, is known as the virgin curve. As the magnetizing force increases, the flux density increases until it reaches the saturation point, at which any increase in magnetizing force results in no increase in flux density. After that, the current will be reduced, and the flux density will slowly decrease until the magnetizing force reaches zero, which occurs at point B. This is the retentivity point, where there is a residual magnetism with a value ranging from zero to point B. After that, there will be a corrective force, which is the reverse magnetizing force required to remove the residual magnetism from the material, as shown in the figure at point C, where the flux density is zero. This point is called coercivity. It should be noted that hardened steel would require a stronger reverse magnetizing force to remove the residual magnetism. After point C, the reverse magnetizing force increases until it reaches the saturation point at point D, and then the magnetizing force is reduced to decrease the flux density until it reaches another retentivity point at point E. With the same principle the required force to remove this residual field is as shown between points zero and F.

From the hysteresis loop, a number of primary magnetic properties of a material can be determined.

1. Retentivity: A measure of the residual flux density corresponding to the saturation induction of a magnetic material. In other words, it is a material's ability to retain a certain amount of residual magnetic field when the magnetizing force is removed after achieving saturation. (The value of **B** at point B on the hysteresis curve.)

2. Residual Magnetism or Residual Flux: The magnetic flux density that remains in a material when the magnetizing force is zero. Note that residual magnetism and retentivity are the same when the material has been magnetized to the saturation point. However, the level of residual magnetism may be lower than the retentivity value when the magnetizing force did not reach the saturation level.
3. Coercive Force: The amount of reverse magnetic field that must be applied to a magnetic material to reduce the magnetic flux to zero. (The value of **H** at point c on the hysteresis curve.)
4. Permeability, μ: A property of a material that describes the ease with which a magnetic flux is established within the component.
5. Reluctance: The opposition that a ferromagnetic material shows to the establishment of a magnetic field. Reluctance is analogous to resistance in an electrical circuit.

In the case of hard steel, the following qualities produce a wide hysteresis loop:

- Low permeability, making it hard to magnetize.
- High retentivity, so it retains a strong residual magnetic field.
- High coercive force, requiring a high reverse magnetizing force to remove the residual magnetism.
- High reluctance, resulting in significant resistance to magnetizing force.
- High residual magnetism, allowing it to retain a strong residual magnetic field.

On the other hand, soft or low carbon steel will provide the following characteristics:

- High permeability, making it easy to magnetize.
- Low retentivity, so it retains a weak residual magnetic field.
- Low coercive force, requiring a low reverse magnetizing force to remove the residual magnetism.
- ·Low reluctance, resulting in minimal resistance to magnetizing force.
- Low residual magnetism, so it retains a weak residual magnetic field.

10.6.7 Examination Medium

The iron particles used to indicate the location and shape of the discontinuities should meet special specifications and precautionsfor selection, as stated in ASME V. Ferromagnetic particles used for examination must meet the following requirements.

Particle types should be treated to impart colour, such as fluorescent pigments, non-fluorescent pigments, or both, to enhance visibility against the background of the surface being examined.

The particles shall be used within the temperature range limitations specified by the manufacturer.

REFERENCES

ASNT (1980a). *Dry Penetrate Training Manual*, American Society of Nondestructive Testing, Columbus, OH.

ASNT (1980b). *Magnetic Particle Test Training Manual*, American Society of Nondestructive Testing, Columbus, OH.

ASNT (1980c). *Radiographic Method Manual*, American Society of Nondestructive Testing, Columbus, OH.

ASNT (1980d). *Ultrasonic Method Training Manual*, American Society of Nondestructive Testing, Columbus, OH.

11 Construction for Steel Structure

11.1 INTRODUCTION

This chapter presents the method of constructing a steel structure by presenting a case study of constructing a steel structure for an office building and cases of construction a workshop. The case presents project by AISC and EC3, detailed as follows and also a steel structure workshop with crane capacity 10 ton which is follow ECP that is match with the British standard.

The building is intended for office use and consists of four storeys, covering an area of approximately 7200 m^2 (80,000 square feet). The total amount of steel required for this building is approximately 400 tonnes of structural steel, equating to 56 kg/m^2 (10 Ib/ft^2)t.

This building was constructed in 2003, and the approximate costs for fabrication and erection were 9 dollars per square foot.

It features a standard bay size of 9 m by 9 m (30 ft by 30 ft) and consists of three by seven bays with around 964 pieces of structural steel.

This project is finished in 140 days considering that the performance rate is different from country to another. The schedule plan level is presented in Figure 11.1 to identify the relation between the activates for these types of projects.

As shown in the schedule, material procurement occurs early in the process, even before the shop drawings are generated. This timing ensures that the steel mill has adequate time to produce the required steel. Material procurement involves quantity surveying, placing the manufacturing order, delivery of steel from the manufacturer, and storing the steel in the fabrication shop. Detailing includes creating shop drawings for every piece in the structural frame. These shop drawings are submitted to the contractor and structural engineer to ensure compliance with the engineer's design. Upon approval of the shop drawings and the arrival of steel from the mill, the steel pieces are precisely fabricated. Once structural steel members are fabricated, they are transported to the yard, loaded onto trucks, and delivered to the job site for erection.

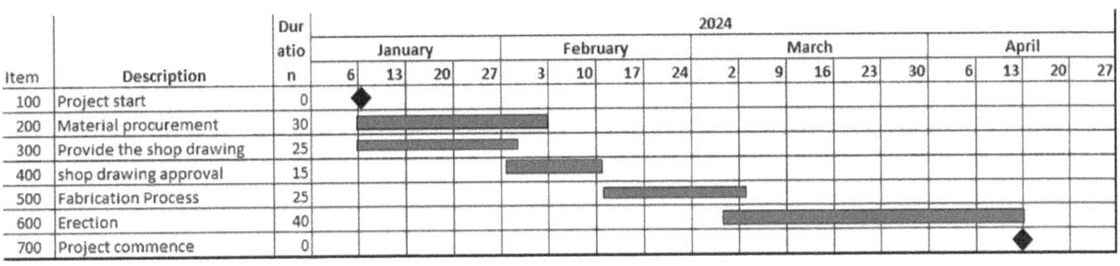

FIGURE 11.1 Project schedule plan.

DOI: 10.1201/9781003307303-11

11.2 DETAILING PROCESS

To start the detailing process, the fabricator used specialized detailing software to create a three-dimensional model of the steel frame based on the design prepared by the structural engineer. This model can be developed from data input by the detailer or from data imported from the designer's design software. The model aids the detailer in visualizing connections during the development of shop drawings.

During the detailing process, shop drawings are developed by the fabricator's detailer to convey precise information on how to fabricate each steel member in the structure. Each member shall be assigned a unique piece mark for identification and tracking. The detailer also prepares a list describing the structural steel members as presented in Figure 11.2. Shop drawings for beams P No.1, No.2,.., No.100 from the case study are shown above.

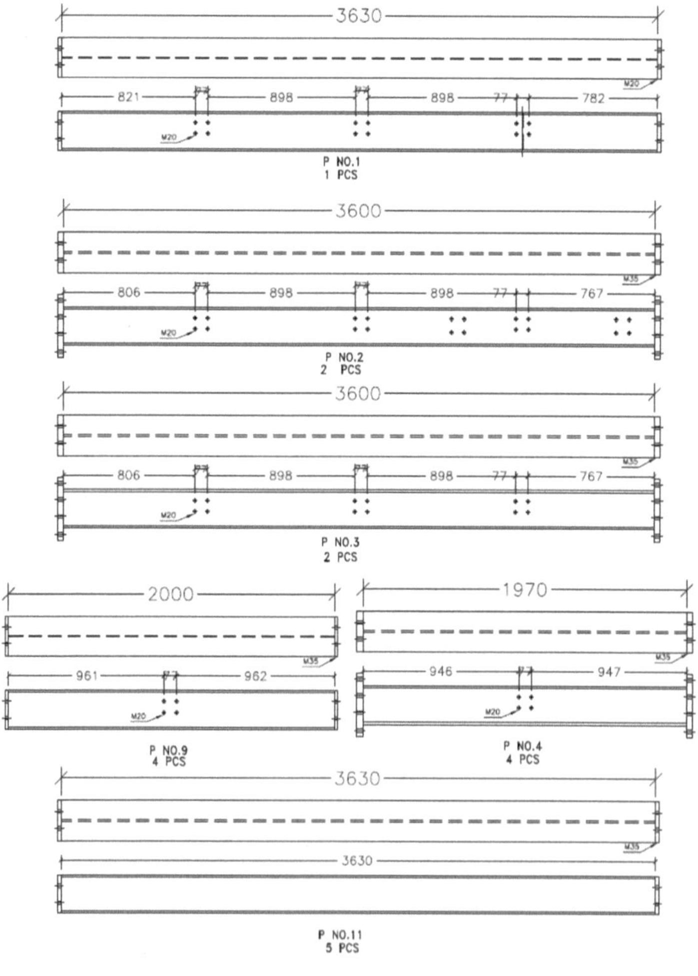

FIGURE 11.2 Steel detailing drawing sample.

Structural steel members are fabricated by precisely cutting, shearing, punching, drilling, fitting, and welding as presented in Figure 11.3 and Figure 11.4 respectively, to produce the configurations detailed in the shop drawings. Each member is labelled with a piece mark, length, and job number for identification (Figure 11.6).

In case of long span of the steel member, cambering is an effective method used in some buildings to offset deflections of steel floor beams and girders under the weight of concrete floors. Without cambering, the depth of the floor concrete will need to be varied to achieve level floors. Cambering is performed during the fabrication process by the fabricator and introduces a slight arch in the beams. Cambering was specified for most of the beams as per the project requirements. $C = 3/20$ means that the beams should be cambered 20 mm at the centre. Cambering can be achieved by applying heat to certain points of the beam or by mechanically applying pressure to the beam using specialized machinery.

Once fabricated, each finished structural steel member is labelled. In this case, the fabricator labelled each steel member with its piece mark, sequence number, and the last two digits of the job number. Specific labelling practices may vary by fabricator. The labelling of the steel pieces is crucial, as it helps

FIGURE 11.3 Punching the holes.

FIGURE 11.4 Welding in fabrication workshop.

FIGURE 11.5 Steel pieces after fabrication with their numbers.

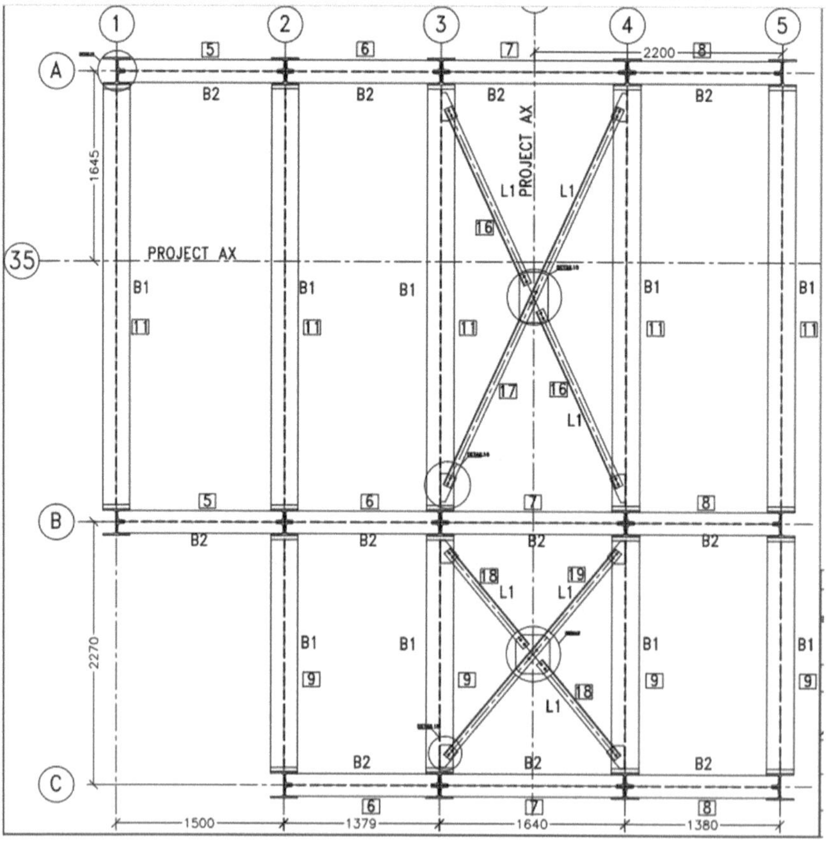

FIGURE 11.6 Steel number marking on the drawings.

identify members on the job site As an example Figure 11.5 presented the steel marks identify the beam location by the mark number. Figure 11.6 is an example for marking the steel column and these numbers presents the axis number in two direction and the floor.

One of the very important drawings and need to prepare early to start on site is the anchor bolts and base plate detailing in addition to the pedestal drawings as presented in Figures 11.7 and 11.8. This anchor bolt and base plate detailing guides the foundation subcontractor in the placement of anchor bolts. It contains horizontal and vertical information for anchor bolt placement, as well as dimensions and elevations of levelling plates (if used).

The drawing in figure 11.7 presents the non-shrink grout underneath the steel base plate and the shape of the anchor bolts are presented in Figure 11.8. In many projects at the center of the base plate can define north and east (N,E) in addition to the height that can be precise in adjusting during the erection by using total station as presented in Figures 11.10 and 11.11.

11.3 LEVELLING PLATES

A levelling plate is a steel plate fixed on top of a foundation to support a structural column. Levelling plates for the case study project have a thickness of 7 mm (1/4 inch) and conform to ASTM A36. They were installed after the anchor bolts were set.

The levelling plates were levelled and grouted prior to the installation of the columns. This is one method for setting columns; however, practices may vary with other erectors (Figures 11.10–11.14).

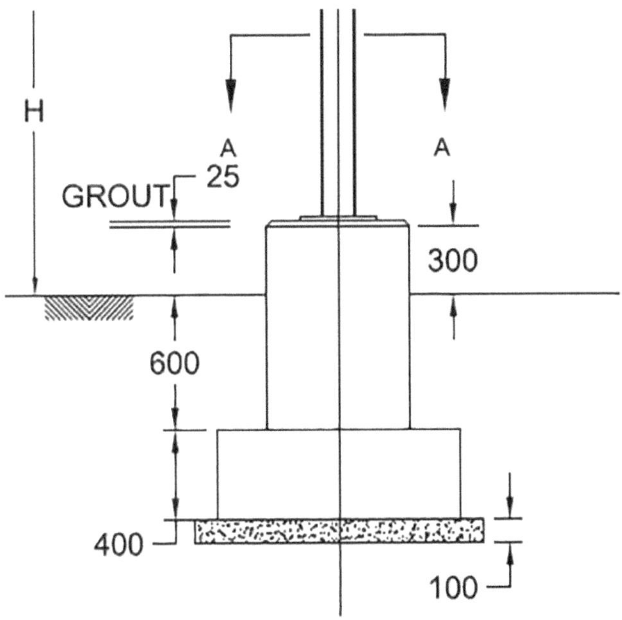

FIGURE 11.7 Typical pedestal cross section.

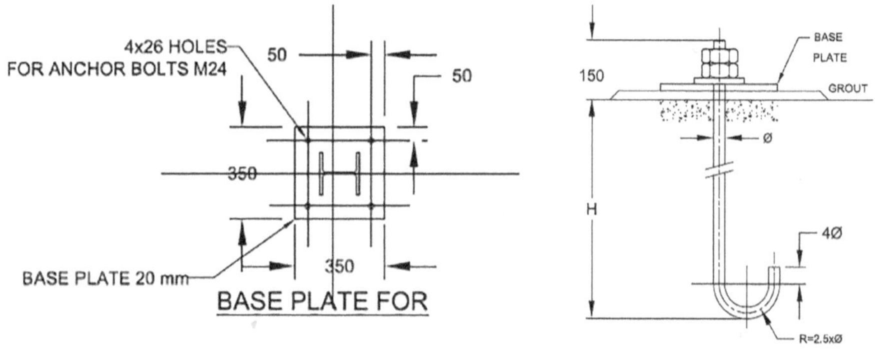

FIGURE 11.8 Typical base plate and anchor bolts.

FIGURE 11.9 Base plate erection

Before the erection process started, anchor rods and levelling plates were surveyed by the erector's surveyor to check their correct positioning. If anchor bolts are improperly located, the foundation subcontractor must rectify them before the erection process can start.

The surveyor used a total station (centre) and prisms (left and right) to confirm the horizontal and vertical location of each levelling plate and anchor bolt.

The survey provided information about any deviations of anchor rods and levelling plates from the plans. Tolerances for the location of anchor rods and levelling plates are specified in the AISC Code of Standard Practice . As shown in Figure 11.9, some times they are using a wire for the axis and adjust the elevation by the height scale equipment as shown in Figure 11.9a and b.

11.4 ERECTION SEQUENCE PROCESS

During detailing, the steel frame was divided into six steps illustrated in the erection drawings. This is done after receiving the drawings in general and should have a constructibility review session to discuss with the team about the project sequence

FIGURE 11.10 Survey equipment on plate.

FIGURE 11.11 Total station adjusted.

process. The sequences represent the order in which a zone or section of the frame will be erected. They are designed to improve the efficiency of the erection process. Planning the sequences allowed for parallel construction operations to take place. As per figure 11.12 presents the sequence of construction in six steps. The first step is to construct the column on the left pats and the second step is to erect the columns on the right parts and in step 3 erect the left parts the two floors. In step 4, erect the

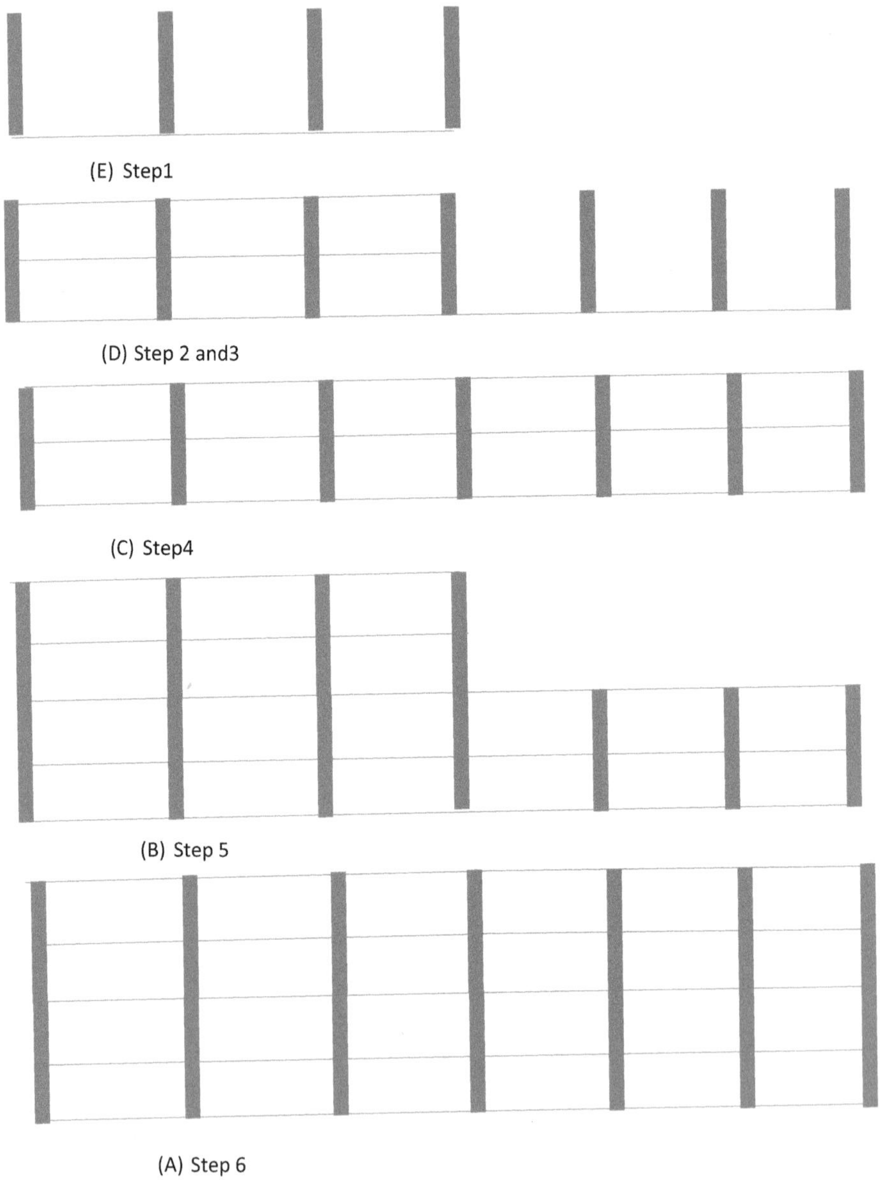

FIGURE 11.12 Construction consequence per steps.

slab on the two floors on the right parts and then in step 5 erect the columns in left parts and then their two slabs and the last step erect the column on the right part then erect its two slabs.

The last images to appear are of the masonry stair and elevator shafts, which were built after the frame was erected. The steel frame was connected to the vertical shafts to provide permanent lateral stability to the frame (Figure 11.13).

As per the AISC, case study for the administration building there are only seven workers formed the erection crew for this project, consisting of one foreman, two ironworkers hooking steel, two ironworkers connecting steel, a crane operator, and an oiler. The beam lifting for the office building in Figures 11.14 and 11.15 presents the lifting for the crane track girder. The erection by using scaffolding for workshop is presented in Figure 11.16.

Common tools used by ironworkers include spud wrenches, pins, and correction bars of various sizes. These tools are used to align steel members, and spud wrenches are also used to tighten bolts.

The crane used for the project was a 100-tonne crawler-mounted crane with a 160-foot boom and a 40-foot jib. The boom (upper right) is the main projecting structure of the crane, while the jib (lower right) is the smaller structure attached to the boom. There are many configurations and sizes of cranes available, which are selected based on cost, availability, speed, reach, and capacity.

Upon delivery to the job site, fabricated steel was unloaded and placed on wood or steel blocking. The blocking allowed for chokers to be attached to each member for subsequent hoisting and erection.

Shakeout of structural steel occurred after unloading. Shakeout is the process of sorting steel pieces on site so that they can be efficiently erected, involving the organization and spacing of the steel members.

This is a special column hoisting device used on the case study project. It has a release attached to a rope to facilitate its detachment from a column. After the column is secured with the required bolts, an ironworker actuates the release directly from the ground to disengage the hoisting cable (Figure 11.17).

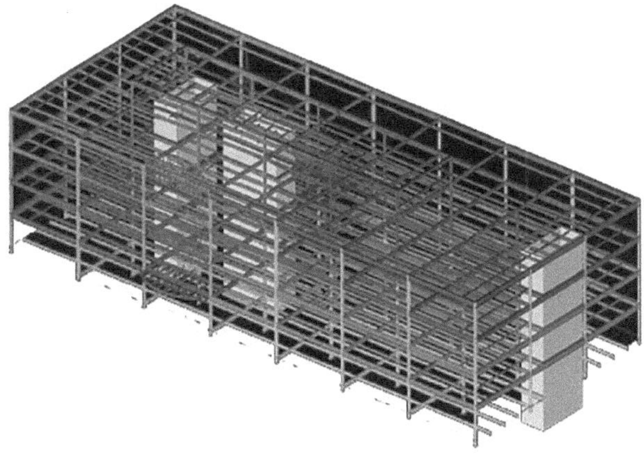

FIGURE 11.13 Structure isometric drawing.

FIGURE 11.14 Erection with the labour.

FIGURE 11.15 Lifting for the crane.

This is the first column being hoisted into position. Using ladders, ironworkers temporarily bolted the column. Prior to erection, the erector must consider column stability in accordance with safety standards and the AISC Code of Standard Practice Figure 11.18.

The first structural steel members to be hoisted were the columns. This is a view of sequence 1 after the columns were set. Note that all columns have a splice on the top.

The case study project columns were spliced at an elevation of 131′–3″ (between the third and fourth floors). A column splice is used to connect two columns vertically.

FIGURE 11.16 Erection for the steel frame.

FIGURE 11.17 Special lifting equipment tools.

The column splices consisted of two shop-bolted steel plates attached to the column flanges. The bolts were loosened before hoisting the upper column into place to facilitate the connection (Figure 11.19).

FIGURE 11.18 Photo of column after erection.

FIGURE 11.19 Column splice.

11.5 COLUMN BASE PLATE

The column base plate as shown in Figure 11.20 considering that this is in case of fixed support may be another configuration presents depends on the support design. Levelling plates were detailed slightly larger than the column base plates to ensure complete support of the base. This practice may vary by fabricator.

FIGURE 11.20 Base plate shape.

A choker is a rigging assembly used to attach a load to a hoisting device. It is usually made of wire rope or synthetic fibre. Chokers are used to hoist beams into position. They are positioned at the centre of gravity of the beam to ensure that it remains level when hoisted (Figure 11.20).

Multiple lift rigging is a process used to hoist several pieces of steel simultaneously. This is accomplished by utilizing a rigging assembly that facilitates the attachment of up to five independent loads to the hoist rigging of a crane. In this project, up to four beams were hoisted at a time. Multiple lift rigging speeds up the lifting process by reducing the total number of lifts required for the frame. Figure 11.21 starts bolting fixation after adjusting the lifting piece in its location.

FIGURE 11.21 Align holes for bolt fixation.

11.5.1 Preliminary Bolting

This is a view of the completed hoisting activity for sequences 1 (left) and 2 (right). It took approximately 5 days to hoist the 166 pieces of steel that are part of the two sequences.

Metal decking was placed and attached using a series of welds. Controlled deck zones were used during installation. A controlled deck zone is an area in which the initial installation and placement of metal decking may take place without the use of guardrail systems, personal fall arrest systems, fall restraint systems, or safety net systems, and where access to the zone is controlled.

11.5.2 Perimeter Safety Cable

On multi-storey structures, perimeter safety cables are installed at the interior and exterior perimeters of floors and openings as soon as the metal decking has been installed.

Snow was removed from the locations where shear studs were to be placed (upper left). Using a stud welding gun (an automatic welding device), the shear studs were then welded to the structural members below the metal decking. Shear studs are used to ensure composite action between the concrete slab and structural steel beams (upper right). Whether shear studs are used depends on the structural engineer's design.

This is a view of the welded studs. Shear studs were tested to ensure that they were correctly welded to the structural steel members in compliance with the bend test of the AWS D 1.1 Structural Welding Code.

This is a view of the completed hoisting activity for sequences 3 and 4. Hoisting these two sequences took approximately 9 days, in contrast to the 5 days it took to hoist sequences 1 and 2. Sequences 3 and 4 consisted of two floors (365 pieces of steel), while sequences 1 and 2 consisted of one floor (166 pieces of steel).

11.5.3 Temporary Bracing and Plumbing Up

Temporary bracing was installed before plumbing up the structural frame. It is used to provide temporary lateral stability to the structure and to ensure that the frame is plumb. The temporary bracing was removed by the erector once the structural steel frame was secured to the vertical masonry elevator and stair shafts, which were installed after the completion of the steel frame. This temporary bracing are from steel wire and in X bracing type to facilitate the plumbing up.

Plumbing up refers to the vertical alignment of the structure. A theodolite was used to verify the alignment of the columns around the perimeter of the building. The foreman directed ironworkers on the structure to shift the frame using the temporary bracing.

Turnbuckles (lower left) are a common component of the bracing system. A turnbuckle is a device that usually consists of a link with screw threads at both ends, which is turned to bring the ends closer together. Using several turnbuckles connected to the temporary bracing, the steel frame was adjusted until it was vertically aligned Figure 11.22.

11.5.3.1 Final Bolt-Up

After a section of the structural frame was assembled and vertically aligned, ironworkers permanently fastened the structural steel members with additional bolts and welds. Tension-controlled bolts (F1852) were used. Final bolt-up was performed

FIGURE 11.22 Turn Buckle.

using a torque wrench, which stops automatically when the bolt shears at a tension control groove, indicating that the required tension has been reached.

This is a view of the completed hoisting activity for sequences 5 (left) and 6 (right). Hoisting both sequences took approximately 9 days, similar to the time taken for sequences 3 and 4.

FIGURE 11.23 Start the wall.

11.5.3.2 Lateral Stability

One of the final activities in the steel erection process was to secure the structural steel frame to the elevator and stair masonry shafts. On the left is a plan view of a stair shaft as depicted in the first-floor framing plan. Figure 11.23 illustrates how perimeter beams were attached to the masonry. Angles were connected to the masonry shaft using 18 mm (3/4") diameter bolts every 200 mm(8"), which were set by the mason. The angles were then welded to the steel beams.

In this case, the elevator shafts provide lateral stability to the structural steel frame of the building.

11.6 START UP

The structural steel frame was inspected upon completion by the local jurisdiction, allowing work to proceed accordingly. Depending on the complexity of the frame, additional periodic inspections and testing may also be required by the local jurisdiction. The design engineer conducts contract administration and field observation activities to ensure quality.

Index

Note: **Bold** page numbers refer to tables and *italic* page numbers refer to figures.

Kolibri 360 GmbH
Ferdinand-Jühlke-Straße 7
99095 Erfurt, Deutschland
produktsicherheit@kolibri360.de